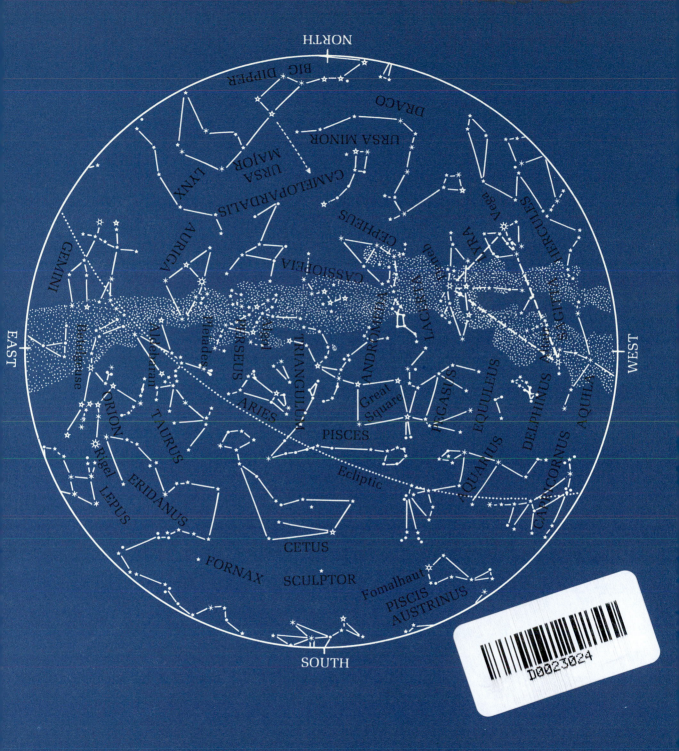

The late autumn sky, on November 15 at 9 P.M.; December 1 at 8 P.M.; or December 15 at 7 P.M., standard time.
Adapted from H. A. Rey, *The Stars*, published by Houghton Mifflin Company. Copyright © 1952, 1962, 1967, 1970 by H. A. Rey.

★ THE RESTLESS UNIVERSE ★

★ THE RESTLESS UNIVERSE ★

AN INTRODUCTION TO ASTRONOMY

HARRY L. SHIPMAN

UNIVERSITY OF DELAWARE

HOUGHTON MIFFLIN COMPANY   BOSTON
DALLAS   GENEVA, ILLINOIS
HOPEWELL, NEW JERSEY   PALO ALTO   LONDON

Library of Congress Catalog Card No.: 77−78584
ISBN: 0−395−25392−6

★ CONTENTS ★

★ PREFACE ★

The Viking landers have reached the surface of Mars, returning data about the weather conditions, soil conditions, and possible presence of life on the red planet. Evidence is rapidly accumulating to prove that black holes, enigmatic compressed objects that do not let light escape, do exist in the real world. The construction of several large telescopes has at last provided extragalactic astronomy with a firm observational foundation, shedding light on the physics of galaxies and the evolution of the universe.

In the last two decades astronomy has been an explosive discipline, with new insights emerging every year. This introductory astronomy text is addressed to the nonscientist, describing recent advances in this field in non-mathematical terms. This book differs from similar ones in its emphasis on contemporary astronomy—events that readers have probably heard about from the media, the material they expect to see in an introductory astronomy course.

The mysteries of starbirth, the violence of stardeath, and extragalactic astronomy are given as much space as possible. A particular difference between this text and others is the treatment of planetary science. Thanks to data returned by space probes, our understanding now extends beyond the first-look preliminary interpretations and spectacular, awe-inspiring pictures. This book, not presuming that either the student or the course instructor has any background in geology, describes the scientific payoff from the recent spectacular decades of planetary research. We are now beginning to understand the forces that shape planetary surfaces; remarkable insights have come from a comparison of similar geological processes at work on different planets. A simple description of planetary environments does not do justice to the field of planetary science.

A minimal quantity of mathematics is used in the text itself, although a few tables, graphs, and boxes demonstrate the use of numerical data without requiring the reader to go through algebraic manipulations or calculations. The extensive treatment of the Copernican revolution, found in some texts, is shortened to make room for the contemporary material. There is still a thorough qualitative description of the appearance of the night sky without an extensive treatment of coordinate systems. Lengthy discourses on stellar properties are shortened to make room for an expanded treatment of stellar evolution.

I thank John Warner of the University of Minnesota, Darrel Hoff of the University of Northern Iowa, Joseph Gould of St. Petersburg Junior College, David Morrison of NASA Headquarters, and Peter Foukal of the Harvard-Smithsonian Center for Astrophysics for their very helpful comments in reviewing the book in manuscript. All of us in the astronomical community are grateful to those federal and private agencies and foundations that support astronomy; most of the exciting research reported in this book has been supported by them. I personally thank the National Science Foundation, the Research Corporation, the University of Delaware Research

Foundation, and the National Aeronautics and Space Administration for the support of my own research.

This book is dedicated to my parents.

H.L.S.

★ THE RESTLESS UNIVERSE ★

★ PART ONE ★

BASICS

Next time it's clear, go out and look at the sky. Depending on the time of day or night, you will see the sun, the moon, a planet or two, some stars, and perhaps a galaxy. What are these astronomical bodies? How distant are they? How do they evolve? Is there life elsewhere? These are questions that the astronomer tries to answer.

Astronomers have to live with a limited ability to conduct research, since almost all our information about the universe comes from light. Celestial objects beam electromagnetic radiation of all types—light, radio waves, x rays, and other forms—to our waiting telescopes. To decode the mysterious message of starlight, we need a knowledge of the basic properties of electromagnetic radiation. Also, to begin to understand our position in the universe, we must first understand our position in the solar system. These two topics are the focus of Part One.

★ CHAPTER ONE ★

THE ASTRONOMICAL UNIVERSE

Figure 1.1 shows a small group of stars. The business of astronomy is the understanding of various types of objects in the universe: planets, stars, galaxies, and the universe itself. We ask questions about the nature of these objects, and the distinguishing characteristic of successful scientific research is the development of productive answers to well-posed questions. This chapter discusses the overall organization of the universe, the kinds of questions asked in astronomical research, and the laws of physics and pieces of equipment used in astronomical research.

★ MAIN IDEAS OF THIS CHAPTER ★

Astronomers ask three types of questions about celestial bodies: Where is it? What is it? How does it evolve?

Virtually all the observational evidence used in astronomical research comes from measurements of the radiation that celestial objects emit.

The laws of physics, describing how matter interacts and how it emits radiation, play an essential role in the interpretation of astronomical observations.

FIGURE 1.1 Stars comprise one of the types of objects that an astronomer investigates. Stars are often found in groups or star clusters. The photograph shows the star cluster Messier 67. (Hale Observatories)

★ 1.1 THE NATURE OF ASTRONOMICAL ★ RESEARCH

☆ OVERVIEW OF THE UNIVERSE ☆

Astronomical research consists of asking and answering questions about the universe. While answering the questions is not easy, knowing the right questions to ask is the heart of successful scientific research. Good questions have realistic alternatives for answers. For example, a negative answer to the question, "Is the moon made of green cheese?" does not provide much insight into the moon's nature.

An astronomer begins the series of questions by addressing the overall organization of the universe. The first big question we ask about an object is "Where is it?" Is this object something small and near, or large and far away? Go outside on some clear night, look up, and ask, "How far away are the stars?" It is difficult to escape the impression that they are little light bulbs suspended a few kilometers away on a crystal sphere (see Figure 1.2). You must stretch your imagination to visualize each of these twinkling points as a giant sun like our own, more than a million kilometers across and some trillions of kilometers from the earth.

FIGURE 1.2 From *But We Love You, Charlie Brown,* by Charles M. Schulz, Holt, Rinehart & Winston, 1959. © 1959 United Features Syndicate, Inc.

BOX 1.1 POWERS-OF-TEN NOTATION

The immense scale of the universe makes it rather inconvenient to write cosmic distances in conventional ways. The nearest known star, apart from the sun, is 4,103,000,000,000,000,000 centimeters (cm) distant. It would be difficult to keep track of the zeros if you wanted to use this number in a calculation. It is far simpler to write it as 4.103×10^{18} cm, where the 10^{18} is just 1 followed by 18 zeros, or one quintillion (American style). Powers-of-ten notation can also be used to write out small numbers. The density of matter in interstellar space is about 0.1 atom per cubic centimeter, or 0.000,000,000,000,000,000,000,000,166 gram per cubic centimeter, more conveniently written as 1.66×10^{-25} g/cm^3.

Powers-of-ten notation uses an exponent to indicate the overall size or *order of magnitude* of a quantity. A quantity is written as

$$(number) \times 10^{(exponent)}$$

where the exponent indicates how many places to shift the decimal point. Shift the decimal point to the right if the exponent is positive, and to the left if the exponent is negative. Thus $1 \times 10^{-2} = 0.01$; $1 \times 10^3 = 1000$; and 6×10^6 is 6,000,000.

Powers of ten become quite useful in calculations. To multiply, add exponents; to divide, subtract exponents; and to raise a number to a power, multiply the exponent by the power. Thus,

$$10^3 \times 10^2 = 10^5$$
$$10^3 \div 10^4 = 10^{-1}$$
$$(10^6)^2 = 10^{6 \times 2} = 10^{12}$$
$$2 \times 10^3 \times 4 \times 10^{10} = 8 \times 10^{13}$$

In the metric system, prefixes are often used to indicate the order of magnitude of a unit of measurement. For example, 1 micrometer (also known as a micron) is 10^{-6} meters (m). Commonly used prefixes are micro = 10^{-6}, milli = 10^{-3}, kilo = 10^3, mega = 10^6, and giga = 10^9. These prefixes can also be applied to other units. A person's annual salary can be expressed in kilobucks, a university budget in megabucks, and the national debt in gigabucks.

NOTE TO THE READER: These boxes present supplementary material that would break up the flow of ideas in the text were they not set off. A few boxes, like this one, deal with ideas that are essential to the understanding of later material. Since some readers have probably seen this material before, it is separated. Most boxes cover the ideas expressed in the text at a somewhat deeper mathematical level than the text itself. The mathematically sophisticated boxes can be skipped or pondered over, depending on the reader's (and the course instructor's) desires.

However, some boxes, like this one, should not be skipped unless the reader feels comfortable with the material in them.

The overall organization of the universe, the scale of cosmic distances, is sketched in Figure 1.3. Each panel of the figure shows a view of the universe, and each panel shows some 100,000 (or 10^5) times as much space as the panel above it. (Powers-of-ten notation is explained further in Box 1.1 for those who have not seen it before.) The earth comfortably fills the top panel.

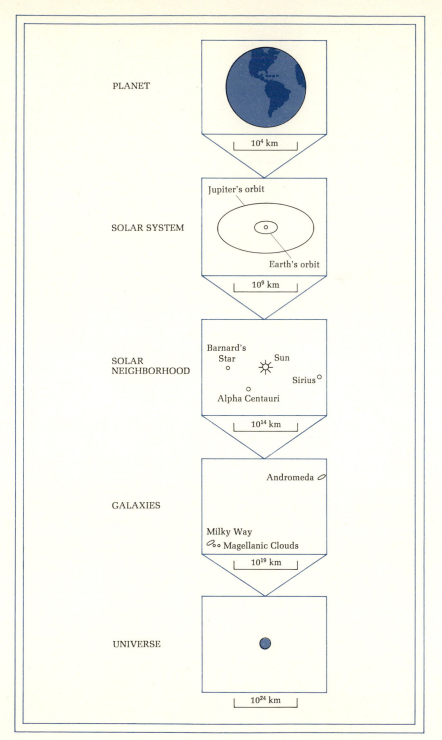

PLANET

Jupiter's orbit

SOLAR SYSTEM

Earth's orbit

10^4 km

10^9 km

Barnard's
Star

Sun

SOLAR
NEIGHBORHOOD

Sirius

Alpha Centauri

10^{14} km

Andromeda

GALAXIES

Milky Way
Magellanic Clouds

10^{19} km

UNIVERSE

10^{24} km

FIGURE 1.3 Five views of different parts of the universe. Each panel shows 10^5 times as much space as the panel above it.

9

To bring the sun into view, we must look at 10^5 times as much space, shrinking the earth to 10^{-5} (one one-hundred-thousandth) of its size, as shown in the second panel. The solar system is an empty place. Shrink the view again and a few of the nearest stars become visible. Interstellar space, too, is quite empty; *stars* are very large balls of gas wandering through the even larger expanses that separate them. Shrinking the picture again, we can see that stars are organized in large swarms called *galaxies*, each containing 10^{11} stars. The nearest large galaxy is the Andromeda galaxy, illustrated in Color Photo 17. Shrink again and our cosmic view only now encompasses the entire observable universe, stretching as far as our telescopes can see—about 20 billion light years. (One light year is the distance light travels in a year. All units of measurement are summarized in section 1.2.)

A comparison of planets and stars can illustrate the importance of knowing the distance to an astronomical object. If it's clear, go out tonight and look at the sky. Find a planet; it won't twinkle as the stars do. (Newspapers often list the locations of visible planets.) Compare that planet to a bright star. In appearance, these two objects are rather similar, but one is about 10^5 times as far away as the other (recall Figure 1.3). The planet is a world of its own, comparable in size to the earth, and it shines by the reflected light of our own sun. The star is a ball of glowing gas similar to the sun, producing about 10^{10} times as much light as the planet does. The fundamentally different characteristics of these two types of objects only become apparent when their relative distances are known. In the sky, they just look like two lights that move in slightly different ways, which is the way they were regarded by the ancients.

☆ THE NATURE OF STARS ☆

With a rough knowledge of the distance to an object, an astronomer can begin analyzing radiation from the object in an attempt to determine its nature. It is now that questions become arranged in a natural order or hierarchy. The second basic question, "What is it?" is too vague to answer at first. The working scientist needs to know what observations to make, what calculations to perform. Which calculations are mathematical exercises and which provide some insight? Good, probing questions can guide research into productive channels. Consider how we came to understand the nature of stars, for example.

Although astronomy is the oldest of all the sciences, the study of the physical nature of stars did not begin until the nineteenth century. Earlier watchers of the skies were primarily interested in the motions of the planets; the star catalogues that then existed were primarily used by observers needing a backdrop for measuring planetary positions. Mere lists of stars were not very helpful to people trying to analyze them.

The nineteenth century marked the beginnings of stellar spectroscopy, the science of stellar analysis. The *spectrum* of a star is a record of the intensities of the different colors or *wavelengths* in its light. Light from stars is spread out into the rainbow colors of the spectrum, and it is this spectrum that is photographed and preserved for further study. Pioneer spectroscopists real-ized that the spectra of different stars varied in appearance; they hoped that the spectra might provide clues to the nature of the different types of stars. In a monumental effort, Annie Jump Cannon of the Harvard Observatory led a group that photographed and catalogued the spectra of more than 200,000 stars early in the twentieth century. This catalogue brought the science of spectroscopy to a mature stage. The big question, "What is it?" was replaced by the more specific question, "What does a stellar spectrum look like?"

The connection between the appearance of a stellar spectrum and condi-tions on the star came later, when a group of theoreticians was able to inter-pret the observational work of the early spectroscopists. They asked, "What kind of a spectrum would a stellar surface with a given temperature and pressure produce?" This question can be answered by a calculation, and a comparison of the results of the calculation with the spectra of real stars made the determination of stellar temperatures a reality. Spectroscopic tech-niques also made it possible to determine the chemical composition of stars, although any kind of direct sampling of the gas in a star is now just a wild dream. The theoretical work, combined with the observations, answered the second big question, "What is a star?" The detailed interpretation of the message of starlight will be discussed more fully later in this book.

But the analysis of stars did not stop with a catalogue of stellar tempera-tures, pressures, and chemical compositions. With a partial understanding of the nature of stars, astronomers could ask the third big question, "How do stars evolve?" Analysis of stellar spectra showed that most stars were roughly the same size. Examination of stars in star clusters (Figure 1.1), where all the stars are the same age, provided additional insight into the in-trinsic differences among stars. Theorists would calculate what stars should look like at various evolutionary stages, and observers would then go out and look at real stars to check the calculations. The agreement between theory and observation is good enough so that we understand many, but not all, of the stages of a star's life cycle.

☆　A HIERARCHY OF QUESTIONS　☆

Human progress in understanding the nature of stars and stellar evolution was marked by a change in the character of the questions that were asked as research progressed. People sometimes wonder how a science can progress at all, when each new experiment or calculation leads to additional ques-tions. Does the endless parade of questions go on forever? The questions may continue, but their character changes as the science works its way down a hierarchy. As a science matures, the questions become more specific and pointed as they focus on more detailed properties of the objects of study.

In a science's early days, scientists attempt to sort out the overall proper-ties of the objects they study. Planetary science is still in this stage, since space probes and ground-based investigations continually uncover new and unexpected features on the surfaces and in the atmospheres of the planetary members of the solar system. The usual scientific acitvity in these early stages is the cataloguing and classifying of objects under study. Just as Annie Cannon catalogued stellar spectra in the early days of stellar astronomy, so are teams of planetary scientists naming, mapping, and cataloguing the different surface features of the inner planets of the solar system. The chief challenge here is to come up with a classification scheme that has some relationship to the intrinsic nature of the objects being studied.

A science matures when the questions proceed down the hierarchy toward more analytical ones. Theorists, after randomly searching for some pattern to the observations, come up with a *model* for the type of object they study. A scientific model is the centerpiece of a mature science. A model is a theoretical picture of what an object is made of and how its constituents interact with each other. The laws of physics can then be applied to the model, and some of its properties can be calculated theoretically. Theorists calculate what an observer should measure; if the model is a successful one, the theoretical calculations should be reasonably close to the observations.

When a mature science becomes more sophisticated, the analytical questions become even more detailed. Once the gross differences in stellar spectra were explained, attention could focus on some of the finer details. The model is refined when more and more physical effects are included in it. Better observations permit more precise comparisons of model and reality. This painstaking pursuit of details provides additional insight into the broad questions asked earlier.

It might seem that the hierarchy of questions could continue indefinitely, for researchers could seek to dot every *i* and cross every *t* in the astronomical landscape. However, when answers to questions cease to provide insight into broader issues, the level of interest in the microscopically fine details falls off, and researchers work on something else. The study of planetary motion is one example of a fully mature field of science. By 1930, the model for planetary motion had been checked quite thoroughly. Additional routine comparison of planetary positions with calculated positions no longer provided any new insights, and activity in the field dwindled away. Observations of planetary motion go on, but these investigations are generally designed to provide additional information bearing on other fields of science. The model for planetary motion has been checked out.

Once the right questions are asked, how are they answered? The key step in scientific research is the quantitative comparison of model with observation. Someone seeking to predict the future evolution of the sun, for example, must work with a believable model of the solar interior. This model is checked by quantitative comparison with observations. Is the sun, for example, as large as the model says it should be? Does the sun shine as brightly as the model dictates? Only when model and observation agree can the model be used to predict the future evolution of the sun.

The unique feature of science is that both the model and the measurements used to check it can be described in quantitative terms. We do not simply ask whether the sun is bright or dim, large or small; the model describes how bright and how large the sun should be, and the observer makes measurements to check the model. Any kind of quantitative measurement requires some kind of universally acceptable unit by which to measure it. The next section briefly reviews units of measurement and introduces a concept that is ubiquitous in the astronomical world, the concept of energy.

☆ **LENGTH** ☆

In this book I use the metric system, where the fundamental unit of length is the *meter* (m), a length of a little more than a yard. The prefixes of the metric system allow a number of other units to be introduced naturally: the *micron* (μ) or *micrometer* (μm), which is 10^{-6} m; the *centimeter* (cm), which is 10^{-2} m; and the *kilometer* (km), which is 10^{3} m, equivalent to 1.6 miles.

Some other units of length are used in astronomical work. Wavelengths of light are often expressed in *angstroms* or angstrom units (Å); 1 Å = 10^{-10} m. Astronomers measure distances between stars in *parsecs* (pc); 1 pc = 3.085×10^{18} cm. Stars in our part of the galaxy are, on the average, roughly 1 pc apart. One *astronomical unit* (AU) is the average distance between the sun and the earth, 1.495979×10^{13} cm.

Astronomers are at a disadvantage when it comes to measuring sizes, because most of the objects we are concerned with are too remote to measure with measuring sticks. The *angular diameter* of something in the sky, or the angle between light rays coming from either end of it, is determined by its *linear diameter* and its distance from the observer. Consider the sun and the moon, for example. Both are about half a degree across in the sky (Figure 1.4), but the sun really is much larger and farther away. If the angular size of an object is measured in degrees, one can express the numerical relationship between angular diameter, linear diameter, and distance as

$$\frac{\text{Linear diameter}}{\text{Distance}} = \frac{\text{angular diameter}}{57.295 \text{ degrees}} \qquad (1.1)$$

FIGURE 1.4 An illustration of the relationship between linear diameter and angular diameter. Although the sun and moon have the same angular diameter (0.5°), the moon's linear diameter is much smaller, and the moon is closer to the earth. The relative distances of the sun and moon are incorrect in the drawing; if they were shown to scale, the sun would be almost 10 meters from the earth.

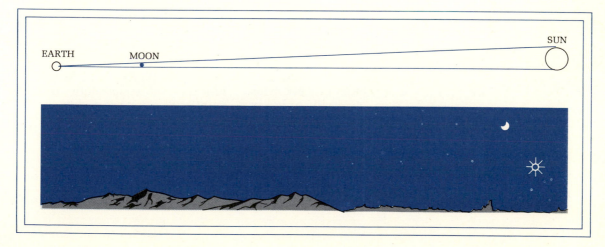

Thus an object with a large angular diameter can be either a modest-sized object close by (at a small distance) or a very large object far away (with a large linear diameter). Once the distance of an object is known, its diameter can be determined by measuring its size and plugging the numbers into equation 1.1.

☆   MASS   ☆

The metric unit of mass measurement is the *gram* (g); 1 gram is $\frac{1}{28}$ of an ounce, and 1000 grams or 1 kilogram (kg) is 2.2 pounds.

☆   ENERGY   ☆

Astronomical objects don't just sit quietly in the universe; they shine, giving off light that we later detect in our telescopes. *Energy* flows from the sun to the earth in the form of sunshine. This energy can be collected and made to do useful work; in one type of solar power plant, sunlight is used to heat water, creating steam, which can then be used to make a turbine turn and generate electricity. Much of cosmic evolution occurs when energy is transformed from one form into another.

Box 1.2 TEMPERATURE SCALES	CELSIUS	KELVIN	FAHRENHEIT
Boiling point of water	100	373	212
Room temperature	20	293	68
Freezing point of water	0	273	32
Absolute zero	−273	0	−459

Temperatures may be measured in degrees Celsius (°C) or Kelvins (K). The Fahrenheit scale, in common use in the United States will gradually be phased out as we go metric. (In many parts of the country, temperatures are now being given in both Fahrenheit and Celsius.) Zero degrees Celsius is the freezing point of water; zero Kelvins is the absolute zero of temperature, the point at which atoms and molecules are motionless. A comparison of various temperature scales is shown in the table above.

NOTE: This box should not be skipped unless you feel you understand temperature scales.

Numerous units have been used to measure energy; the one most as-tronomers use is the *erg*. An erg is a very small amount of energy; a 2-g in-sect crawling along at a speed of 1 centimeter per second (cm/sec) has an en-ergy of motion of 1 erg. Energies at the atomic level are often expressed in *electron volts* (eV); 1 eV = 1.60×10^{-12} ergs. Ergs and electron volts are the only two units of energy used in this book.

An object like the sun puts out a certain amount of energy every second. The *luminosity* of an astronomical object, measured in ergs per second, indi-cates the rate at which this object puts out energy. A 100-watt light bulb has a luminosity of 10^9 ergs/sec, miniscule when compared to the solar lumi-nosity of 3.9×10^{33} ergs/sec.

These units of measurement allow us to describe in quantitative terms what happens to a particular type of astronomical object. The sun can be described as a large ball of 1.989×10^{33} g of gas, 6.95×10^{10} cm in radius, emitting 3.9×10^{33} ergs of energy every second. But to analyze the physical conditions within this ball of gas, to try to model the various physical proc-esses taking place in an object like the sun, a theorist must understand how matter interacts with other matter. Since astronomers observe radiation emitted from planets, the sun, the stars, and the distant galaxies, the physi-cal principles governing electromagnetic radiation are also a part of our basic tools. These physical laws are the subject of the next two sections of this chapter.

★ 1.3 THE BEHAVIOR OF GASES ★

Most astronomical objects are gaseous. A gas consists of a bunch of isolated atoms flying around and colliding with each other. We can't study these atoms in detail because they are too small; but the collective effects of some 10^{23} atoms colliding determine the behavior of a liter of gas. What overall properties does a gas have?

The *temperature* of a gas is simply a measure of the energy that the atoms in a gas have when they move around. In a hot gas, the atoms zip around rapidly and collide frequently; in a cool gas, they move more sluggishly and collide less often. An absolute way to measure temperature is in *Kelvins* (K) or degrees *Celsius* (°C) above the absolute zero (Box 1.2). The energy of the atoms in a gas is proportional to the absolute temperature of the gas. Thus the atoms in the atmosphere of the brightest star Sirius (with a temperature of 9900 K) are almost twice as energetic as the atoms in the sun's atmosphere (with a temperature of 5760 K). At absolute zero, atoms would not move at all.

When atoms in a gas move around, they collide. If a gas is confined in a container, these atoms collide with the walls of the container and exert a *pressure*. A pressure of 75 pounds per square inch (psi) in the tire of a ten-speed bicycle comes from the relentless pounding of the air molecules on the walls of the inner tube. The amount of pressure depends on how many atoms are in a given volume of gas, on their speed, and on the density and temperature of the gas. The gas *density* is the amount of mass per unit volume; this property should not be confused with the total mass of the gas, which reflects how much stuff there is.

★   1.4 RADIATION   ★

Most of the energy flow from one astronomical object to another occurs in the form of radiation. Such an energy transfer occurs, for example, when someone is sunburned at the beach on a hot summer day. Energy generated deep within the heart of the sun moves across the 1.5×10^8 (150 million) km of space separating the sun and the earth as sunlight, a form of *electromagnetic radiation*. Electromagnetic radiation transfers energy from one place to another via the wavelike motion of electric and magnetic forces through space. Visible light, x rays, gamma rays, and radio waves are all forms of electromagnetic radiation (Table 1.1).

The energy contained in electromagnetic radiation comes in little packets called *photons*. Each photon carries a certain amount of energy. A high-energy x-ray photon carries far more energy than a low-energy radio photon, but both photons are forms of electromagnetic radiation. Photons of visible light have an intermediate amount of energy. The energy per photon in visible light can be related to the color of the light; red light has less energetic photons than blue light.

Thinking of radiation as a beam of bulletlike photons is only considering half the picture. Electromagnetic radiation also has wavelike properties,

TABLE 1.1   FORMS OF ELECTROMAGNETIC RADIATION

RADIATION TYPE	WAVELENGTH (Meters)	FREQUENCY (Hertz)	PHOTON ENERGY[a] (Electron volts)
Gamma rays	1.2×10^{-13}	2.5×10^{21}	10^7
X rays	1.2×10^{-9}	2.5×10^{17}	10^4
Ultraviolet	1.5×10^{-7} (1500 Å)	2×10^{15}	8.3
Blue	4×10^{-7} (4000 Å)	7.5×10^{14}	3.1
Red	6×10^{-7} (6000 Å)	5×10^{14}	2.0
Infrared	2×10^{-6}	1.5×10^{14}	0.62
Microwave	3×10^{-4}	10^{12}	0.004
Radio	1	3×10^8	1.2×10^{-6}

[a] Units of energy are discussed in section 1.2.

Source: Adapted from Harry L. Shipman, *Black Holes, Quasars, and the Universe.* Copyright © 1976 by Houghton Mifflin Company, p. 6.

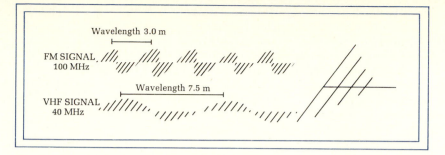

FIGURE 1.5   Electromagnetic waves of different wavelengths travel to the right toward a television antenna. The shorter-wavelength signal reverses direction more often as it passes by the antenna, and so it has a higher frequency.

which are best visualized by thinking of long-wavelength radio and television signals. Electrons in a metal television antenna—electrically charged particles in the metal making up the antenna—are pushed back and forth along the antenna by the electric and magnetic forces carried by the incoming electromagnetic radiation. The motion of the electrons creates an electric current in the antenna that is then amplified by the television set and converted into a picture with accompanying sound (Figure 1.5).

Different forms of electromagnetic radiation differ in the *frequency*, or number of times per second, that electric and magnetic fields reverse direction as the radiation passes by an antenna. (Frequency is measured in *hertz* or cycles per second.) A high-frequency radio wave, such as that used for FM signals, with a frequency of 100 megahertz (100 MHz or 10^8 cycles per second), reverses direction two and one-half times faster than a low-frequency VHF television signal of 40 MHz (4×10^7 cycles per second). Further, each form of electromagnetic radiation has a wavelength, equal to the distance between two places where the electromagnetic forces point in the same direction (Figure 1.5). The higher the frequency, the shorter the wavelength, because shorter waves pass by a standing observer more often (Figure 1.5 and Box 1.3). Thus a high-frequency FM signal has a shorter wavelength than a low-frequency television signal.

The frequency of an electromagnetic wave is also related to the energy carried by its photons. High-energy photons are characteristic of high-frequency radiation, and low-energy photons are characteristic of low-frequency radiation. The electromagnetic spectrum extends over a vast range of energy. It has been divided into a number of regions because different types of equipment detect and measure different types of radiation. The names, wavelengths, frequencies, and energies of these regions of the electromagnetic spectrum are shown in Table 1.1.

Most, but not all, of the electromagnetic radiation in the universe is produced thermally by hot objects. The nature of the radiation produced by an object can depend on the properties of the object itself, but it is useful to ask what kind of radiation is produced by an ideal radiator. Such an object is called a *black body* because it has no intrinsic color. A hot black body contains a number of atoms moving around at random, and the energy with which these atoms move around depends on the object's temperature. In hot objects atoms jitter around rapidly, while in cooler objects atoms lumber around slowly. Predictably, the photons emitted by hot objects are more energetic or bluer than the photons emitted by cold objects. Any black body has a characteristic spectrum of radiation that it emits, dependent only on the body's temperature. Figure 1.6 illustrates some representative black-body spectra. A black body emits a greater quantity of radiation, and the radiation becomes more energetic or bluer, as the body's temperature increases. While astronomical bodies are not black bodies, real astronomical objects share the same general properties by emitting more energetic radiation at high temperatures. Box 1.3 summarizes the radiation laws in equation form.

★ 1.5 A CASE HISTORY: THE PLANET ★ SATURN

The previous sections of this chapter have described, in rather general terms, the approach that astronomers take when attempting to find out what's out there. While the rest of this book applies this general approach to

FIGURE 1.6 Radiation emitted by black bodies, objects without intrinsic color, becomes both more intense and bluer with increasing temperature.

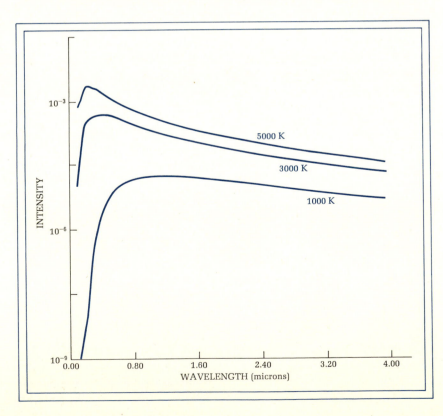

the entire universe, it is perhaps useful to apply these principles to one particular object in an effort to understand how astronomers use them. One of the most beautiful objects in the sky is the planet Saturn with its intriguing ring system (Figure 1.7, Color Photo 8). While Chapter 6 will discuss the giant planets in more detail and will describe some of the generalities of this section in more specific terms, it is useful to see how far the scientific approach (plus intuition) can go. We start with the first big question.

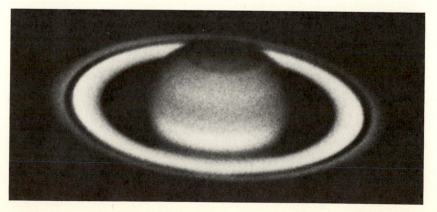

FIGURE 1.7   The planet Saturn. (Lick Observatory photograph)

Look, sometime, at the planet Saturn in the evening sky. It looks like a little light bulb, suspended in space, a few thousand kilometers from the earth at most. For over three millennia the human race viewed Saturn as just one of many celestial light bulbs, suspended in the great planetarium dome of the sky to amuse us. It was only the Copernican revolution of the sixteenth century, described in more detail in Chapter 3, which showed that, in fact, nontwinkling celestial objects like Saturn were worlds like our own. Careful observation showed that planets such as Saturn wander with respect to the starry background; they do not remain fixed in any one constellation. Careful observation of those motions indicated that the simplest, most believable model of planetary motion is one in which the sun lies at the center of the solar system and the planets orbit around it, held in their paths by gravity.

How was this model of the solar system, depicted in the second panel of Figure 1.3, proved to be correct? While its role in the Copernican revolution is not completely clear, the telescope made possible precise observations of planetary positions, and allowed the theorists and observers of the eighteenth century to check the predictions of a model based on Isaac Newton's law of gravity. Furthermore, the telescope allowed us to begin attacking the second big question, "What is it?" for the planet Saturn; it is thus worthwhile to take a few paragraphs to describe the telescope and its workings.

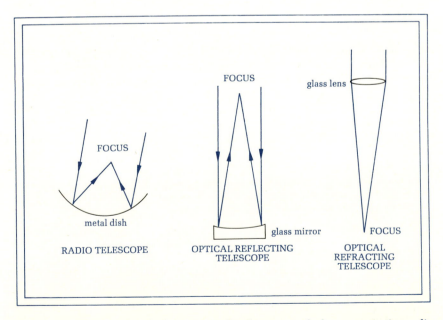

FIGURE 1.8 Three types of telescopes. Each telescope works by concentrating radiation at a point called the *focus*.

The primary function of a telescope is the collection and focusing of electro-magnetic radiation at a single point, called the *focus*, where some instrument, be it the eyeball or some other detector, can observe and record the position and intensity of the source of radiation. All telescopes—radio, optical, ultraviolet, x-ray, or whatever—work on the same general principles. Most optical telescopes use reflection to concentrate radiation, as Figure 1.8 illustrates. A large metal dish or mirror collects the incoming radiation from some celestial object and reflects it toward the focus of the telescope. The curvature of the reflector must be precisely figured so that all radiation hitting any part of the reflector is aimed at the same point. Because the focus of a *reflecting telescope* is in the telescope's beam, additional secondary mirrors are often used to reflect the beam of focused radiation off to one side. *Refracting telescopes* use a lens to concentrate radiation toward a focus; the word "refractor" is used because lenses bend (or refract) light. The human eyeball is a very small refracting telescope. The size of a telescope determines the total amount of radiation that the telescope can gather in; a large telescope can deposit more radiation at the focus than a small one can. Figure 1.9 shows the 90-inch telescope of the Steward Observatory, one of the big telescopes used in contemporary astronomical research.

The application of the telescope to astronomical research was one of the decisive factors in the confirmation of the Copernican model of the solar system. Copernicus hypothesized that Saturn was a giant sphere about $1\frac{1}{2}$ billion km from the sun, rather than a little light in the sky a few million kilometers away at most. The Copernican model of the solar system made some rather detailed predictions about the appearance of heavenly bodies. For example, the model stated that Venus should look like a crescent when it is bright and near the sun. Galileo used his telescope to observe Venus and verified that the model was correct. It was many observations like this that proved that the Copernican model was correct and that Saturn is $1\frac{1}{2}$ billion km away from both the sun and a planet near the sun, Earth.

☆ **WHAT IS IT?** ☆

This second big question also requires the use of the telescope if it is to be answered. In a telescope the planet Saturn shows up as a disk, the angular diameter of which can be measured. When Saturn is exactly $1\frac{1}{2}$ billion km from Earth, its angular diameter is 0.0046 degrees (°). Application of equation 1.1, the relationship between angular diameter and linear diameter, shows that Saturn is a huge ball with a radius of 60,400 km, some nine and one-half times the radius of Earth. A photographic plate placed at the focus of a telescope can record the image of Saturn in full detail and reveal it as a world with bands and rings (Figure 1.7). Yet this beautiful photograph by itself only tells us that Saturn is a huge striped ball with rings around it. What is this ball made of? We must appeal to another tool of observational astronomy.

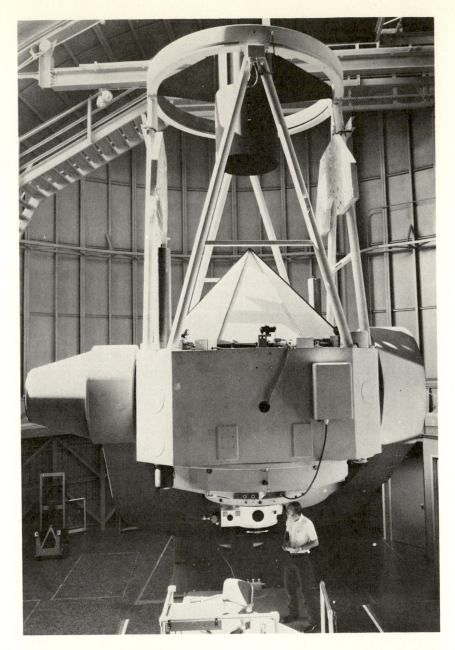

FIGURE 1.9 A large reflecting telescope, the 90-inch telescope of the Steward Observatory, University of Arizona. The mirror of this reflecting telescope is just above the observer's head; it is protected during the daytime by the conical cover visible at the bottom of the telescope. Light reflected from the main mirror can be brought to a focus at the top of the telescope, or it can be reflected back down to the bottom. (Steward Observatory photograph)

Telescopes can do more than just magnify objects and produce pretty pictures. Since they bring all the light hitting the telescope mirror to a focus, this light can then be analyzed. A *spectrograph* attached to the business end of a telescope focuses the light from the planet or star onto a slit and then photographs that slit in all the wavelengths of the electromagnetic spectrum. The spectrum of Saturn is shown in Figure 1.10.

The appearance of the spectrum of Figure 1.10 shows that the atmosphere of Saturn contains considerable quantities of methane gas. The spectrum of Saturn is crossed by numerous dark lines or *absorption lines*. Since the spectrum is simply a photograph of a vertical slitlike section of Saturn in various wavelengths, these absorption bands show that Saturn is reflecting very little radiation at certain wavelengths. A recollection of the properties of electromagnetic radiation (section 1.4) indicates that Saturn is trapping photons of very specific energies. How does this prove Saturn has methane in its atmosphere?

When sunlight impinges on the Saturnian atmosphere, it includes photons of many different wavelengths. Methane molecules in the Saturnian atmosphere are capable of absorbing certain energies by moving from a lower to a higher *energy state*. If a photon of the correct energy hits them, the methane molecules capture it and it cannot escape. Other photons are reflected off the cloud decks in Saturn's atmosphere and are reflected back to waiting observers on Earth. Since the methane molecules are only able to absorb certain energies, they leave gaps in the spectrum that are as recognizable as fingerprints. The evidence of the spectrograph shows that Saturn contains methane gas (Figure 1.11).

The spectrograph has proved to be an invaluable tool in analyzing the chemical compositions of astronomical objects. The energies that particular molecules absorb are unique, and the spectrograph allows us to determine the composition of distant celestial objects almost as accurately as we could if we were able to bring a piece of each into a laboratory for analysis.

FIGURE 1.10   The spectrum of Saturn. Shown is a photograph of a slit cutting from a piece of the image of Saturn in various wavelengths in the infrared part of the spectrum. The dark *absorption lines* cutting across this image indicate that less radiation in particular spectrum lines is reaching the telescope. The lines marked "O_2" are caused by the absorption of particular energies or wavelengths of radiation by the earth's atmosphere. The lines marked "CH_4" show where Saturn's atmosphere absorbs radiation. The faint trace above the Saturnian spectrum is the spectrum of Saturn's rings; no absorption lines from methane are present in this spectrum. (Yerkes Observatory photograph)

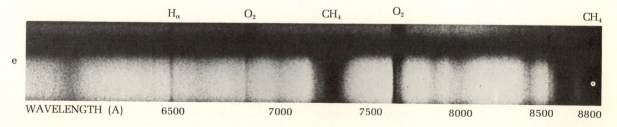

FIGURE 1.11 This schematic drawing illustrates how absorption lines are formed in the atmosphere of Saturn. Photons from the sun (black and color) strike the atmosphere of Saturn. Methane molecules in Saturn's atmosphere absorb photons of specific energies (color) because they can jump from certain energy states to others, soaking up specific amounts of energy. Photons with the wrong amounts of energy (black) are not absorbed and are reflected by the cloud deck. The absence of photons of specific energies in Saturn's spectrum (Figure 1.10) indicates that methane gas is present in Saturn's atmosphere.

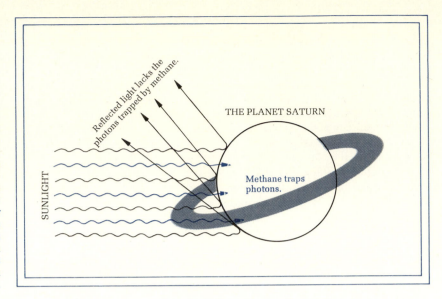

Reflected light lacks the photons trapped by methane.

THE PLANET SATURN

SUNLIGHT

Methane traps photons.

Spectroscopy can work its wonders throughout the electromagnetic spectrum. Atoms generally show their spectral signatures in visible light, which can penetrate the earth's atmosphere. Solids have spectral signatures that lie in the infrared part of the spectrum, and infrared radiation is blocked from some ground-based telescopes by the absorption pattern of the earth's atmosphere (Figure 1.12). Other parts of the electromagnetic spectrum must be observed from space, but they too have their own unique messages. High-energy x-ray photons, for example, provide information about hot million-degree masses of gas in the universe. Spectroscopy is one of the key tools used in answering the second big question, "What is it?" It tells us that Saturn is a giant ball of gas.

☆   THE RINGS OF SATURN   ☆

While spectroscopy has determined the nature of the planet Saturn, by itself spectroscopy cannot say much about the nature of the rings. The spectrum of the rings is shown as the faint spectrum above the planetary spectrum in Figure 1.10. No methane absorption lines are visible in the spectrum of the rings; the only spectral features in Figure 1.10 are absorption lines produced by the atmosphere of the earth. Another astronomical tool, theoretical modeling, provided the key step toward determining the physical constitution of the rings.

Saturn's rings were first discovered by Galileo, the first astronomer to use a telescope. His discovery is ambiguous, for he only wrote that Saturn seemed to be a "triple" planet, not a disk like the others. Christiaan

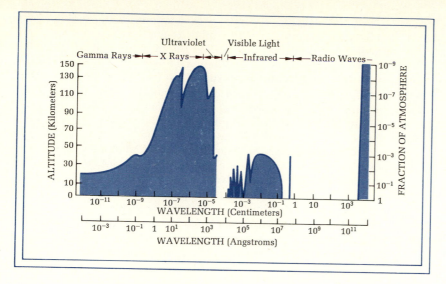

Huygens, working in the seventeenth century, constructed a better telescope and discovered the true nature of the rings as thin disks. But were they solid or were they swarms of small particles? Saturn is so far away that telescopes could not separate the individual ring particles; theoretical calculation was the only method of attack. James Clerk Maxwell, a Scottish scientist who later produced the complete theory of light as an electromagnetic wave, asked a critical question: If the rings were a solid disk, could they remain coherent or would they be broken up by the gravitational forces of the planet? Maxwell's calculations showed that the answer to this question was no, the rings must be swarms of small particles. Observations confirmed this conclusion, and infrared and radio observations of the 1960s and 1970s showed that the rings consist of small ice particles.

☆ HOW DID IT EVOLVE? ☆

Spectroscopy and theoretical calculations can provide information about the surface and rings of Saturn, but the bulk of its material is contained in the interior. Investigations of the surface only partially attack the problem of evolution. Most evolutionary questions are attacked by developing a theoretical model of an object, testing it by comparing it with observation, and seeing how it did and will evolve.

A theoretical model of Saturn as a ball of gas with the same composition as the sun fits the observations. If this model is correct, it indicates that Saturn will change very little in the future, and so the title of this section indicates that Saturn's past evolution is the interesting theme. One unique feature of Saturn's evolution that the model builders can understand is the existence of a ring system. Theoretical calculations of the gravitational forces on a Saturnian satellite in a close orbit around the planet indicate that if such a satellite were as close to Saturn as the rings are now, it would be pulled apart by the overwhelming gravitational forces of the planet. Perhaps an eleventh Saturnian satellite once existed and its orbit evolved into one that came too close to the planet, so that the satellite broke down into a swarm of moonlets that we now see as rings. More likely, the gravitational forces of Saturn were too strong to allow any satellite to form that close in, and the rings of small particles formed instead.

To a large extent, knowledge of the evolution of giant planets like Saturn is a matter for the future. These planets are very, very distant from the earth; a cloud the size of the state of Pennsylvania would be barely distinguishable on Saturn. In 1979, the Pioneer 11 spacecraft that flew by Jupiter in December 1974 will pass by Saturn, and if its cameras are still working we will obtain a closer view of this intriguing ringed planet. Two Voyager spacecraft, launched in late 1977, will rendezvous with Saturn in 1981. Planetary science is a young field; the investigation of planetary evolution is a matter for the future, especially for planets as distant as Saturn.

SUMMARY

Astronomical research proceeds by asking questions about the universe. While a researcher's curiosity can lead to many questions, detailed investigations are usually related to one of three big questions about an astronomical object: How far away is it? What is it? How does it evolve? A model of the object, based on a general idea of its physical makeup and the laws of physics, is compared to the observational evidence of the real world. If the model resembles reality, it provides some insight into the nature of the astronomical body under study.

This chapter has covered:

The main types of questions astronomers ask;

The laws of physics describing the interaction of matter and radiation in astronomical bodies;

The methods astronomers use for measuring the intensity of the radiation emitted by stars, planets, and galaxies; and

The application of these methods to the study of the planet Saturn.

KEY CONCEPTS

Absorption lines	Black body	Electron volt
Angstrom	Celsius	Energy
Angular diameter	Centimeter	Erg
Astronomical unit	Density	Focus
Atomic energy levels	Electromagnetic radiation	Frequency

Galaxies	Meter	Reflecting telescope
Gram	Micron	Refracting telescope
Hertz	Model	Spectrograph
Kelvins	Order of magnitude	Spectrum
Kilometer	Parsec	Stars
Linear diameter	Photon	Temperature
Luminosity	Pressure	Wavelength

REVIEW QUESTIONS

1. An astronomer seeking to learn as much as possible about the planet Jupiter might ask the following questions. Place them in the order in which they would be asked.

a. Is Jupiter made of gas or rock?

b. How far away is Jupiter?

c. What is responsible for the formation of clouds in Jupiter's atmosphere?

2. Can you think of an explanation for the maturity of the scientific field studying planetary motion and the relative youth of the study of planetary surfaces and atmospheres?

3. A Boeing 747 airplane is larger than a small Piper Navajo. If the two planes have the same angular size in the sky, which one is closer to the observer?

4. Blue stars have temperatures exceeding 10,000 K, red stars have temperatures of 2000 to 3000 K, and planets have temperatures of 40 to 800 K. Which of these objects would be observed by an ultraviolet astronomer? An optical astronomer? An infrared astronomer?

5. Why is the pressure in automobile tires higher after the car has traveled on an interstate highway for an hour, heating the air in the tires?

6. Explain why spectroscopic analysis of the radiation from the planet Saturn shows that methane is present in its atmosphere.

7. Why are large telescopes superior to small telescopes?

8. Would you need to use a large telescope to study radiation from the sun?

FURTHER READING

Some books describing recent astronomical research are:

Friedman, H. *The Amazing Universe.* Washington: National Geographic Society, 1975.

Golden, F. *Quasars, Pulsars, and Black Holes.* New York: Scribners, 1975.

Shipman, Harry L. *Black Holes, Quasars, and the Universe.* Boston: Houghton Mifflin, 1976.

★ CHAPTER TWO ★

THE CELESTIAL BALLET

KEY QUESTION: *How do the motion of the sun in the sky and the appearance of the various groups of stars change with the daily and yearly astronomical cycles?*

The aspects of astronomy that are closest to our daily lives are the motions of the sun from sunrise to sunset and from year to year and, perhaps, the appearance of different groups of stars or constellations (Figure 2-1) in the sky at different seasons. The sun's motion through the sky is the regulator of our weather patterns, and the duration of time between sunrise and sunset plays an important role in the vocations and avocations of many people. Familiarity with the appearance of the sky, which can provide satisfaction and recreation for a lifetime, comes from an understanding of the daily and seasonal motions of the sun and from an ability to recognize the patterns of the major constellations.

FIGURE 2.1 A photograph showing the constellation Sagittarius rising amidst the trees about 4 A.M. in mid-March. The drawing in the inset shows the appearance of the constellation, which is sometimes visualized as a teapot with a handle, spout, and lid. (Photograph by Helen K. Moncure)

Seasonal phenomena, such as the varying length of daylight and the changing appearance of the constellations, are caused by the rotation of the earth on its axis and its orbital revolution around the sun.

It is possible to visualize these phenomena by placing the earth at the center of a huge celestial sphere on which the stars move, even though the earth is only one small planet in the universe.

★ 2.1 THE SUN AND THE SEASONS ★

Astronomy is the oldest of the sciences. People of different cultures, seeking to understand the world they lived in, usually began by looking up at the sky and trying to understand the movements of the sun, the moon, and the stars. Many early astronomers sought to comprehend the passage of the seasons by constructing an accurate calendar, which required close observation of the motions of celestial bodies. They built huge observatories, where pairs of rocks, archways, hallways, or windows marked the locations in the sky of important celestial events. Such ancient observatories can be found all over the world: Stonehenge in England, the pyramids in Egypt, Cahokia Mounds near St. Louis, Missouri, and the Caracol Observatory in Chichen Itza on the Yucatan peninsula of Central America. The Polynesians wanted to understand the movements of the stars so they could find their way across the Pacific Ocean, navigating the thousand-kilometer stretches of ocean that separated inhabited islands.

The most apparent motion of celestial objects is the daily rising and setting of the sun, moon, and stars. This motion is caused by the rotation of the earth on its axis. Another piece of the celestial ballet is the variation of the sun's path across the sky from one season to the next, resulting in a change from the long days and late sunsets of summer to the short days and early sunsets of winter. The seasonal cycle is perfectly regular, repeating itself every year. The construction of an accurate calendar demands an understanding of the passage of the seasons and a precise measurement of the length of the year.

The earth is a poor vantage point if you seek to visualize the motion of the sun, because the apparent motion of the sun is caused by the motion of the orbiting, rotating earth. It is better, at first, to step back from the earth, this chunk of rock orbiting 150×10^6 km (150 million km or 93 million miles) from the ball of gas called the sun, and look at our solar system from afar. Two such views are shown in Figure 2.2. While the earth orbits the sun, it rotates on its axis. The daily (or diurnal) rotation of the earth causes daylight and darkness, since someone living at any one point on the earth is carried into and out of the sunlight by the earth's rotation. The axis of rotation is tilted relative to the earth's orbit around the sun, $23\frac{1}{2}°$ away from the perpendicular to the orbit.

This tilt of the earth's axis causes the seasons. Look first at the left side of Figure 2.2, showing the earth on December 21, in winter for the northern hemisphere. While the earth rotates, any individual is carried from daylight to darkness along the circles shown. Someone living in Washington, D.C.,

FIGURE 2.2 The earth, orbiting the sun, experiences seasonal variations in the duration of sunlight because of the tilt of its axis. Barrow, Alaska, (B) is north of the Arctic Circle. It suffers from perpetual darkness at the winter solstice and enjoys the midnight sun in summer. Washington, D.C., (W) is in the north temperate zone, having short daylight hours in winter and long ones in summer. The Galapagos Islands (G) are on the equator, and they, like all equatorial locations, experience 12 hours of sunlight every day of the year. (The earth and sun are not drawn to scale in this picture.)

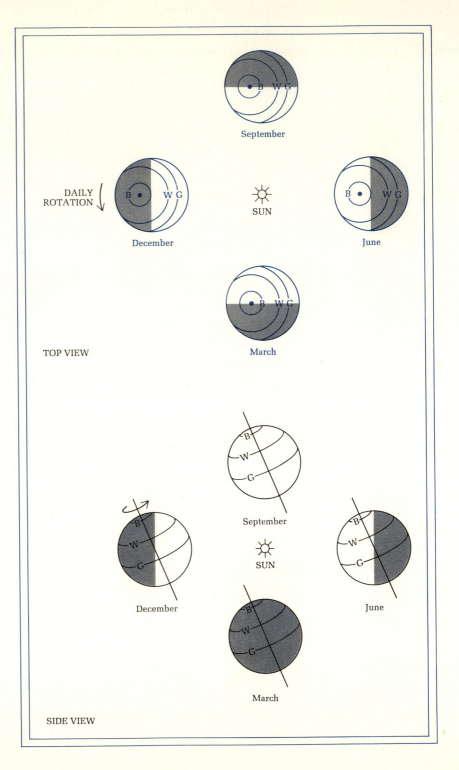

will spend more than half the daily journey around the earth's axis in darkness, so the days will be short and the nights long. Six months later, in summer, the earth will have orbited around to the other side of the sun, but its axis will still point in the same direction. Here, Washington will spend more than half the day in sunlight, and daylight will last into the evening. This variation in the length of the daylight hours is partly responsible for the coldness of winter and the warmth of summer.

The seasonal pattern of sunlight illustrated by Washington, D.C., is shared to a greater or lesser degree by most other places on the earth. There are a few locations that carry this pattern to extremes. Barrow, Alaska, is north of the Arctic Circle. On December 21, its motion around the earth's axis never carries it into daylight, and so it suffers perpetual darkness. The people of Tromsö, Norway, also above the Arctic Circle, refer to the time during which the sun never rises as "the tunnel." In summer, the seasonal cycle makes up for the cruel treatment of winter, and locations like Barrow and Tromsö never rotate out of daylight; the sun remains above the horizon even at midnight. A great contrast to Barrow is the Galapagos Islands, located on the equator, which have 12 hours of daylight every day of the year. South of the equator the seasonal pattern remains, but the daylight hours are long in December and short in June, producing Southern Hemisphere summer while the Northern Hemisphere experiences winter.

The variation in the length of the daylight hours is not abrupt; the long nights of winter gradually shift to the long days of summer. Figure 2.2 illustrates two extremes, called *solstices*. The *winter solstice*, occurring on December 21 or 22, marks the time when the Northern Hemisphere is tilted furthest away from the sun and suffers the longest nights and the least amount of sunshine. The *summer solstice* marks the longest day and occurs on June 21. In between the solstices there are two *equinoxes*, occurring when the daily paths of all points on the earth are half in daylight, half in darkness. The *vernal* (spring) *equinox* occurs on March 21 or 22, and the *autumnal equinox* occurs on September 23. (The terms "equinox" and "solstice" can also refer to the place in the sky that the sun occupies at these special times.)

The astronomical pattern of the seasons is responsible for the coldness of winter and the warmth of summer, but the meteorological extremes lag behind the solstices by about one month. For example, the Northern Hemisphere is generally coldest in January and hottest in July. This lag is caused primarily by the time it takes for the patterns of wind and rainfall (or snowfall) to respond to the changing solar radiation.

★ **2.2 STARS AND CONSTELLATIONS** ★

☆ **THE SEASONAL CYCLE** ☆

The seasons are marked not only by variations in the length of daylight and the weather, but also by a change in the groups of stars or *constellations* that are visible at sunset. Figure 2.3 again shows a view of the solar system; the focus is now on the relationship between the sun and the stars as seen from the earth. (The stars have been drawn much closer to the solar system than

FIGURE 2.3 The earth in orbit around the sun, showing the visibility of different constellations in the evening sky in different seasons. The observer shown at each season is looking overhead at sunset and sees different constellations at different times of year. (Details from H. A. Rey, *The Stars*, published by Houghton Mifflin Company. Copyright © 1952, 1962, 1967, 1970 by H. A. Rey.)

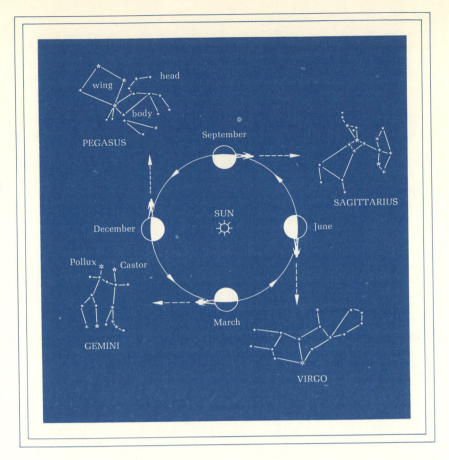

they really are in order to show both sun and stars in the same figure. If the stars were placed at their true locations relative to the scale of the sun and the earth, the figure would be several kilometers across.) Again begin by looking at the earth in its December position. An observer at the equator at sunset would look roughly overhead and see the constellation Pegasus, the Flying Horse. (Connect the stars as shown in the figure, and the group looks somewhat like a horse with a large, square wing. The Great Square of Pegasus is the constellation's most distinguishing feature.) Three months later, in March, the same person would see a different group of stars in the overhead position, the bright twin stars of Gemini, Castor and Pollux. As we follow the earth around in its orbit, it becomes apparent that Virgo is overhead at sunset in June, and Sagittarius is overhead in September.

Another way to visualize the changing pattern of the constellations is to imagine where the sun would be if it and stars could be seen at the same time. In March, at the vernal equinox, the Sun would be near the constellation Pegasus. The sun's location at this time is also called the equinox, and its position in the sky is quite close to Pegasus, in the rather obscure constellation Pisces. In June the sun is in Gemini, in September it is in Virgo, and in December it is in Sagittarius.

While the year passes, the sun traces out a path through the sky because our vantage point changes as the earth orbits the sun. This path is called the *ecliptic,* because eclipses of the sun and moon occur in that part of the sky. In the vicinity of the ecliptic are a group of twelve constellations, called the *zodiac:* Aries, Taurus, Gemini, Cancer, Leo, Virgo, Libra, Scorpius, Sagittarius, Capricornus, Aquarius, and Pisces. Some of these constellations contain quite a few bright stars and are easy to locate, while others contain only faint stars and can be quite difficult to see. The zodiacal constellations, particularly the brighter ones, are useful ones to know because the moon and planets are generally found in or near them. They are also significant in the pseudoscience of astrology (see the appendix to Chapter 3).

BOX 2.1 STAR AND CONSTELLATION NOMENCLATURE

The sky contains 88 recognized constellations (see Appendix B). Many of these groups of stars were first delineated in ancient times, and the fainter ones were named by eighteenth- and nineteenth-century astronomers. The naming of new constellations was sometimes controversial. For example, in 1790 the German astronomer Johann Bode formed the small constellation of *Friedrichs Ehre* (Frederick's Glory) in honor of Frederick II (the Great), the king of Prussia, by pilfering some stars from Andromeda's northern hand and moving her hand to a new celestial location. Most astronomers did not acknowledge the theft, and Friedrichs Ehre eventually disappeared from the star charts. Andromeda's hand, however, remained in the new position, and the small constellation Lacerta (Lizard) took over the contested place in the sky. In 1930 the International Astronomical Union put an end to these disputes of nomenclature and divided the sky into 88 constellations with carefully defined boundaries. Many small star groups disappeared in this division. A star grouping that is not one of the recognized constellations is called an *asterism.* Surprisingly, the Big Dipper is an asterism since it is part of the larger constellation Ursa Major, the Great Bear.

While the names of the 88 constellations can be dictated by international fiat, the names of individual stars are not as universally accepted. The brightest individual stars have names derived from Arabic star catalogues, and these names often refer to some anatomical detail of the constellation in which they are located. For example, Zubenelgenubi and Zubeneschemali are two stars that, in Arabic times, were the northern and southern claws of Scorpius. Since then, the constellation Libra has been formed, and they are now both part of it. All stars visible to the naked eye are given Greek letter designations in rough order of brightness; thus Zubenelgenubi and Zubeneschemali are known as Alpha and Beta Librae. Unambiguous star designations are given as numbers in the bright star (HR), Henry Draper (HD), and Smithsonian (SAO) star catalogues.

REMINDER: Almost all the boxes in the rest of this book present more technical details and, like this one, can be skipped on a first reading or in a fast-paced course.

☆ THE CONSTELLATIONS ☆

Many of us find knowing one's way around the nighttime sky to be a very satisfying skill. On a spring or summer evening, in the company of good friends, I have waited many times for the bright stars to come out at dusk, picking them out one by one and recognizing them when they appear. Later, the patterns of the major constellations emerge from the glow of twilight. It is very pleasant to recognize these patterns, seeing them as familiar stars rather than as little lights in the sky. Just a few minutes spent with the sky and a star chart can enable anyone to recognize the four or five bright constellations seen in each season's sky. There is no need to know all 88 constellations in order to know the sky, just as there is no need to memorize the location of every tiny street in order to travel comfortably around a city. The star charts in this book are intended to help you recognize the bright constellations, such as Sagittarius in Figure 2.1 or Orion in Figure 2.4.

Good starting points for constellation identification are the *circumpolar constellations*, those that never set from the view of Northern Hemisphere observers, located in the temperate zone between latitudes 30° north and 50° north. As Figure 2.5 indicates, the earth's rotation carries an observer's horizon around the sky, causing most stars to rise and set. The circumpolar stars just wheel around the sky. At 6 P.M. on December 1, for example, an ob-

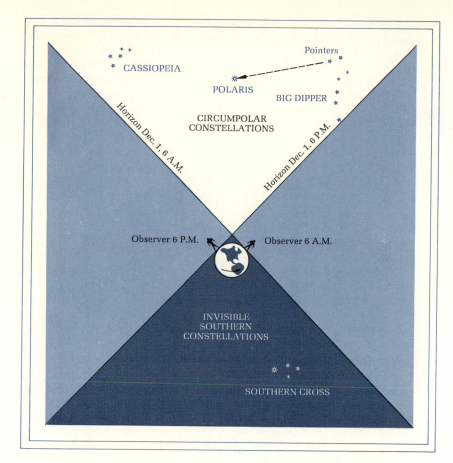

server at latitude 40° north will see the Big Dipper near the horizon. The Dipper, containing seven fairly bright stars in the shape of a dipper, is easy to find if it is not behind a tree or building or lost in the haze near the horizon. The Dipper never sets while the earth rotates; 12 hours later, at 6 A.M., it is high above the horizon. Thus the Dipper is circumpolar. There are also a number of stars in the southern sky that never rise for observers in the Northern Hemisphere; among these invisible southern constellations is the famed Southern Cross (Crux).

The two stars at the end of the Dipper's bowl can guide the eye to another landmark of the northern sky. Draw a line through them, follow that line halfway up the sky, and you will encounter Polaris, a moderately bright star with no other bright ones near by. The earth's axis points almost precisely at Polaris so that, to the unaided eye, it remains fixed in the sky while the earth rotates. Polaris, the pole star, always sits above the north point on the horizon, and is a useful directional indicator.

It is not difficult to identify the brighter northern circumpolar constellations. Any one of the star charts (Figures 2.6 through 2.10) will illustrate

37

where they are. Someone observing the after-dinner sky, at about 8 or 9 P.M., can usually find the Big Dipper, although it might be lost near the horizon in autumn. The Big Dipper's two end stars indicate Polaris and, under favorable conditions, the Little Dipper or Little Bear (Ursa Minor) may be visible. Opposite the Big Dipper is the W or M of Cassiopeia, another easily located constellation. Cassiopeia and the Big Dipper are landmarks that can indicate the position of the fainter circumpolar constellations.

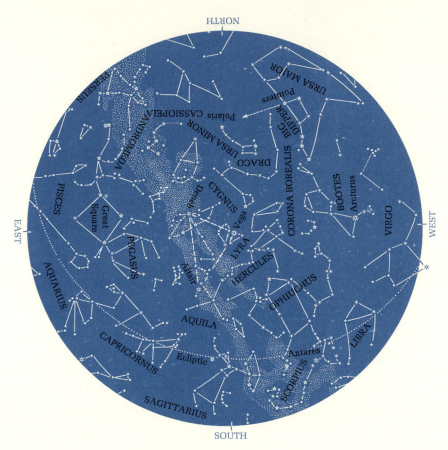

FIGURE 2.6   The late summer sky, on August 16 at 9 P.M.; September 1 at 8 P.M.; or September 15 at 7 P.M., standard time (or one hour later on each date by daylight-saving time).

Highlights Summer Triangle (Vega, Deneb, Altair), Sagittarius (low in the south), Scorpius (very low in the southwest)

Begin by looking straight overhead, where two very bright stars, one 30° to the east of the other, are visible. The eastern star is *Deneb*, the tail of Cygnus, the Swan; the western star is *Vega* in the inconspicuous constellation Lyra. These two stars, together with *Altair*, a bright star high in the southern sky, form the *Summer Triangle*. The collection of bright stars in the far south is *Sagittarius* (Figure 2.1), and low in the southwest is *Scorpius*. Scorpius, a spectacular constellation and one of the few that looks very much like its name, is fading fast into the southwestern haze at this time of year. It is best seen from the southern United States, where it rises quite high in the sky earlier in the summer and is well worth observing. In the west is *Arcturus,* one of the landmarks of the spring sky, and in the northeast are the precursors of autumn. (Adapted from H. A. Rey, *The Stars,* published by Houghton Mifflin Company. Copyright © 1952, 1962, 1967, 1970 by H. A. Rey.)

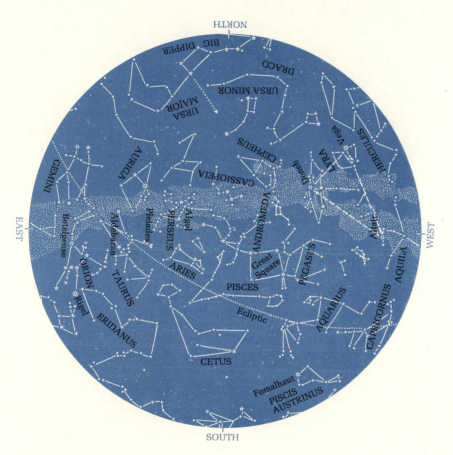

FIGURE 2.7   The late autumn sky, on November 15 at 9 P.M.; December 1 at 8 P.M.; or December 15 at 7 P.M., standard time.

Highlights Great Square of Pegasus (high, south), Perseus (high, east), Fomalhaut (low, southwest), Cassiopeia (overhead)

The autumn sky has a few bright stars, though the spectacular display of winter begins to rise in the east. The easiest group to locate is *Cassiopeia,* in the shape of a W or M overhead. East of Cassiopeia is *Perseus,* a loose jumble of rather bright stars that can, with some effort, be recognized as the Greek champion. High in the southern sky is the *Great Square of Pegasus,* four medium-bright stars set in a square with 15° sides. The summer stars, including the Summer Triangle, set in the west as the winter stars rise in the east. (Adapted from H. A. Rey, *The Stars,* published by Houghton Mifflin Company. Copyright © 1952, 1962, 1967, 1970 by H. A. Rey.)

FIGURE 2.8 The mid-winter sky, containing the best collection of bright stars, on January 16 at 9 P.M.; February 1 at 8 P.M.; or February 15 at 7 P.M., standard time.

Highlights Orion, the Winter Hexagon (Rigel, Aldebaran, Capella, Castor and Pollux, Procyon, and Sirius)

Orion, the Mighty Hunter, is the centerpiece of the winter sky, with Betelgeuse (pronounced "beetle juice") in his left shoulder and Rigel in his right foot. Orion's belt is easily recognized, containing the stars Alnitak, Alnilam, and Mintaka.

The other bright stars of the sky form the *Winter Hexagon,* starting with Rigel in Orion's right foot. Aldebaran, upward and to the right, is the brightest star in Taurus, the Bull. H. A. Rey's connection of the stars in this group forms a real bull, with the Pleiades representing one of the horns. The other horn can be connected to Auriga, the Charioteer, to form a pentagonal grouping of bright stars containing the bright star Capella. East of Auriga, northeast of Orion are the twin stars of *Gemini,* Castor and Pollux. South of Gemini is Procyon in Canis Minor, the Little Dog. Further south still is Sirius, the brightest star in the sky, in Canis Major, the Big Dog. Southern observers can see Canopus south of Sirius.

Sirius, Procyon, and Betelgeuse form an equilateral triangle with beautifully contrasting colors. Betelgeuse is red, Procyon is yellow, and Sirius is white.

Gemini and Taurus are both in the zodiac, so planets in the vicinity of these constellations can disrupt the patterns shown in this chart. (Adapted from H. A. Rey, *The Stars,* published by Houghton Mifflin Company. Copyright © 1952, 1962, 1967, 1970 by H. A. Rey.)

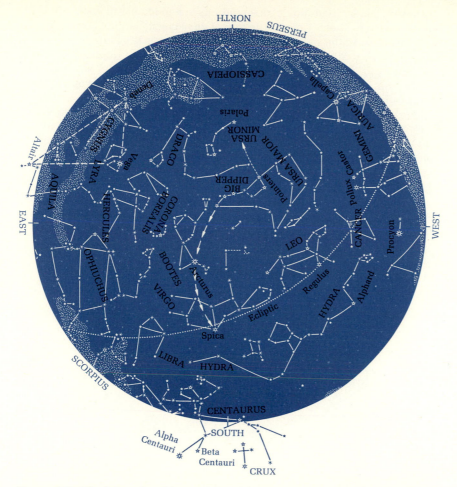

FIGURE 2.9   The late spring sky, on May 1 at 10 P.M.; May 15 at 9 P.M.; or June 1 at 8 P.M., standard time.

Highlights Big Dipper, Regulus (Leo), Arcturus (Bootes), Spica (Virgo), Scorpius rising (low southeast)

A good starting point is the *Big Dipper*, overhead. The pointers can easily be followed to Polaris, and can be followed backward to the bright star *Regulus* in the foot of *Leo*, the Lion. Leo is often difficult to locate completely under poor observing conditions, but the sickle shape of the stars north of Regulus is generally apparent. The *arc* or the Big Dipper's handle can be followed to *Arcturus*, a very bright star located in Bootes, the Herdsman. Bootes' head is rather amorphous, but east of the head is *Corona Borealis*, a notable circlet of bright stars. The sweep of the Dipper's handle can be carried beyond Arcturus to *Spica*, the main star of Virgo, the Virgin. Ruby red Antares, the Scorpion's heart, is rising in the southeast; later in the evening Scorpius will be a spectacular object.

Southern observers can take advantage of this season to note one of the landmarks of the southern skies, the *Southern Cross* (Crux). It is only in the tropics that this surprisingly small constellation is seen well. Alpha Centauri and Beta Centauri are visible southeast of the Cross; Alpha Centauri is the nearest star to us other than the sun. (Adapted from H. A. Rey, *The Stars*, published by Houghton Mifflin Company. Copyright © 1952, 1962, 1967, 1970 by H. A. Rey.)

These brighter circumpolar constellations are good ones to know because they are visible at any time of the year. Other constellations come and go with the seasons; the star charts on the last few pages illustrate this procession. Looking at the pictures in a book gives only a feeble impression of the true beauty of these constellations; a real knowledge of the sky and the satisfaction of feeling at home with the stars come only when you observe the stars in the night sky.

Figures 2.6 through 2.9 show the appearance of the sky at various seasons. They are drawn so that if you hold this book upside down over your head, and look up at a chart, the chart represents the sky at one of the times indicated. The brightness of the stars in the real sky varies considerably more than can be shown in a star chart, and many of the faint stars shown on the charts will not be visible from the light-polluted skies of some suburbs. The caption for each chart describes the landmarks of each seasonal sky, the first constellations to locate. Box 2.2 provides more technical details on the star charts.

FIGURE 2.10 The southern stars. Bright groups are *Crux* (the famous Southern Cross) and *Centaurus* (the Centaur), which are highest in the sky in June. Magellan's navigators placed a ship in a large part of the southern sky, but modern astronomers divided the ship into *Vela* (Sail), *Puppis* (Stern), and *Carina* (Keel). The remainder of the southern sky has few easily recognized bright constellations, but it does contain a couple of bright first-magnitude stars in dark regions of the sky: Achernar at the end of *Eridanus* (the River) and Fomalhaut in the Southern Fish. The two fuzzy spots outside the Milky Way are the two Clouds of Magellan, two irregular galaxies quite close to the Milky Way galaxy. (Adapted from H. A. Rey, *The Stars,* published by Houghton Mifflin Company. Copyright © 1952, 1962, 1967, 1970 by H. A. Rey.)

The star charts, representing the appearance of the entire sky, are meant to be held over your head (Figure B2.2A). It is best to try to

90° spans the distance from the horizon to the overhead point or *zenith*. It is easy to visualize how large 10° is by holding your fist out at arm's length (as shown in Figure B2.2C).

BOX 2.2 THE STAR CHARTS: TECHNICAL DETAILS

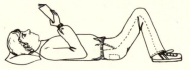

FIGURE B2.2A How to use the star charts.

pick out one constellation at a time. Proportions near the horizon, however, are distorted; it is impossible to represent proportions accurately when transferring a hemispherical sky onto a flat sheet of paper.

Different symbols are used to indicate the brightness of different stars in the sky. The magnitude scale used to describe stellar brightnesses numerically is described further in Chapter 9; here the largest symbols represent the brightest stars, with magnitude 0 the brightest and magnitude 5 the faintest (Figure B2.2B).

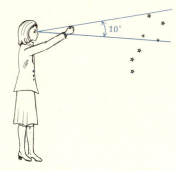

FIGURE B2.2C Distances between stars in the sky are measured in angles, and the degree is the unit for angular measurement. A clenched fist held at arm's length spans a 10° angle in the sky, equal to the separation between the two stars at the top of the Big Dipper's bowl.

FIGURE B2.2B The magnitude scale used in Figures 2.6 to 2.10. (From H. A. Rey, *The Stars*, published by Houghton Mifflin Company. Copyright © 1952, 1962, 1967, 1970 by H. A. Rey.)

Angular distances in the sky are measured in degrees. Thus,

The visibility of stars near the horizon depends on several factors. The horizon shown in these charts is the geometrical horizon at latitude 40° north; southern observers will see the southern stars higher up, and northern observers will see the northern stars higher up. Trees and buildings are, of course, appreciable obstructions. Dust particles in the air can brighten the sky near the horizon and obscure even the brightest stars from some locations. City lights and moonlight can impair the visibility of the fainter stars shown on these charts (see the discussion of light pollution in Chapter 7).

The preceding sections describe the appearance of the sun and the stars from the earth. Our vantage point was that of an omniscient observer looking down on the solar system from a great distance and determining what someone on this ball of rock, revolving around the sun and rotating on its axis, would see. While such a viewpoint can provide an adequate qualitative description of the motions of the sun and stars in the sky, it cannot provide quantitative details. An understanding of the operation of ancient astronomical observatories such as Stonehenge requires a more detailed description of celestial motions. It is difficult to make things precise when the situation is viewed from afar, and so we now shift to the vantage point of the earth.

Yet how can we eliminate the complexities of the earth's motion: its rotation on its axis and the revolution of the earth around the sun? As Galileo, the great philosopher-scientist of the 1600s, first realized, all motion is relative. Is the sky at rest above a spinning earth, or is the earth at rest under a rotating sky? It is possible to adopt either viewpoint and correctly describe the celestial motions. Although it is far easier to explain a rotating earth than to explain a rotating sky, during most of human history people thought the earth was stationary while the sky rotated. In the remainder of this section, we shall translate the rotation of the earth to a rotation of the sky.

The principal step in eliminating the earth's motion is to place the stars, planets, and the sun on a giant sphere surrounding the earth. This sphere is quite analogous to a planetarium dome (Figure 2.11), and it is called the *celestial sphere*. We now visualize the rotation of the earth as a rotation of the sphere itself; instead of having the horizon sweep around the sky, we pivot the sphere on the points where the earth's rotation axis intersects the sphere. These pivot points are the two *celestial poles*, north and south. The North Celestial Pole is within half a degree of Polaris, and so it is almost correct to visualize the entire sky wheeling around Polaris while the day passes. The line on the celestial sphere halfway between the celestial poles is called the *celestial equator*.

Spend some time with Figure 2.11. Different regions of the celestial sphere are shaded to highlight the differing behaviors of various stars. Objects near the North Celestial Pole, in the white region of the sphere, are circumpolar and never set. Objects in the region shown in color, near the South Celestial Pole, never rise above the horizon. Compare this view, centered on the earth, with Figure 2.5, which demonstrates the same ideas from the viewpoint of someone looking at the rotating earth.

The appearance of the celestial sphere is different at different latitudes. The bottom panel of Figure 2.11 shows a cross section of the celestial sphere. The height of the North Celestial Pole above the northern horizon is equal to the observer's latitude; someone at the North Pole will see Polaris directly overhead, while someone on the equator will see Polaris on the horizon.

The behavior of objects that are not circumpolar still depends on their location relative to the celestial equator. From the viewpoint of a Northern Hemisphere observer, an object north of the celestial equator will spend

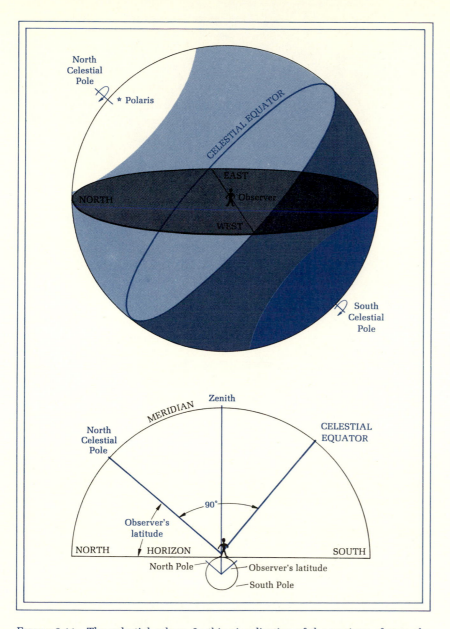

FIGURE 2.11 The celestial sphere. In this visualization of the motions of stars, the observer is at the center of a giant sphere that rotates around the earth once a day, pivoting around the North and South Celestial Poles. Midway between the poles is the celestial equator. Objects in the white region are circumpolar and never set; objects in the color region never rise (compare Figure 2.5).

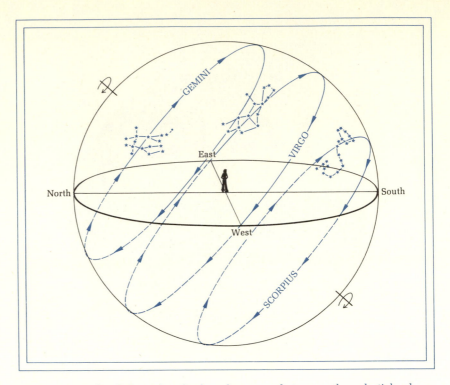

FIGURE 2.12 The daily paths of selected groups of stars on the celestial sphere. Gemini is north of the celestial equator, with north declination; it rises in the northeast, sets in the northwest, and spends more than 12 hours a day above the horizon (the solid part of an object's daily path indicates when it is above the horizon.) Scorpius, south of the celestial equator, rises in the southeast, sets in the southwest, and is above the horizon for less than half of its daily path. Virgo is on the equator, rises in the east, sets in the west, and is above the horizon for half of a 24-hour period. (Details from H. A. Rey, *The Stars,* published by Houghton Mifflin Company. Copyright © 1952, 1962, 1967, 1970 by H. A. Rey.)

more than half of its daily track above the horizon, remaining in the sky for more than 12 hours in the day. An object south of the celestial equator will remain in the sky for less than 12 hours each day.

An object's position in the sky relative to the celestial equator determines its daily behavior. A precise measurement of this position is given by an object's *declination*, its angular distance from the celestial equator, measured in degrees. Figure 2.12 explores this relation in more detail. Objects with north declination, such as the constellation Gemini at declination 25° north, rise in the general northeast direction, pass high in the northern sky, and set in the general northwest direction. Objects with zero declination are on the celestial equator, and they follow the same path as the constellation Virgo, rising in the east and setting in the west. Objects south of the celestial equator, such as the constellation Scorpius at declination 30° south, rise in the

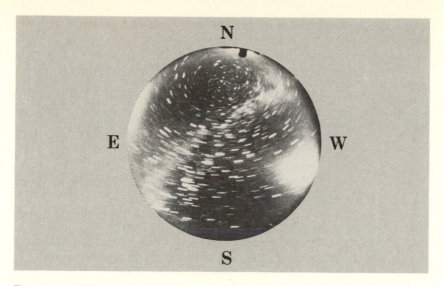

FIGURE 2.13   The Kitt Peak sky. A camera with a fish-eye lens recorded the star trails made while the stars wheeled around the sky for the 20 minutes of the exposure. The Milky Way is visible overhead. The patch of light in the west is the zodiacal light, sunlight reflected from dust particles circling the sun between the planetary orbits (see Chapter 7). In the east and northeast are reflections from the city lights of Tucson and Phoenix, Arizona. (© by the Association of Universities for Research in Astronomy, Inc. The Kitt Peak National Observatory.)

southeast, pass low in the southern sky, and set in the southwest. Objects with very large declinations are near the poles and are either circumpolar or invisible, depending on which pole they are near. The Big Dipper, with a declination of 60° north, is circumpolar; the Southern Cross, at 60° south, is permanently invisible to someone living at latitude 40° north.

Figure 2.12 presents another way of understanding the motions of the sun at various times of the year. While the sun travels around the ecliptic during the year, it passes through different constellations, some of which are shown in Figure 2.12. When the sun is in Gemini, in the month of June, it follows the same path in the sky as Gemini, rising in the northeast, passing high overhead, setting in the northwest, and remaining in the sky for more than 12 hours. June thus is a summer month. When the sun is in Virgo, in September, it rises due east, sets due west, and remains in the sky for exactly 12 hours. The autumnal equinox occurs in September. In winter, when the sun is in Scorpius, it follows the same path in the sky as Scorpius, rising in the southeast, following a low arc across the southern sky, and setting in the southwest. Note that, through the year, the sun rises and sets at different points on the horizon.

Figure 2.13 is a time-exposure photograph of the sky as seen from the Kitt Peak National Observatory; it was taken with a fish-eye lens that shows the entire sky. The image of any star is seen as a short line that defines the path of the star during the 20 minutes of the time exposure. A good review of the major points of this section is provided by Figure 2.13. A close examination of that photograph and a solid review of this section should result in an understanding of the motions of the stars shown in the photograph. Polaris scarcely moves; circumpolar stars describe paths that keep them above the horizon; southern objects just barely peep above the horizon. A visit to a planetarium will also reinforce the explanations of celestial motions presented in this section.

<h2>★ 2.4 ANCIENT ASTRONOMY ★</h2>

The motions of the sun and stars reflect the seasonal cycle in two ways. At different times of the year, different stars are visible in the evening sky. The declination of the sun changes as it moves along the ecliptic and as the year passes, and the points where the sun rises and sets change accordingly. An ancient astronomer could follow either or both of these changes to keep track of the passing of a year. Did ancient peoples keep track of the sky in this manner? The mute evidence of ancient observatories indicates that they did.

Perhaps the most famous ancient observatory is Stonehenge, located in the flat, dry Salisbury Plain in southern England (Figure 2.14). Stonehenge

FIGURE 2.14 Stonehenge, an ancient astronomical observatory. (The British Tourist Authority)

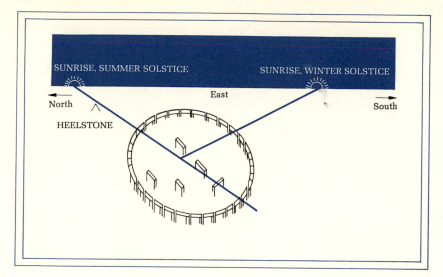

FIGURE 2.15 Schematic illustration of the use of Stonehenge as an astronomical observatory. On the summer solstice, when the sun rises as far to the northeast as it ever does, someone standing at the center of the monument will see the sun rise in the center of an archway, over an outlying stone called the "heelstone." It is possible that two of the archways are aligned wtih sunrise on the winter solstice.

was built in several stages, in a period that lasted a thousand years, by various cultures in the late Stone and early Bronze Ages. Even the earliest stage of Stonehenge, referred to as "Stonehenge I," has considerable astronomical significance. On the summer solstice, and only at that time of year, the sun rose over an outlying stone called the "heelstone" when viewed from the center of the monument (Figure 2.15). The Bronze Age Wessex people, who erected the huge stones of Stonehenge about B.C. 1900, preserved this alignment by centering the ring of huge stone archways on the center of the old monument, the circular ditch of Stonehenge I. It is also possible that one of the five inner Stonehenge archways was aligned with sunrise on the winter solstice. Unfortunately, the Stonehenge builders did not write. They left no instructions for using their observatory, and it is impossible to tell whether the winter-solstice alignment is just coincidence.

Many other ancient observatories were aligned with sunrise at noteworthy times of the year. Only the cultures of the Near East left any written records attesting to their understanding of astronomical phenomena, but the alignments of buildings or other monuments with significant sunrises and sunsets indicate that other cultures were also interested in the seasonal cycle. American Indians built a stone circle in the Big Horn Mountains of Wyoming and a circle of wooden posts at Cahokia Mounds, a settlement east of St. Louis, Missouri, and aligned both circles with sunrise at the solstices. Central American natives developed extremely complex calendars and equally elaborate observatories.

An indefatigable British investigator, Alexander Thom, has uncovered a number of megalithic observatories on the west coast of Britain that were used to track the movements of the moon as well as the simpler movements of the sun. These structures of standing stones, built before Stonehenge, marked the alignment of significant astronomical events with distant points on the horizon. Such an alignment allowed the declination of the moon to be measured extremely precisely by noting its rising point on the horizon. There is considerable evidence that these very early cultures, still in the Stone Age, understood the complex motions of the moon in such a way that they could predict lunar eclipses. Although they could neither read nor write, these people were able to understand the motions of celestial objects.

<h2>★ 2.5 PRECESSION OF THE EQUINOXES ★</h2>

Most of the material in this chapter has described the astronomical manifestations of the daily and yearly motions of the earth. These cycles repeat, day after day and year after year; the spring sky 100 years from now will look much the same as the spring sky of the 1970s, the sky shown in Figure 2.9. It would seem that the heavens are truly everlasting, that the astronomical clock will keep on cycling forever. Even apart from the overall effects of the evolution of the universe, there is a third cycle in the earth's motion that takes 25,700 years to complete. This cycle, the *precession of the equinoxes*, is so slow that it is scarcely perceptible during a human lifetime.

FIGURE 2.16 Over a 26,000-year period, the orientation of the earth's axis in space changes slowly, so that Polaris will not always be the pole star. This slow shift also results in a shift in the position of the celestial equator and the equinoxes.

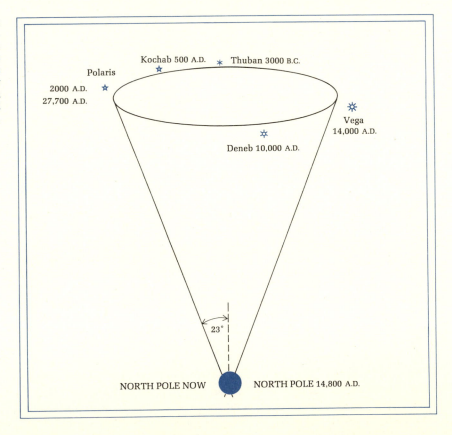

50

Precession occurs because the earth's pole very slowly changes its orientation in space (Figure 2.16). At present, the pole points to Polaris, making it the pole star; but this state of affairs will not last forever. The pole describes a very slow circle through the sky, and it will move away from Polaris in several hundred years. There will be no bright star near the pole until 10,000 A.D., when the North Celestial Pole passes within 2° of Deneb, the bright tail of Cygnus. Vega, another bright member of the Summer Triangle, will become the pole star in 14,000 A.D., when the cycle is half completed. The cycle will be completed 25,700 years from now, and a thousand generations hence stargazers will again have Polaris as their pole star.

Precession affects the position of some other celestial checkpoints—the solstices and the equinoxes. Recall that the equinox is the intersection of the ecliptic with the celestial equator, and the solstices are the two points on the ecliptic that are the farthest away from the celestial equator. The changing position of the celestial poles results in a change in the position of the celestial equator, the line in the sky 90° from the poles. Since the ecliptic remains fixed, the equinoxes slide along the ecliptic, following the 25,700-year cycle of precession. As a result the position in the sky that the sun occupies on the first day of spring, the vernal equinox, shifts very slowly. When Hipparchus made the first star catalogue in B.C. 130, the vernal equinox was in Aries; now it has slipped two-thirds of the way through Pisces. In 2600 A.D. it will enter the constellation Aquarius, defined by the constellation boundaries set down in 1930, and we shall enter the celebrated Age of Aquarius. (Astrologers define the constellation boundaries differently, and their Age of Aquarius may begin sooner than 2600 A.D.)

SUMMARY

The motions of the stars can be visualized in two ways. A distant view of the earth, rotating on its axis and revolving around the sun, can illuminate the gross features of the motions of the sun and the stars: Why are the nights long in winter and short in summer? Why are different constellations visible at different times of the year? However, it is necessary to adopt the viewpoint of an earthbound observer, seeing the celestial ballet played out on the stage of the celestial sphere, to understand the details of the celestial motions. The stars are the background, circling the earth once a day. The sun, progressing along the ecliptic, takes our seasons from winter solstice, through the vernal equinox and the hot days of summer, back to the autumnal equinox and to the chilly days of winter.

This chapter has covered:

The consequences of the earth's rotation on its axis and revolution around the sun in its orbit;

The phenomena connected with the seasonal cycle, including the variation in the length of the day and the different constellations visible at different times of year;

The appearance of the brighter constellations;

The use of the celestial sphere to describe these phenomena in detail;

The principle underlying ancient observatories such as Stonehenge; and

The precession of the equinoxes.

KEY CONCEPTS

Asterism	Circumpolar constellations	Precession
Autumnal equinox	Constellation	Summer solstice
Celestial equator	Declination	Vernal equinox
Celestial pole	Ecliptic	Winter solstice
Celestial sphere		Zodiac

REVIEW QUESTIONS

1. On December 21, El Centro, California, has 10 hours of daylight, and Yreka, California, has 8.4 hours of daylight. Which town is further north? In which town will daylight last longer in summer? Explain your answers.

2. If you were living in the Galapagos Islands (on the equator), would any stars remain above the horizon all the time?

3. If you wanted to choose a site for a telescope designed to survey as much of the sky as possible, would the best location be near the equator or near the pole?

4. Name the *best* time of year to observe the following constellations: Big Dipper, Orion, Pegasus, Cygnus, Gemini.

5. Redraw Figure 2.12, the celestial sphere, from the viewpoint of someone living at the North Pole.

6. Redraw Figure 2.12 from the viewpoint of someone living at the equator.

7. Sketch the probable appearance of an all-sky time-exposure photograph like Figure 2.13 if the camera were located near the equator.

8. Sketch an all-sky time-exposure photograph as it would appear from the North Pole.

9. Write a brief instruction manual explaining the use of Stonehenge as an observatory in determining the date of the summer solstice.

10. Will the Big Dipper always be circumpolar from latitude 40° north?

11. Suppose you were asked to verify the existence of precession using no equipment other than that available to pretelescopic astronomers: your eyes. You would have to observe the stars for a long period of time to see any observable effect. How would you go about this task?

Three good books on constellation identification are:

Menzel, D. H. *A Field Guide to the Stars and Planets*. Boston: Houghton Mifflin, 1964.

Rey, H. A. *The Stars*. Boston: Houghton Mifflin, 1970.

Whitney, C. A. *Whitney's Starfinder*. New York: Knopf, 1974.

★ CHAPTER THREE ★

PLANETARY MOTION

Not all objects in the sky share the slow seasonal motion of the sun and the stars. Someone observing the moon night after night will notice that it zips quickly from one zodiacal constellation to the next, remaining in each for two or three days while it passes through its phases. The starlike wandering planets are much less conspicuous than the moon, but careful examination of the sky will show that these, too, move with respect to the starry background. The sun, moon, and the planets form our solar system. Close analysis of the movements of the planets led sixteenth- and seventeenth-century astronomers to analyze the structure of the solar system and to show that the earth, too, is a planet in orbit about the sun.

FIGURE 3.1 Jupiter has four satellites that orbit around it in much the same way as the moon orbits the earth. Galileo Galilei, the first person to turn the power of the telescope on the heavens, discovered the four satellites of Jupiter and, in his notes, drew their motion about Jupiter. These four objects are found on different sides of Jupiter as they circle the planet. (Yerkes Observatory photograph.)

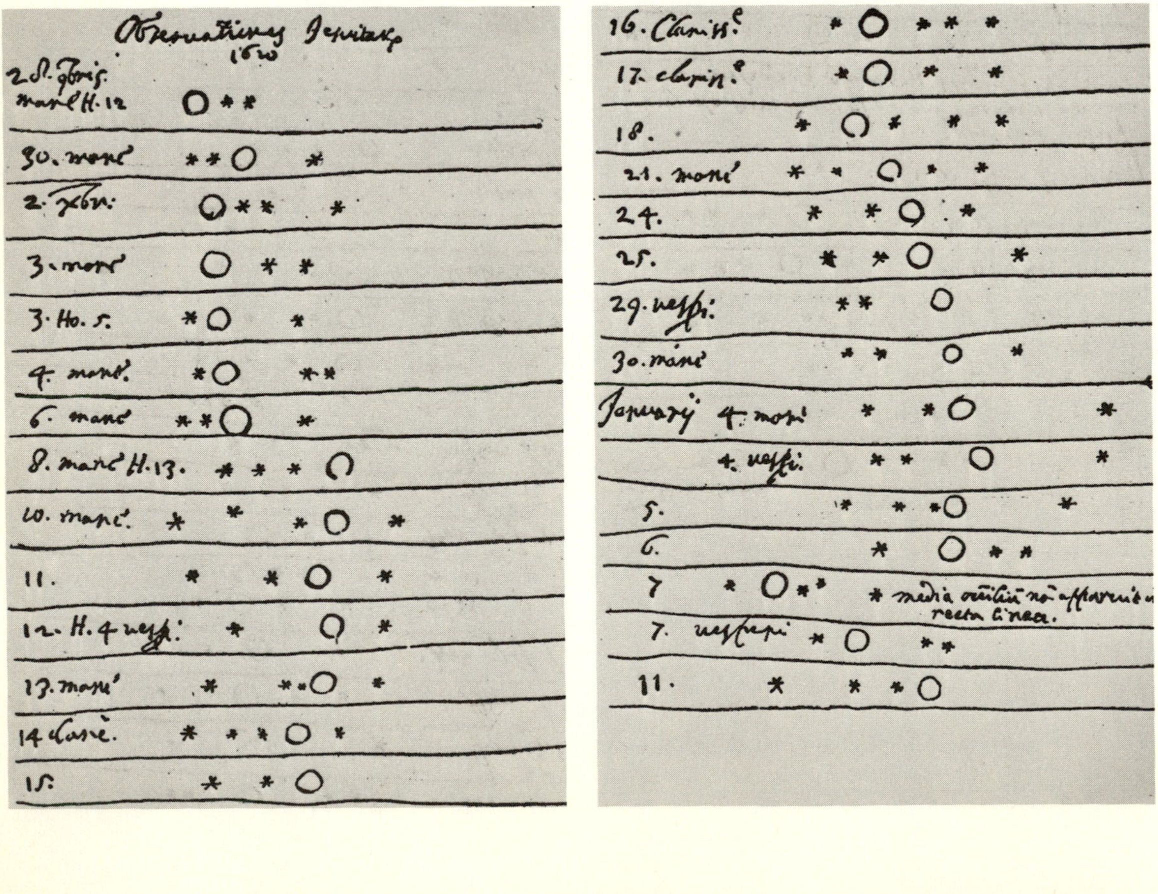

★ MAIN IDEAS OF THIS CHAPTER ★

The moon orbits the earth. Its phases and the tides are a consequence of its changing position in space.

Planets in the solar system orbit the sun; the major irregularities in their motion as viewed from the earth result from the motion of the earth around the sun.

Gravity is the force that keeps the planets in their orbits.

★ 3.1 THE MOON ★

The moon is one of the most poetic celestial objects. It does not overpower us with its brilliance as the sun does; it just softly illuminates the night, allowing people to take moonlit walks without stumbling around in the dark. For people who live in the far north, the moon provides a source of light for the long, dark, and cold winter nights. But the moon's motion is not simple. It goes through *phases;* sometimes it is a thin crescent, sometimes it is a full disk (Figure 3.2). Why?

FIGURE 3.2 The full moon rising among some saguaro cacti in Arizona. (Photograph by Tom C. Cooper from *Arizona Highways,* September 1974, p. 32.)

The moon is a *satellite* orbiting around the earth; the moon's changing position relative to the earth and sun generates its phases (Figure 3.3). The lunar cycle or month starts when the moon is in the same direction as the sun when viewed from the earth. When the moon is *new*, it is invisible since we on earth see only the dark side. A few days later, the moon has progressed in its orbit to the point where we can see a little sliver, or *crescent*, of the sunlit side of the moon. At this point in the cycle the moon is still close to the sun in the sky, and sets shortly after sunset. When the crescent moon is very young, shortly after new moon, the dark side can sometimes be seen because earthlight, sunlight reflected off the earth, illuminates the moon.

One week after new moon, the moon is at a right angle to the sun in the sky. Half of the part of the moon that we see from the earth is now sunlit, and half is in darkness. We see a half moon at this *first-quarter* phase, named because the moon is one-quarter of the way around its orbit. The first-quarter moon is a prominent feature of the late afternoon sky, since it rises at noon and sets at midnight.

A few days after first quarter, the moon has moved around in its orbit so that all but a slim crescent, in the part of the moon farthest from the sun, is illuminated. In this *gibbous* phase, the almost complete disk of the moon is quite bright, dominating the southeastern part of the evening sky. A few

FIGURE 3.3 The moon goes through phases when its position relative to the earth and sun changes during the course of a month. The sun is on the left, far away from the picture.

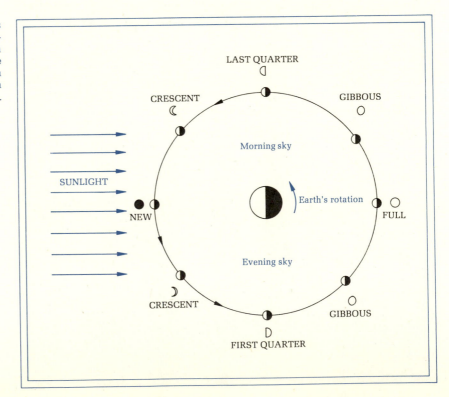

LAST QUARTER

CRESCENT

GIBBOUS

Morning sky

SUNLIGHT

Earth's rotation

NEW   FULL

Evening sky

CRESCENT

GIBBOUS

FIRST QUARTER

days after gibbous phase, the moon has reached the point in its orbit opposite to the sun. The entire sunlit side of the moon now faces the earth. This *full* moon rises in the east at sunset (Figure 3.2).

During the last two weeks of the lunar cycle, the moon retraces its phases in reverse order—gibbous, last quarter, crescent, and back to new moon again. The moon completes the cycle 29.53 days after new moon and is once again invisible. In the second part of the lunar cycle, the moon is primarily seen in the morning sky.

☆ **ECLIPSES** ☆

At new and full moon, it is quite possible that the moon, earth, and sun will fall directly in line; at such times an *eclipse* occurs. If it is precisely in line with the sun and the earth, the new moon will block the sun's rays from the earth; the sun will then be invisible, or eclipsed, to someone on earth located in the moon's shadow. Because the lunar shadow is at most only a few hundred kilometers across at the earth's surface, only a few people see the sun completely blocked out in a *total solar eclipse*. At other points on earth, the moon only partially blocks the sun's rays, and a *partial eclipse*, showing the sun with a bite taken out of it, is visible. At full moon, a *lunar eclipse* will occur if the earth's shadow falls on the moon, preventing sunlight from reaching the moon and illuminating the lunar surface. In most lunar eclipses, the earth's atmosphere refracts or bends enough sunlight so that the moon appears to be a dull red.

Eclipses can only occur when the earth, moon, and sun are all in the same plane (Figure 3.4). The moon's orbit is tilted relative to the plane of the earth's orbit and intersects it at two places called *nodes*. Therefore, the new or full moon must be at or near a node if an eclipse is to occur. Since the locations of the nodes do not change very much in a year's time, the new and

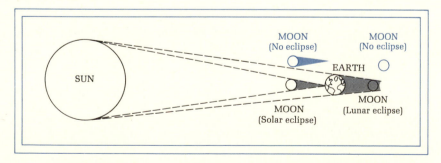

FIGURE 3.4 Solar eclipses occur when the earth, moon, and sun are lined up, with the moon in the middle. Lunar eclipses occur when the earth is between the sun and the moon. In this view from the plane of the earth's orbit, the color illustrations show the moon's position when the alignment is not precise and no eclipse occurs.

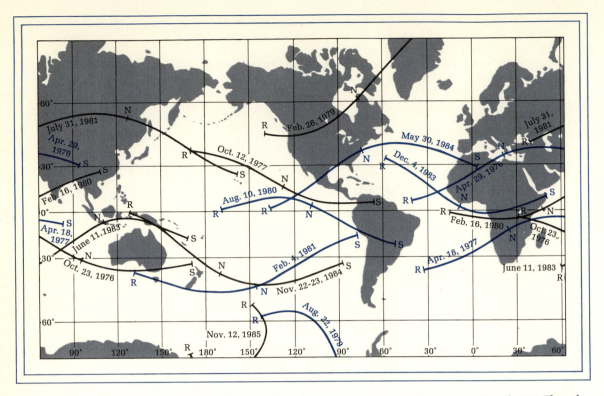

FIGURE 3.5 Paths of 15 total and annular eclipses between 1976 and 1985. The color tracks indicate annular eclipses, in which the moon's disk is not quite large enough to cover the sun completely and a ring of sunlight is visible around the moon's edge. On each track, the letters R, S, and N indicate sunrise, sunset, and noon positions. (The central lines have been plotted from data in the *Canon of Solar Eclipses*, by J. Meeus, C. C. Grosjean, and W. Vanderleen, 1966. Adapted from *Sky and Telescope*, August 1976, p. 96. Used by permission.)

full moons occur near nodes only twice in any single year—in eclipse seasons a little less than 6 months apart.

Eclipse seasons do not always occur at the same time of the year, because the nodes do shift very slowly around the moon's orbit, making a complete cycle every 18.61 years. The cyclical nature of the movement of the nodes, combined with the regular periodicity of the moon's motion, causes any individual eclipse to recur 18 years and $11\frac{1}{3}$ days later. The second eclipse will have approximately the same duration as the first one; if it is a solar eclipse the track of the moon's shadow on the earth's surface will have a similar shape, This period was named the "Saros" by the Babylonians, and their knowledge of the eclipse cycle allowed them to predict eclipses from past eclipse records. Our understanding of the motion of the moon is sufficiently complete that we can now prepare detailed maps showing just where on the earth the moon's shadow will fall during future solar eclipses (Figure 3.5). Observers in the shadow track will see a total eclipse; others will see only a partial eclipse.

Thanks to the predictability of eclipses, they are regarded as spectacular but routine events in the twentieth century. The aesthetic beauty of total solar eclipses moves people to travel thousands of miles to be at the small spot on the earth's surface where the sun is totally blotted out by the moon's shadow. At one time, professional astronomers made great efforts to be at those points on the earth's surface where total eclipses were visible, to photograph the outer solar atmosphere. Now there are other ways to observe the solar corona, and eclipse expeditions, though still undertaken, are not as critical for solar research. Figure 3.6 shows the solar corona during an eclipse; the black disk of the moon blocks out the searing light of the solar surface, allowing astronomers to take photographs of the outer atmosphere.

Before the lunar motion was understood as well as it is now, though, eclipses were regarded as terrifying events. The sun is the energy source for all life on earth; its permanent disappearance would mean disaster for the human race. Primitive and not-so-primitive cultures feared that the eclipse was a forerunner of doom; the French thought that an eclipse of the sun in

FIGURE 3.6 Photograph of a total solar eclipse that occurred on June 8, 1918, at Green River, Wyoming. The black disk is the back side of the moon, which covers the disk of the sun. Because the overpowering light from the solar disk is blocked out, the fainter outer atmosphere of the sun, or *corona*, is visible. (Hale Observatories)

August 1560 predicted a universal insurrection or a worldwide flood. A few people shut themselves up in fumigated caves for protection. Occasionally, eclipses have produced beneficial results, according to Herodotus' description of the eclipse of B.C. 585:

> The Lydians and the Medes had been at war for five years, with the fortunes of war alternating between the two camps, when during a battle in the sixth year the sky grew dark and night came abnormally. The combatants saw a celestial warning in this, put down their arms and made peace.[1]

There is growing evidence that many cultures, even those that had not developed a written language, understood the art of eclipse prediction. While it takes modern mathematics to predict the precise path of a solar eclipse, we can determine when the eclipse seasons will occur in a given year if we know the locations of the nodes. A number of megaliths, rings of large standing stones, on the west coast of the British Isles are built so that pairs of stones, a distant mountain or island, and the setting point of the moon at significant times in its orbit are aligned. The megalith builders left nothing to confirm that these stone rings were in fact used to observe the moon's motion and to predict eclipses. However, they are so accurately constructed that many people conclude they were used as observatories.

★ **3.2 THE MOON'S ORBIT AND GRAVITATION** ★

The moon has orbited the earth, following the same path, century after century. Romantic couples of classical Greece enjoyed the same moon as amorous twentieth-century Americans. What holds the moon in orbit? If something is dropped out of a second-story window, it falls; why doesn't the moon fall down?

Perhaps the question should be asked somewhat differently. Why doesn't the moon speed off into the depths of interplanetary space? The orbiting moon travels at a speed of 3700 km/h, about four times as fast as a typical airplane. What keeps it from flying off into the great beyond?

The stability of the moon's orbit is due to a delicate balance between two opposing tendencies: the moon's orbital motion and the gravitational attraction between the moon and the earth. The moon, instead of falling toward or flying away from the earth, falls around the earth, maintaining this balance between gravity and motion while it circles. There is no friction at the moon's altitude, no air resistance to slow the orbiting moon, so this balance will last forever. The moon will continue to circle the earth indefinitely along a trajectory much like its present one.

[1] Quoted in *The Flammarion Book of Astronomy*, ed. G. C. Flammarion and A. Danjon, trans. A. Pagel and B. Pagel (New York: Simon and Schuster, 1964).

Isaac Newton capped the Copernican revolution of the sixteenth and seventeenth centuries when he showed that gravitation kept the moon in orbit. The glue that holds the earth-moon system together is the same force that holds people, books, houses, and other objects to the earth's surface—the force of gravity. Newton described *gravitation* in a precise, scientific form by writing a quantitative description of the gravitational force: The force F between any two objects in the universe equals a constant G times the product of their masses M_1 and M_2, divided by the square of the distance D between them. Or, more concisely,

$$F = G \frac{M_1 M_2}{D^2}$$

(3.1)

This equation allows one to express the gravitational force between two objects as a number. The constant G is just a numerical conversion factor; if the two masses M_1 and M_2 are measured in grams and the separation D is measured in centimeters, then setting $G = 6.67 \times 10^{-8}$ results in a force F computed in dynes.

Some definitions are necessary. What's a dyne? A dyne is just a unit of measurement used for forces. What's a force? A *force* is an influence that tends to set an object in motion, a push or a pull. The Newtonian synthesis included a full description of the science of mechanics, a description of the motion of objects. A force F was defined to equal the mass M of an object times its acceleration a, so Newton's second law of motion reads

$$F = Ma$$

(3.2)

Acceleration a measures the change in an object's motion. If an automobile accelerates from 0 to 100 km/h (about 0 to 60 miles per hour) in 10 seconds, its motion has changed by 100 km/h in 10 seconds, for an average acceleration of 100 km/h per 10 seconds or 10 km/h/sec. This acceleration is produced by the force of the engine transmitted through the wheels.

The chain of definitions ends, and the Newtonian synthesis of celestial motion is complete. Choose two orbiting objects in the solar system. Equation 3.1 makes it possible to calculate the force between them. Equation 3.2 uses the force to produce an acceleration—the response of the two objects to the force between them. We can then use the tools of mathematics to calculate the resulting motion of the objects, their future orbits. The computers that calculate planetary orbits, showing where planets are in their orbits so that space probes can be aimed to arrive on target, work in just this manner: calculating forces, calculating accelerations, and then calculating orbital paths.

The theory seems complete, but is it correct? Will gravity actually work to keep planets in their orbits? Newton was able to verify that the same force that causes an apple to fall to earth holds the moon in orbit. Given the gravity we know on earth and the moon's distance from earth, the law of gravitation, equation 3.1, tells us that that force is just the force needed to keep the moon circling endlessly. Verification of this statement requires a numerical calculation, given in Box 3.1. You may or may not want to work through the details of the calculation, but the general trend of the argument is worth

**Box 3.1   The Moon
and the Apple**

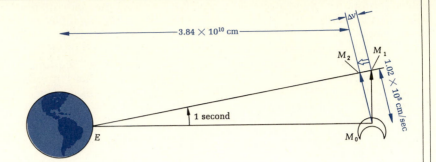

CONJECTURE: The force (gravity) that causes an apple to fall to earth is the same force that keeps the moon in orbit.

To prove this conjecture, apply Newton's law of gravitation, equation 3.1, to calculate the gravitational accelerations of the moon and an apple and see if the different accelerations agree with observation.

THEORY: A falling object with a mass M responds to gravity by accelerating. The gravitational force equals Ma, so we can write

$$F_{gravitational} = G \frac{\cancel{M}M_{earth}}{D^2} = \cancel{M}a$$

where D is the object's distance from the earth's center,[1] or

$$a = \frac{GM_{earth}}{D^2}$$

Thus the theory gives the ratio a_{moon}/a_{apple} as

[1] For this calculation we assume that the mass of the earth is sufficiently larger than the mass of the moon so that the moon can be considered as orbiting around the center of a fixed earth.

$$\frac{a_{moon}}{a_{apple}} = \frac{\cancel{G}M_{earth}/D^2_{moon}}{\cancel{G}M_{earth}/D^2_{apple}}$$

$$= \frac{1/D^2_{moon}}{1/D^2_{apple}}$$

The apple is 6378 km from the earth's center, and the moon is 384,000 km from the earth's center, so the ratio of accelerations should be

$$\frac{a_{moon}}{a_{apple}} = \frac{1/(384,000)^2}{1/(6378)^2}$$

$$= 2.76 \times 10^{-4}$$

OBSERVATION: If the theory is any good, the observed ratio of accelerations should equal the value just calculated. The acceleration of an object falling at the earth's surface is 980 cm/sec/sec. The illustration shows the moon moving around the earth. In 1 sec, it changes its direction of motion, as it falls toward the earth, by a small amount Δv. This Δv can be geometrically calculated, because the large triangles EM_0M_1 is similar to the small triangle $M_0M_1M_2$, and the ratio (short side)/(long side) is the same in the two triangles. In 1 sec, the moon travels through 1/(27.322 days × 86,400 sec/day) of its orbit. Its whole orbit

cont'd on next page

covers $2\pi \times (3.84 \times 10^{10}$ cm), so the moon's speed is

$$\frac{1}{27.322 \times 86{,}400 \text{ sec}} \times 2\pi \times 3.84 \times 10^{10} \text{ cm} = 1.02 \times 10^5 \text{ cm/sec}$$

Taking the ratio (short side)/(long side) of each triangle gives

$$\frac{\Delta v}{1.02 \times 10^5} = \frac{1.02 \times 10^5}{3.84 \times 10^{10}}$$

The change in velocity Δv is 0.271 cm/sec in 1 sec. Algebraically, $\Delta v/v = v/D$, so $\Delta v = v^2/D$. This change is just the acceleration a_{moon}. The ratio of the accelerations, $a_{\text{moon}}/a_{\text{apple}}$, that is observed is

$$\frac{a_{\text{moon}}}{a_{\text{apple}}} = \frac{0.271}{980}$$
$$= 2.76 \times 10^{-4}$$

in good agreement with the theoretical value.

some attention. A theoretical result is calculated and compared with an observational result. Quantitative agreement between the two indicates that the theory works, in this case at least. Similar agreement for a variety of calculations made with different objects in the solar system shows that the theory is a valid one.

★ **3.3 TIDES** ★

The *tides*, a familiar part of life along the seashore, are another manifestation of the force of universal gravitation. The moon pulls on all the constituent parts of the earth according to equation 3.1, pulling harder on the parts that are nearer to it. Most of the earth is a solid ball of rock; the different gravitational forces on it average out, so that the gravitational action on the whole mass of the earth is similar to the action on a point mass located at the earth's center. But the oceans are not attached to the rock, and the difference between the gravitational forces acting on the earth's center and the force acting on the ocean water produces the tides.

Figure 3.7 shows how the tide-producing gravitational forces work. The water in the Pacific Ocean is closer to the moon than the center of the earth is, and it is thus pulled toward the moon. The water in the Atlantic Ocean is further away from the moon than the earth's center, and consequently the earth is pulled away from it. If the continental land masses did not obstruct the free flow of water around the earth, there would be two high tides, at A and at P. As the earth rotated, these high tides would change position; at any particular location there would be two tides per day, one when the moon was overhead and one about 12 hours later.

But land masses do exist, interfering with the tidal forces. Instead of setting up a bulge like that shown by the dotted line of Figure 3.7, the tidal forces create currents that slosh around the ocean basins. These currents,

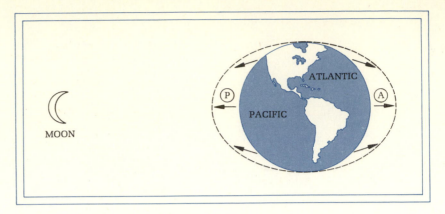

FIGURE 3.7   Tides are raised in the earth's oceans by the action of the moon. Ocean waters at *P* are pulled away from the earth's center because they are closer to the moon; the earth's center is pulled away from the Atlantic Ocean waters at *A*. If here were no land masses, high tide would always occur at the points on the earth nearest to and farthest from the moon. The presence of land masses, however, makes the tides more complicated.

whirling around bays and gulfs and traveling up rivers, produce the rise and fall of the water level at the seashore. Because the shorelines are irregularly shaped, the tides can be quite complicated. At the inner ends of deep bays or rivers, the high tides can lag behind the overhead passage of the moon by more than 7 hours. Funneling actions in places such as the Bay of Fundy can produce tides as high as 15 m (about 50 feet). The complex swirling currents in the Gulf of Mexico produce only one high tide per day, rather than the two that would normally be expected, and the two daily tides on the California coast are of rather unequal heights.

★   3.4 PLANETARY MOTION   ★

Universal gravitation has been remarkably successful in explaining the motion of the moon and its action in producing ocean tides. The moon's phases and eclipses are a consequence of its changing position in its orbit. Can gravitation explain the motion of the planets as well?

The planets, like the moon, do not remain in one place relative to the starry background. Each wanders through the zodiac while it orbits the sun, confounding stargazers who try to identify a zodiacal constellation but can't when a bright planet confuses the usual pattern of bright stars. Although there are nine planets in the solar system, naked-eye observers can see only five of them in the sky. Earth is not seen in the sky because we live on it, and Uranus, Neptune, and Pluto can be seen only with telescopes. The other naked-eye planets are conspicuous objects.

Two of the visible planets, Mercury and Venus, are *inferior planets*; their orbits are inside the earth's orbit (Figure 3.8). Because of their locations, an observer on the earth always sees them near the sun. Mercury is always within 28° of the sun, and Venus strays no further than 46° away. Conse-

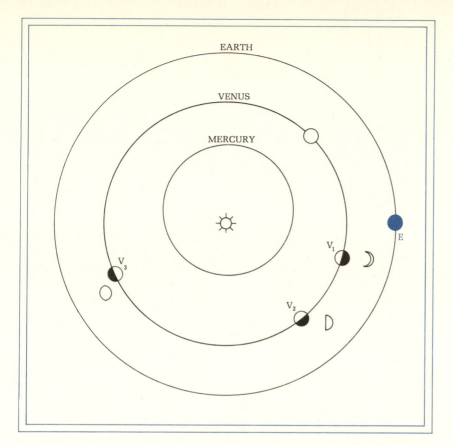

FIGURE 3.8 Orbits of the two inferior planets, Venus and Mercury, and Earth. These planets show phases and are never too far from the sun in the sky. With Earth at E, only a small crescent of the sunlit side of Venus is visible at V_1. Half the disk is visible at V_2, and almost all the disk is visible at V_3. (The planetary orbits are drawn to scale, but the planets are drawn far too large for their orbit sizes.)

quently, they either follow or precede the sun in its motion across the sky; they are seen in the eastern sky in the morning or the western sky in the evening. Mercury is fairly faint and can be quite difficult to spot, especially from urban areas where dust and haze enhance the brightness of twilight. Venus, in contrast, is quite difficult to miss when it is in a good location. At its brightest, it is the brightest object in the sky other than the sun and moon, and it is often far enough away from the sun that it is still high in the sky when twilight ends. Venus and Mercury, as seen through a telescope, both show phases because different fractions of the illuminated disk are turned toward the earth in different parts of the planets' orbits.

The other three naked-eye planets—Mars, Jupiter, and Saturn—move quite differently. Because their orbits around the sun are outside that of the earth, they are called *superior planets* (Figure 3.9). All superior planets travel eastward in the sky, wandering through the zodiac, and they can be

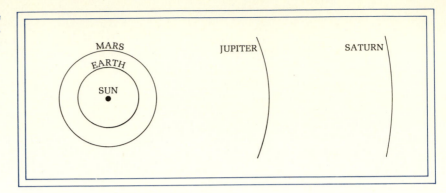

FIGURE 3.9 Orbits of some of the superior planets, drawn to scale.

found anywhere in the sky relative to the sun. Planets nearest the sun move fastest in their orbits. Mars, the speediest of the superior planets, takes less than 2 years to go around, while sluggish Saturn takes 30 years, an appreciable fraction of a lifetime.

For an interval of a month or so during the earth's year, a superior planet's eastward motion through the zodiac is interrupted when the earth's motion carries the earth past it (Figure 3.10). The fast-moving earth overtakes the superior planet, and for a while the planet appears to move backward, or westward, in the zodiac. This phenomenon is called *retrograde motion*. Retrograde motion is quite difficult to explain if the planets are visualized as bodies circling the earth, but it is a logical consequence of the architecture of the solar system with a central sun.

A picture of the solar system as nine planets orbiting one central body and held in orbit by gravitation produces the gross features of planetary motion that are apparent to a casual observer of the nighttime sky. The great difference in the behavior of inferior and superior planets, with one group always staying close to the sun and the other group moving through the zodiac with no concern for the sun's location, is readily explained by the position of the earth in the planetary system. A scientific theory must do more than just provide qualitative explanations, though. Detailed eclipse predictions require a precise analysis of the lunar orbit, and detailed calculations of planetary positions require a similarly precise application of the theory of gravitation to the planets. The detailed agreement between calculations and observations of planetary motion indicates that gravitation is a proper explanation of planetary motion.

★ 3.5 GRAVITATION IN THE SOLAR SYSTEM ★

☆ KEPLER'S LAWS ☆

Reflecting on the material of the last section for a moment, it becomes apparent that the planets move in a regular manner. Figures 3.8 and 3.9 show that the planets all move in orbits that are roughly circular. The time a planet

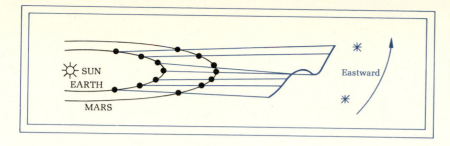

FIGURE 3.10 When Earth overtakes Mars in its orbit, Mars's eastward progression through the zodiac is temporarily interrupted; Mars moves backward, and retrograde motion occurs.

takes to make one complete revolution around the sun, or its *period*, is related to its distance from the sun. Speedy Mercury, the closest to the sun, only takes 88 days to go around once, while Saturn takes 30 years.

The seventeenth-century astronomer Johann Kepler realized that these regularities of planetary motion might be meaningful. He sought to analyze these regularities quantitatively, because a scientific explanation of planetary motion could not come from a general description of overall trends. He sought out the Danish astronomer Tycho Brahe so that he could make use of Tycho's life work, the most complete and accurate collection of planetary observations then in existence. He analyzed these observations and derived three empirical laws describing the way in which planets move:

1. Planets move in elliptical orbits with the sun at one focus of the ellipse.

2. Planets move in such a way that, in any given amount of time, the area swept out by a line joining the planet to the sun is constant. Planets thus move fastest in their orbits when they are closest to the sun.

3. The square of a planet's orbital period is proportional to the cube of its distance from the sun.

Some exploration of these laws is given in Box 3.2. The first two laws describe the shape of a planet's orbit and its orbital speed. Although the orbits are elliptical, on the scale of Figures 3.8 and 3.9 a circle is a good representation of an ellipse. The third law describes the relationship between the size of a planetary orbit and the planet's period. Anyone armed with *Kepler's laws* could accurately predict planetary positions. However, these empirical laws just systematize *observations* of planetary positions; they provide no understanding of the physical causes of the motion. Kepler succeeded in describing *how* planets move, but Isaac Newton completed the model of the solar system by describing *why* they move as they do.

The crowning achievement of Newtonian gravitation was its precise explanation of the laws of planetary motion. Box 3.1 shows that the moon stays in its orbit because of gravity, and Newton was able to prove that Kepler's three empirical laws are a natural consequence of Newtonian gravity. Newton's theory provided an understanding of planetary motion, a scientific model of the solar system. Such an understanding, usually provided by a complete scientific model, is the goal of scientific research.

69

Box 3.2   Illustration
of Kepler's Laws

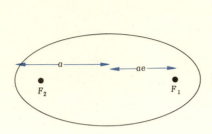
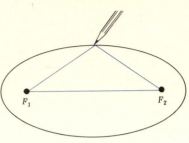

1. Planets move in elliptical orbits with the sun at one focus.
An ellipse, as shown above, is a curve drawn so that the sum of the distances from any point on the ellipse to the two *foci* F_1 and F_2 is constant. The distance from the center of the ellipse to each focus is *ae*, where *e* is the *eccentricity* of the ellipse and *a* is the *semimajor axis.*

2.   The law of areas
The diagram shows the area swept out by a planet in an elliptical orbit when it moves for a short period of time. If the area is

to be the same no matter where the planet is in its orbit, the planet must move more slowly when it is further away from the sun.

3.   Period² is proportional to (Orbit size)³
Some data on the orbits of three planets is shown in the . table below.
   Kepler's third law states the number in the third column, the square of the period, should be proportional to the number in the fifth column, the cube of the semimajor axis. It is, which shows that the theory is correct. The slight discrepancy for Jupiter arises because this simple formulation of Kepler's third law presumes that the planetary mass is negligible compared to the solar mass. This is not strictly true for Jupiter if agreement to five significant figures is required.

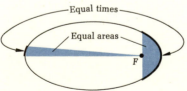

PLANET	PERIOD (years)	PERIOD²	SEMIMAJOR AXIS a (AU)[a]	a³
Mercury	0.24085	0.05800	0.38710	0.05800
Earth	1.00000	1.00000	1.00000	1.00000
Jupiter	11.862	140.71	5.2028	140.83

[a] An astronomical unit (AU) is the distance from the earth to the sun (section 1.2).

Box 3.3 Kepler's Third Law

Newton's law of gravity says that

$$F_{\text{gravitational}} = G\,\frac{M_{\text{planet}}M_{\text{sun}}}{D^2}$$

and by his definition of force

$$F = M_{\text{planet}} \times \text{acceleration}$$

The acceleration needed to keep a planet in a circular orbit is v^2/D (Box 3.1), where the radius of the orbit is D. The orbital velocity v is related to the planet's orbital period P by distance = rate × time, or (Orbital circumference) = $2\pi D$ = vP, giving $v = 2\pi D/P$.

Putting it all together gives

$$F_{\text{gravitational}} = G\,\frac{\cancel{M}_{\text{planet}}M_{\text{sun}}}{D^2}$$

$$= \cancel{M}_{\text{planet}}\,\frac{v^2}{D}$$

$$= \frac{\cancel{M}_{\text{planet}}}{D}\left(\frac{2\pi D}{P}\right)^2$$

Use a little algebra and multiply each side by P^2D^2 to get

$$[P^2\cancel{D^2}]\left[\frac{GM_{\text{sun}}}{\cancel{D^2}}\right]$$

$$= \left[\frac{(2\pi)^2D^2}{\cancel{D}\cancel{P^2}}\right][\cancel{P^2}\cancel{D}D]$$

Cancel as shown, and Kepler's third law is the result:

$$P^2GM_{\text{sun}} = (2\pi)^2D^3$$

(Here P is in seconds, M in grams, and D in centimeters.)

The mathematics involved in deriving some of Kepler's laws from the law of gravity is rather intricate, but the third law can be explained at least qualitatively by intuition. A planet moves in its orbit because it falls around, rather than toward, the center of attraction. A planet close to the sun moves faster because gravity is stronger, and the planet must zip around in its orbit to avoid falling into the sun. A faraway planet such as Saturn experiences a much weaker gravitational force than a close planet such as Mercury, and it can move more slowly in its orbit. If it were to move too fast, the sun's gravity would not be strong enough to hold onto it, and it would fly off into space. Box 3.3 provides a more quantitative derivation of Kepler's third law for the case of circular orbits.

The model of the solar system is essentially complete. The sun sits in the center, and the planets, a number of small balls of rock and sometimes gas, circle around it in their orbits. The shapes of the planetary orbits and their orbital speeds are all determined by one simple law: Newton's law of gravitation. The complex motions of the sun, moon, stars, and planets, as we see them from the earth, disappear; what remains is the elegance of Newtonian gravitation working on a system of nine planets, one of which we live on.

The process of confirming the Newtonian model of the solar system did not stop with the confirmation of Kepler's laws but rather continued through the eighteenth and nineteenth centuries. So far, in our model, we have assumed that the sun is the only source of gravity in the whole system. This approximation was a good starting point, one that allowed Newton to verify his theory without becoming lost in a mathematical morass. But each planet is attracted to all the others by gravity, too. After Newton, mathematicians developed tools that could describe these other planetary gravitational forces; once again, the refinements of the model agreed with observations. Most scientific work consists of such theoretical refinements. On rare occasions, such comparisons are persistently unfavorable, and a continuing mismatch between theory and observation can be a cause of a fundamental change in the theory, or a scientific revolution.

Perhaps the most notable result of the calculation of detailed effects of Newtonian gravity was the discovery of the planet Neptune in 1846. Uranus, the first "telescopic" planet, had been accidentally discovered by William Herschel in 1781 while he was searching for something completely different—faint companions to bright stars. The new planet was carefully followed for some years, and the French astronomer Alexis Bouvard noticed that observations of Uranus did not match calculations. While the years passed, the planet began to deviate more and more from its calculated path. A few people thought it was time to throw out the theory, arguing that Newton's simple law of gravity described by equation 3.1 was not valid in the outer reaches of the solar system, where Uranus orbits. However, the anomaly did not require that the theory be eliminated. J. C. Adams and Urbain Leverrier simultaneously followed up several suggestions that an undiscovered planet was supplying the extra force pulling Uranus out of its orbit. Leverrier, by good fortune, communicated his results to the Berlin Observatory just after that observatory had completed a chart of the region of the sky containing Uranus. J. Galle, at Berlin, examined the sky and found a new planet, Neptune, right where Leverrier and Adams had calculated that it should be. The discovery of Neptune illustrates that even a significant addition of our model of the solar system does not necessarily upset the underlying theory.

Newton's theory of gravity was changed in a more fundamental way by the work of Albert Einstein in the early twentieth century. Einstein's theory is significantly different from Newton's only when very strong gravitational forces come into play. The two theories differ only slightly in their description of the motion of objects in the solar system, and these minor differences are continually being used to test Einstein's theory against its rivals.

Our model of planetary motion in the solar system is now not only complete but also well-tested. Most astronomers do not bother with routine measurements of planetary positions, since further comparison of theory and observation is not likely to provide any fundamentally new insights. A few special, high-precision observational programs do continue in an effort to test Einstein's theory and assist in the aiming of space probes, but these projects are not directed toward the overall confirmation of the model of planetary motion.

A comparison of the Newtonian model of the solar system with its predecessor, the geocentric universe of Ptolemy (section 3.6), illustrates the value of a scientific analysis of the make-up of the universe. Newton transformed a mysterious heaven into something that people could understand and analyze. In the century after Newton, the philosophers of the Enlightenment explored the implications of the major discovery of the Copernican revolution—the fact that we live on one of nine planets circling one of many thousands of stars in the universe.

★ **3.6 THE COPERNICAN REVOLUTION** ★

A glance at the sky presents the very clear impression that the sky is a giant bowl, and that the sun, moon, stars, and planets are all circling around the earth, located at the center of this bowl. This very natural conception governed the first human attempt to decipher our place in the universe: the geocentric universe of classical civilization. This theory is usually associated with the name of Claudius *Ptolemy*, an ancient intellect who lived in Alexandria in the second century after Christ. Ptolemy, summarizing the ancient world view in his astronomy textbook *The Almagest*, actually added a number of refinements to an earlier theory. Most of the model had been worked out several centuries earlier by Greek thinkers, notably Aristotle and Hipparchus.

The Ptolemaic world view saw the universe as a gigantic planetarium.[2] The motions of celestial objects were produced by a system of wheels within wheels in the sky. These wheels were necessary to explain irregularities such as retrograde motion; by Ptolemy's time, after centuries of accumulated observations, the system's insistence on circular motion meant that well over 30 wheels were needed to explain the observations. Ptolemy himself did not believe that there were in fact 30-plus wheels up there in the sky. Rather, the whole system was regarded as a computing device that human beings could use to predict planetary positions. The reality of planetary motion was something known only to God, and not for mere mortals to try to understand. Ptolemy states:

> Let no one, seeing the difficulty of our devices, find troublesome such hypotheses. For it is not proper to apply human things to divine things nor to get beliefs concerning such great things from such dissimilar examples. . . . But it is proper to try and *fit as far as possible* the simpler hypotheses to the movements in the heavens, and if this does not succeed, then any hypotheses possible. Once *all appearances are saved* by the consequences of the hypotheses, why should it seem strange that such complications can come about in the movements of heavenly things?[3]

[2] A planetarium is a hemispherical theater with a projector at the center. The projector displays the stars as they appear in the real sky. A planetarium should not be confused with an observatory.

[3] C. Ptolemy, *The Almagest*, trans. R. C. Taliaferro, *Great Books of the Western World* (Chicago: Encyclopedia Britannica, vol. 16, 1952), p. 429. (Emphasis added.)

The Ptolemaic universe endured for a thousand years. But in the Renaissance, a Polish nobleman, *Nicolaus Copernicus*, began to observe the stars and realized that the Ptolemaic theory did not provide accurate calculations of planetary positions. His work began the Copernican revolution that ended with Newton, the essential steps of which are summarized in Box 3.4.

The Newtonian universe that emerged from the Copernican revolution was not content with merely "saving appearances." It provided a physical explanation of planetary motion. But only one layer of the mystery had been uncovered, for the nature of gravity itself still remains to be discovered even now. How does the moon know that the earth is there? How is the gravitational force transmitted from one object to the other? Newton expressed both the achievements and the limitations of his theory of gravity:

> Hitherto we have explained the phenomena of the heavens and our sea by the power of gravity, but have not yet assigned the cause of this power. . . . But hitherto I have not been able to discover the cause of gravity from phenomena, and I frame no hypotheses, for whatever is not deduced from phenomena is to be called an hypothesis; and hypotheses, whether metaphysical or physical, whether of occult qualities or mechanical, have no place in experimental philosophy. . . . And to us it is enough that gravity does really exist, and act according to the laws that we have explained, and abundantly serves to account for all the motions of the celestial bodies, and of our sea.[4]

The contrast between Newton and Ptolemy is the contrast between a pre-scientific world view and a scientific one. Ptolemy had a limited vision of the universe because he did not have a tradition of astronomical observations to draw on. He was content to describe appearances as well as he could, and he regarded the real model of the solar system as an idea that mere humans could never comprehend. The distinction between divine things and the realm of human inquiry was very sharply drawn. Newton, in contrast, lived at the beginning of a scientific age. Although not all the problems had been solved, human reason could produce a viable model for the solar system, a model that contained the causes of planetary motions. The scientific process of computation and observation could be applied to this model, and Newton's successors completed the job of matching model and reality.

APPENDIX: ASTROLOGY

One relic of Ptolemy's theory of the universe is astrology, which, unlike Ptolemy's description of planetary motion, is still with us. Astrology claims that the positions of the sun and planets in the heavens at the moment of a person's birth allow an astrologer to predict the person's fortune. Astrology has nothing whatever to do with astronomy, but as far as public exposure is concerned, astrology seems to receive as much attention as astronomy. (Sad

---

[4] I. Newton, *The Principia*, trans. A. Motte, rev. F. Cajori (Berkeley: University of California Press, 1966), pp. 546–547.

Box 3.4 **Personalities of the Copernican Revolution**

Nicolaus Copernicus (1473–1543) was responsible for moving the earth from its position at the center of the geocentric universe. He insisted on circular motion, in the same way as Ptolemy did; as a result, his world system became almost as cumbersome as Ptolemy's, with wheels within wheels explaining minor irregularities in planetary motion. He nevertheless realized and expressed the key fact: Retrograde motion is due to the earth's motion around the sun. His fame is due to those who built on that seminal idea to complete the revolution.

Tycho Brahe (1546–1601) did not accept Copernicus's assertion that the sun was the center of the planetary system. He was the first to realize that precise observations of planetary motions would point the way to the true model for the solar system, and his collection of observations formed the basis for the work of those who followed.

Johannes Kepler (1571–1630) was one of the most enigmatic figures of the Copernican revolution. He began his career at Tycho's observatory, seeking to analyze Tycho's observations of Mars to determine the patterns of planetary motion. He extracted his principal laws of planetary motion from Tycho's observations in a tortuous calculation, but hedged his laws of planetary motion in a forest of mysticism that made it difficult for subsequent workers to distill the essence of his achievement.

Galileo Galilei (1564–1642) first turned the telescope toward the heavens, discovering the phases of Venus, craters on the moon, and the satellites of Jupiter (Figure 3.1). These telescopic observations played a key role in the confirmation of the Copernican model of the solar system. His work in physics led to the development of the concept of inertia, the resistance of a massive body to any change in motion and a key ingredient in the Newtonian synthesis. Galileo is most renowned for his *Dialogue on the Two Chief World Systems— Ptolemaic and Copernican,* the first work that debated the two major theories of planetary motion in a language that everyone could understand—Italian rather than Latin. The publication of this work led to his persecution by the Catholic church.

Isaac Newton (1642–1727), born in the year that Galileo died, capped the Copernican revolution by weaving the work of his predecessors into a coherent whole. The magnitude of his achievement is best realized by considering that he is the source of most of the equations listed in the text of this chapter, the quantitative theory of the tides (section 3.3), the quantitative connection between terrestrial gravity and the force that holds the moon in orbit (Box 3.1), the mathematical proof of Kepler's laws (Box 3.3), and many other calculations that proved the Copernican model of the solar system.

but true: One of my astronomy students at Yale University insisted on turning in examination books marked "Astrology 10a," and in southern California I could not talk to someone at a party for 10 minutes without being asked what my birth sign was.)

Ptolemy wrote several books summarizing the knowledge of his time; his *Almagest* dealt with astronomy, his *Geography* with geography, and his *Tetrabiblos* with astrology. The *Almagest* and the *Geography* are historical documents, representing the intellectual achievements of the time but hopelessly outdated now. However, the *Tetrabiblos* is still the basis for current astrological practice.

Perhaps astrology had some real basis when the Ptolemaic theory of the universe was still considered valid. The quotation from the *Almagest* on page 73 indicates that Ptolemy felt that the real universe was destined to remain forever mysterious, and it was easy to believe that the planets and the sun might influence people on the centrally located earth. Such a theory is much less tenable with a Newtonian universe, in which our earth is just one piece of rock orbiting around a sun that is 150 million km away.

How could the other planets, the other rock balls and gas balls, influence people's lives on the earth? Astrological influences from the planets cannot be transmitted by any of the forces we know about. Modern science recognizes four forces: gravitational, electromagnetic, and strong and weak nuclear forces. The nuclear forces do not act over astronomical distances. The gravitational force of a delivery table acts ten times more strongly on a newborn baby than the gravitational force of the planet Mars, regarded as one of the strongest astrological influences. The electromagnetic radiation from a 100-watt light bulb in the delivery room is tens of thousands of times brighter than the radiation from Venus, the brightest planet. In order to explain the potency of celestial bodies as determinants of human fortunes, astrologers must appeal to mystical and mysterious cosmic influences.

The status of astrology as a pseudoscience is confirmed by consideration of the contents of a scientific model. One characteristic of modern science is its reliance on fundamental physical laws that describe, in a precise mathetical form, the interactions of objects in the universe. The theoretical descriptions of these interactions are then confirmed by comparison with observations. What are the laws of astrology? When have these laws been made into a concrete model which can then be tested against reality? Astrologers have rarely made any statistical tests of their predictions, and those tests that have been made show that there is no evidence that astrological predictions are correct.[5]

The persistence of astrology over 2000 years is quite surprising to me as a scientist. Perhaps it reflects the unconscious desire of many people to believe in magic; perhaps it comes from people's need for guidance and comfort in an uncertain world. But whatever comfort is provided by the horoscopes and fortune telling of the astrologers, any claim that this reassurance is based on scientific knowledge is false.

SUMMARY

The motion of the moon—its 27-day circle around the zodiac with its accompanying phases—can be explained easily as a result of the moon's orbit

[5] See Lawrence E. Jerome, ''Astrology: Magic or Science?'' and Bart J. Bok, ''A Critical Look at Astrology,'' *The Humanist* (September–October 1975).

around the earth and its illumination by the sun. Similar explanations of astronomical phenomena are the result of the model of the solar system presented in this chapter. Newton's law of gravitation can be applied to this model—a collection of nine planets orbiting around a central sun.

This chapter has covered:

Lunar phenomena, including the moon's phases, eclipses, and tides;

The role of gravity in keeping the moon in orbit;

The motions of the inferior and superior planets;

The regularity of the motions of objects in the solar system systematized by Kepler's three laws

The role of Newtonian gravity as an explanation of Kepler's laws; and

A brief treatment of the Copernican revolution that preceded the discovery of gravitation.

KEY CONCEPTS

Copernicus	Kepler's laws	Newton
Eclipses	Lunar phases	Nodes
Lunar	*Crescent*	Period
Partial	*First and last quarters*	Ptolemy
Total solar	*Full*	Retrograde motion
Force	*Gibbous*	Satellite
Gravitation	*New*	Superior planets
Inferior planets		Tides

REVIEW QUESTIONS

1. Figure 3.2 was taken at sunset. In what direction was the camera pointing when it was taken? Explain.

2. Show that the duration of a lunar eclipse depends on how close the full moon is to a node in the moon's orbit.

3. Explain why an eclipse does not occur every time there is a new or full moon.

4. If gravity were suddenly turned off in the universe, what would happen to the moon?

5. The Southern Hemisphere of the earth is mostly water, and the Northern Hemisphere contains many land masses. In which hemisphere will most locations have two high tides per day, occurring when the moon is overhead and again about 12 hours later?

6. If you were an inhabitant of the planet Mars, would the earth appear to be an inferior or superior planet?

Suppose that you could see Venus, Jupiter, and the crescent moon in the west, in the evening sky. Questions 7, 8, and 9 relate to this configuration.

7. Draw a sketch of the solar system showing where each object is.

8. What phase would Venus show at this time?

9. Would Jupiter be going through retrograde motion?

10. Consider a planet orbiting as far away from the star Epsilon Eridani as the earth orbits from the sun (1 AU). Epsilon Eridani has a mass of 0.7 solar masses. Will the orbital period of this planet be more or less than a year? Explain your answer.

FURTHER READING

Kuhn, Thomas S. *The Copernican Revolution.* New York: Vintage Press, 1959.

The motions of the earth and planets have been fairly well understood for about 100 years. Planets are held in their orbits by the force of gravity, and the seemingly complex motions of objects in the sky become simpler once one realizes that much of the movement is caused by the rotation and revolution of the earth, itself a planet. Various phenomena connected with planetary motion—the seasons, the phases of the moon, and the tides—are simple consequences of our position as one of nine planets circling the sun.

In subsequent part summaries, tables will separate solid fact, secure interpretation, working hypotheses, and speculation. No such table is needed here, for the ideas discussed in this section are well founded on an extensive history of observations. Work on planetary motion has not completely ceased, but it is now directed toward answering specific questions related to some other field of science. For example, workers at NASA's Jet Propulsion Laboratory need measurements of planetary positions far more accurate than those needed to check out Newton's laws, if they are to send a spacecraft to Mars and have it land in the right place. The observations and interpretations of planetary motion discussed in this part are secure and solid.

★ PART TWO ★

PLANETS

Perhaps the most dramatic of all human ventures is exploration. Astronomers explore the entire universe, but the pulse quickens when actual physical contact with another planet occurs. The only celestial object people have landed on is the moon, but space probes have given us a close view of nearby planets and have liberated us from the fetters of the earth. While human eyes have not gazed on the deserts of Mars, we have the next best thing—the television cameras of Viking 1, which landed in Chryse Planitia on July 20, 1976, exactly 7 years after the first lunar manned mission landed in Mare Tranquillitatis.

NASA's missions have done more than supply beautiful, breathtaking photographs of planetary landscapes. They have supplied useful information that, combined with careful observation via ground-based telescopes, can provide insight into planetary evolution. This information has prompted planetary astronomers to ask the right questions, and planetary science has grown rapidly in the last decade. David Morrison, an expert on the smaller bodies in the solar system, has called this an "Elizabethan age" for planetary science. In the Elizabethan age of history, indefatigable explorers opened up the earth; in the twentieth century, the vista is our solar system. The difference between a telescopic view of a fuzzy orange blob called Mars and the photographs returned by the Mariner and Viking probes is the difference between a completely unknown world and one that, though still mysterious, has been added to the list of known worlds. In the last 10 years we have mapped the moon, Mercury, Venus, Mars, and Jupiter. In 1979 a Voyager probe will map the four large satellites of Jupiter. In 1980 a Pioneer probe will fly by Saturn. What new horizons remain?

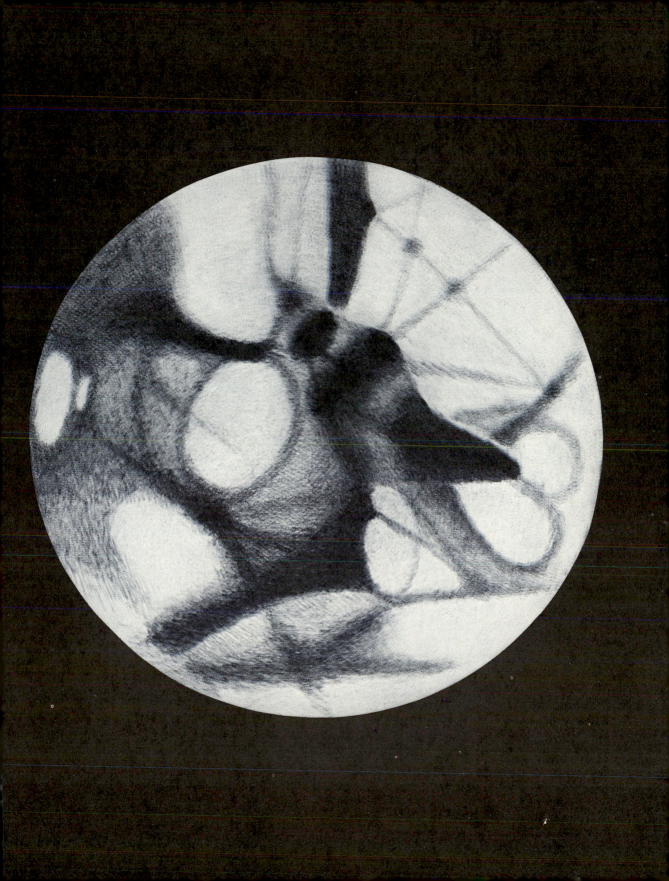

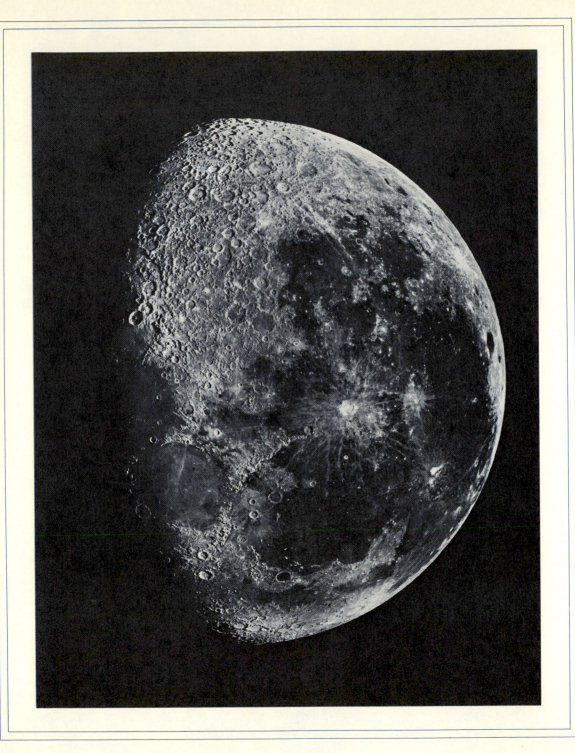

★ CHAPTER FOUR ★

TINY PLANETS: MERCURY AND THE MOON

KEY QUESTION: *What forces shape the landscapes on planets with inactive cores and no atmospheres—Mercury and the moon?*

The study of planets begins with the smallest ones, planets where activity in the core and the erosive forces of an atmosphere do not affect the appearance of the surface. The moon, while not strictly a planet, is the best example; further, six Apollo spacecraft have taken astronauts to the lunar surface and returned over 400 kg of moon rocks. While photographs of the surfaces of Mercury and the moon (Figure 4.1) may be aesthetically intriguing, the real scientific payoff from these missions comes from analysis of the data and a reconstruction of these planets' past history. We can compare these two planets to each other and to the earth, providing considerable insight into the ways in which similar geological processes act under different conditions.

FIGURE 4.1   A photograph of the Marius hills and the surrounding plateau of Oceanus Procellarum by Lunar orbiter 2. The crater at the upper right is named Marius. Giant lava flows created the volcanic plain; the craters are the result of meteorite impacts. (NASA Photo No. 66-H-1632)

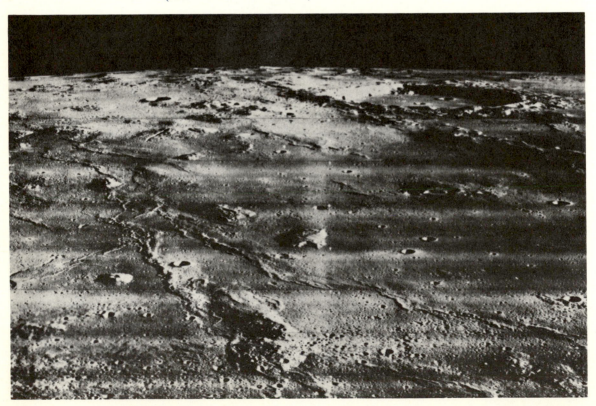

★ **MAIN IDEAS OF THIS CHAPTER** ★

The impact of meteorites creates basins and craters on the moon and on Mercury. Most of these planets' surface features are created by such impacts.

Volcanic activity filled the lunar basins and created some lunar surface features of lesser importance.

Analysis of lunar rocks reinforces the differences between the lunar basins and highlands.

The moon and Mercury, in contrast to Earth, exhibit little geological activity deriving from events in the core.

★ **4.1 SURFACE CONDITIONS** ★

The surfaces of Mercury and the moon (Figure 4.1, Color Photo 1) are quite inhospitable. Both planets rotate extremely slowly, exposing their sunlit sides for a long time and creating very hot temperatures on the surface. More solar radiation reaches Mercury than any other planet, since it is closest to the sun, and its surface temperature reaches 350°C during the 176 days in which the sun remains above the Mercurian horizon. The moon, though cooler, is still hotter than the earth, with a maximum temperature of 117°C (the average terrestrial maximum is 35 to 40°C). Both planets become quite cool during their long nights, with temperatures dropping as low as −170°C just before dawn. Neither planet has an atmosphere to retain heat.

The lack of an atmosphere is the result of the small size of Mercury and the moon. Their surface gravities are small, one-third and one-sixth of the earth's surface gravity, because their relatively small masses result in a low surface gravitational force in spite of their small size. Atmospheres developed on small, rocky planets like the earth and the moon when primeval volcanoes released gases from the planetary interior. These gases were then heated by the action of sunlight, and the molecules in them began to move. The earth, with a strong surface gravity, could hold on to the rapidly moving atmospheric molecules, but the weaker lunar gravity could not pull the wayward molecules back toward the planet. The Mercurian gravity might have been sufficient to retain an atmosphere if Mercury were farther away from the sun but, because of its nearness to the sun, its atmosphere became so hot that the atmospheric gases escaped in the same way as the moon's. Neither of these tiny planets could retain an atmosphere for long.

The absence of an atmosphere, however unwelcome to prospective lunar colonists, is a blessing to geologists who can observe surface features uncomplicated by erosion (Figure 4.2). These planets are geologically inactive, for neither the constructive process of mountain building by continental collision nor the destructive process of erosion by the action of air and water operate to any significant extent on these planets at the present time. The drama of volcanic activity is all a part of the past history of these tiny worlds. Most of our understanding of the surfaces of small, airless planets comes from lunar studies, since the moon has been walked on by six Apollo mission crews, landed on by several automated space probes, and photographed many times from orbit; we have only photographed one-third of

FIGURE 4.2 *Left:* The Oregon trail, blazed a little over 100 years ago by American pioneers, is overgrown in many places and will soon be eroded by the action of wind and rain. (Oregon Historical Society) *Right:* Wheel tracks left by Apollo 14's equipment transporter, brightly illuminated by the sun, will last indefinitely on the lunar surface since there is no air or water to erode moderately deep markings on the surface. (NASA Photo No. AS14-67-9367)

the Mercurian surface in any detail at all. Thus we study the processes that shape the lunar surface in some detail, and compare findings with photographs of Mercury to gain more insight.

★ **4.2 CRATERS AND RUBBLE** ★

☆ **LUNAR SURFACE FEATURES** ☆

Figures 4.1 and 4.2 show the desolate appearance of the lunar landscape, and Figure 4.3 shows an area of Mercury that is similar to the lunar area of Figure 4.1. The moon is far more desolate than the earth, far more forbidding than the driest terrestrial desert. The view is dominated by angular rocks, a grainy, gray soil, and craters dispersed amidst darker volcanic plains.

FIGURE 4.3   On March 29, 1974, the Mariner 10 spacecraft photographed a region of Mercury's northern hemisphere. The volcanic plain at the top of the picture is rather similar to the lunar plain shown in Figure 4.1. Also shown is a more heavily cratered area at the botton of the picture. (NASA/JPL Photo No. P-15427)

The lunar photographs from the Apollo missions are very impressive, even awe-inspiring. The human footsteps on the moon are evidence of a tremendous technological achievement, the crossing of a 384,000-km airless gap that separates the earth from its satellite. But the scientific value of the lunar missions comes only from intelligent analysis of the data returned by various lunar probes. Why is the earth covered with green hills and blue oceans, while the moon is covered with craters and sterile, rubbly plains? The correct answer to this question will unveil at least part of the story of lunar evolution.

FIGURE 4.4 This photograph of the 20.05-day-old moon shows the different types of features visible on the lunar surface. Individual features are identified in Figure 4.5. (Lick Observatory photograph)

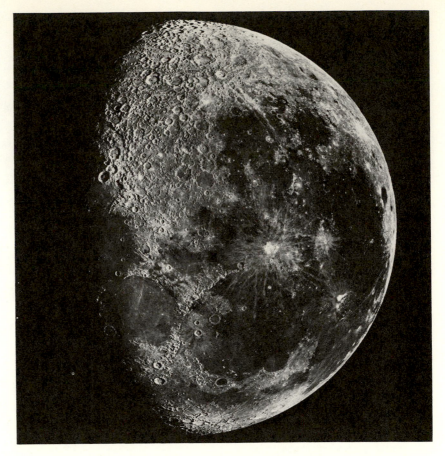

The primary feature of the lunar landscape is the circular crater. An overview of the moon is provided by Figure 4.4, a photograph taken from the earth of the moon just after its full phase. This face of the moon, the near side, is the only one visible from the earth, since the moon rotates on its axis every time it orbits the earth. The lunar surface is divided into two types of terrain: the dark *maria*, unfortunately named "seas" by early lunar mappers, and the lighter colored *highlands* (also called "terrae"; if the maria are seas, the other areas should be terrae, or "earth" in Latin). The dominant feature in both highlands and maria is the crater, which are far more abundant in the highlands. Most lunar craters are named after astronomers; appropriately, three of the most prominent ones are named after stars of the Copernican revolution: Copernicus, Kepler, and Tycho. These craters are identified in Figure 4.5. The maria are given more fanciful names, which may not fit now but which have been used for hundreds of years. The maria Imbrium, Serenitatis, and Tranquillitatis are the seas of Rains, Serenity, and Tranquility. The moon has no atmosphere, and it is difficult to see how rain could fall on the moon. Oceanus Procellarum is the Ocean of Storms.

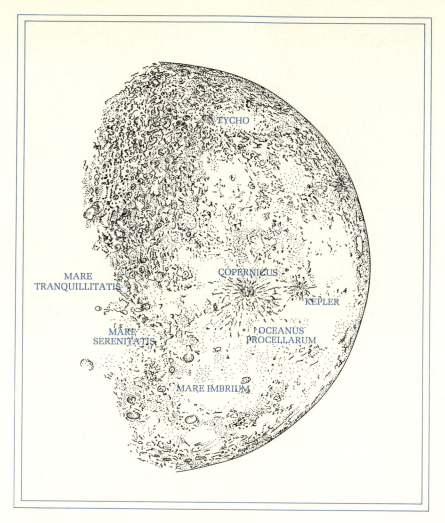

FIGURE 4.5 Identification of some of the features visible in Figure 4.4. The dark *mare* are volcanic plains, as is Oceanus Procellarum ("Ocean of Storms"; an orbiter view of Oceanus Procellarum is shown in Figure 4.1). Copernicus, Tycho, and Kepler are three prominent craters.

Mariner 10 was launched from the earth and flew by Mercury twice in 1974 and once in 1975. This mission has provided us with photographs of about half the Mercurian surface, in about as much detail as telescopic photos of our moon; one is shown in Figure 4.6. The basin at the center of the several rings shown in the photograph is named Caloris (or "Hot") basin. Numerous craters are evident in the photograph, and a recent International Astronomical Union commission has named these craters for prominent figures in the humanities—artists, composers, and authors. Some of the largest

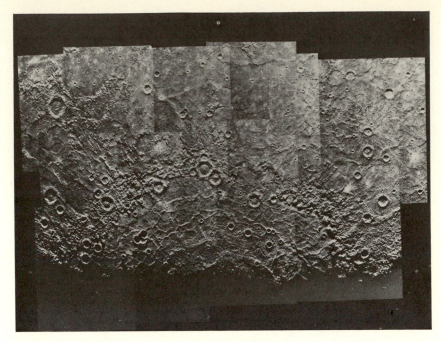

FIGURE 4.6 The left half of this mosaic of pictures from Mariner 10 contains the largest structural feature yet found on Mercury—Caloris basin. This multiringed basin is about 1300 km in diameter. (NASA/JPL Photo No. P-14855)

are named for Michelangelo, Monet, Raphael, Rodin, Beethoven, Bach, Goethe, Homer, Shakespeare, and Tolstoy. Mercurian craters are as common a feature of the surface as lunar craters (named for astronomers).

A close examination of the various lunar and Mercurian photographs shows that craters come in no particular size. Some of the moon rocks brought back by astronauts show impact pits that are tens of microns across, and large basins such as the Caloris basin on Mercury and Mare Imbrium on the moon are about 1000 km across. Currently we believe that almost all these craters were formed by the collision of meteorites with the planet. (The solar system contains a number of wandering rocks, and one of these rocks receives the name "meteorite" when it hits a planetary surface. The properties of meteorites will be described in more detail in Chapter 7.)

☆ **CRATER FORMATION** ☆

What happens when a meteorite strikes the surface of a planet and forms a crater? Figure 4.7 schematically illustrates the ensuing events. The meteorite explodes after penetrating the surface as its energy of motion is transformed into heat by the force of the impact. The exploding meteorite excavates a cavity and blasts material upward and outward, forming a crater. Similar phenomena occur in the formation of bomb craters; the cratering process is perhaps the only geological process that occurs rapidly enough to be reproduced in experiments. Most of this excavated material falls down

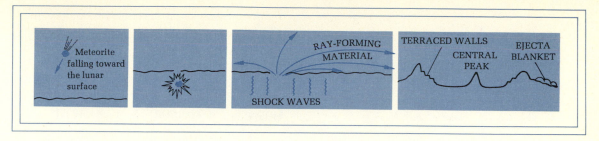

FIGURE 4.7 Schematic illustration of the events taking place during the formation of a large crater. A meteorite falling toward the surface explodes after penetrating the surface, excavating a crater and tossing material across the surface.

fairly near the crater and forms an *ejecta blanket*. Large fragments of ejected material form secondary impact craters where they strike the surface. Some of the material may be tossed large distances, forming *rays* radiating away from the crater center, and some of it may be ejected with such speed that it can escape the moon's gravity entirely. The young craters Tycho and Copernicus both show extensive ray systems in Figure 4.4. In large craters, the inner walls become terraced when the sides of the crater slump downward toward the crater floor. Another characteristic of large craters is the central peaks, whose origins are not understood. Perhaps the shock waves produced by the initial impact cause the central material to rebound.

Figure 4.7 is an attractive, coherent model of crater formation, but how do we know that it represents, in all its details, the way in which real craters are formed? While we can throw rocks at laboratory surfaces and see what happens, or use bomb craters to study the shapes of slightly larger craters, no one can make a 50-km-diameter crater for experimental purposes. The moon is a good laboratory for testing various theories of crater formation, because the absence of an atmosphere means that the appearance of the lunar craters is unaffected by erosion. No lunar winds will blow to upset the appearance of an ejecta blanket. While the lunar crater Copernicus is a billion years old, the lunar surface has scarcely aged since the crater was formed. We can study it almost as though it were made yesterday; on the earth, such a crater would soon have been eroded away by the action of wind and rain, or swallowed up when the continents drifted into each other.

There is one essential difference between the moon and Mercury that makes them an excellent testing ground for theories of crater formation. The moon is much less dense than Mercury; its mean density is only 3.35, while Mercury's is a rather large 5.4. (Box 4.1 explores the concept of density and the difference between density and total mass.) Since the moon and Mercury are roughly the same size, this density difference means that Mercury is over four times as massive as the moon, and consequently its surface gravity is greater (Box 4.2). This stronger gravity means that meteorites will fall faster as they strike the Mercurian surface, because gravity has pulled harder on them. And, something ejected from a crater will travel a shorter distance on Mercury because the higher gravity will pull it back to the surface sooner.

Box 4.1 Density and Mass

The *mass* of a planet measures the total amount of matter contained in the planet, while the *density* measures how closely that mass is packed—the mass per unit volume. Density is expressed in grams per cubic centimeter.

For example, we can compare the moon, Mercury, and Earth:

	MOON	MERCURY	EARTH
Mass (g)	7.3×10^{25}	3.3×10^{26}	5.977×10^{27}
Radius (km)	1700	2400	6378
Volume (cm^3) ($= \frac{4}{3}\pi R^3$)	2.2×10^{25}	6.1×10^{25}	1.087×10^{27}
Density (g/cm^3) ($=$ mass/volume)	3.4	5.40	5.50

The moon and Mercury are quite close in size. Because Mercury has four and one-half times as much mass as the moon does, it has to pack this material more tightly to fit it into a volume only three times larger than the moon's; this tighter packing of material is expressed quantitatively as a higher density. The earth is much larger than the other objects, having a much higher mass, but its density is only slightly higher than Mercury's as it has a proportionately larger volume to pack the mass into.

The density of a planet is related to the kind of material it is made up of. The densities of common materials, in grams per cubic centimeter, are: water, 1; ice, 0.91; common rocks, 3 to 7; iron, 8; lead, 11; platinum, 21.

The size of a crater is determined by the violence of the impact that caused it, so that craters of the same size on Mercury and the moon were produced by equally energetic impacts. It is only the large craters that have central peaks and terraced walls (on both planets), and the size that a crater needs to attain in order to possess these features is about 40 km on each planet.

The debris that is thrown out in the cratering process travels a shorter distance on Mercury than on the moon, since the greater gravity causes material to fall back to the surface sooner. Compare the lunar crater Copernicus (Figure 4.8) with the largest Mercurian craters in Figure 4.6, which are about the same size as Copernicus. Evidently Copernicus has a much more extensive ejecta blanket. Detailed examinations of photographs show that ejecta blankets of similar-sized craters extend one and one-half times further from the crater center on the moon than on Mercury. While most meteorite craters on Earth are so badly eroded that detailed studies are impossible, two cases that have been examined show that ejecta blankets are only half as extensive on Earth as on Mercury, because of the higher surface gravity.

Certainly we do not know all there is to know about crater formation. The Mariner 10 photographs of Mercury, combined with the results of various lunar studies, provide so much information that many insights remain hidden beneath this mound of data. The scientific returns from a space mission often do not appear for many years, because proper interpretation of the

BOX 4.2 SURFACE
GRAVITY OF
PLANETS

At the surface of a planet with mass M_{planet} and radius R, any object (a person, a gas molecule, a meteorite falling toward the surface) with mass M_{object} is attracted toward the planet's center by a force F equal to

$$F = G \frac{M_{planet} M_{object}}{R^2} \quad \text{(B4.1)}$$

where G is the gravitational constant 6.67×10^{-8}, the masses M_{planet} and M_{object} are in grams, and the planetary radius R is in centimeters. The object's response to that force is given by the definition of a force, $F = Ma$ (recall Chapter 3), and so we write

$$M_{object} a = G \frac{M_{planet} M_{object}}{R^2} \quad \text{(B4.2)}$$

Canceling as shown, we find the response of an object to gravity (its acceleration or change in motion under the influence of gravity):

$$a = \frac{G M_{planet}}{R^2} \quad \text{(B4.3)}$$

The gravitational acceleration is a measure of how hard things fall.

One can simplify equation B4.3 by using (Planetary mass M_{planet}) = (volume) × (density d) to obtain

$$a = \frac{G}{R^2} \times \frac{4}{3} \pi R^3 \times d$$

$$= \frac{4}{3} \pi G \times \text{radius} \times \text{density}$$

From the data of Box 4.1, the acceleration of gravity is 162 on the moon, 368 on Mercury, and 980 on Earth, where the units are centimeters per second per second.

FIGURE 4.8 The lunar crater Copernicus, photographed by the 200-inch telescope. This crater is surrounded by an extensive ejecta blanket, unlike the largest craters in Figure 4.6. (Hale Observatories)

photographs takes time. The glamorous phase of a planetary mission like the Mariner 10 flight to Mercury may end when the spacecraft runs out of attitude-control gas, transmits its last picture, and tumbles around help-lessly, unable to point its radio antenna at the earth or to aim its cameras. This phase of the Mariner 10 mission ended in March 1975. Financial support of the mission must continue if scientists are to complete the less visible and more tedious job of sifting through the thousands of pictures, carefully mea-suring such things as the sizes of ejecta blankets. Full realization of the sci-entific potential of a mission like Mariner's comes only when such careful analysis can be done. In these days of dwindling financial support for plane-tary research, the follow-up studies that produce genuine scientific insight can lose out to more glamorous new missions in the battle for inclusion in next year's budget, but if science rather than glory is the objective, the data analysis must continue.

☆ THE LUNAR SOIL ☆

Impacts play another role on the lunar surface, in forming the lunar soil. The moon has no protective atmosphere to shield the surface from the thousands of tiny micrometeorites that pepper any planet every day. These tiny par-ticles, striking the lunar rocks, create very small impact pits that pulverize the lunar surface, creating thin, dusty lunar soil. These small impacts also churn the top layer of the lunar soil, so that any very thin coating of material, such as a ray from a crater, will slowly be buried by the ejecta blankets of these tiny craters. Consequently, old craters do not have extensive ray systems.

★ 4.3 BASINS ★

Perhaps the most spectacular result of impacts on the lunar and Mercurian surfaces is the formation of huge circular basins. Caloris basin (Figure 4.6) is the largest such basin on Mercury, being 1300 km in diameter, but Mariner 10 has mapped 16 other basins, all more than 200 km across. The moon con-tains a comparable number of basins, including Mare Imbrium, 1200 km across, about the same size as Caloris. Figure 4.9 shows Mare Orientale, the lunar basin most similar in appearance to the Caloris basin. This mul-tiringed basin is the youngest of all the lunar basins, since its ejecta overlay the ejecta from the others.

The rings of basins like Mare Orientale and Caloris were presumably created by the impacts that formed them, but the precise nature of the ring-formation process is not understood. The regular pattern seems to sup-port the idea that they are frozen shock waves, the preservation of a ripple pattern by the rocks of the planet. Alternatively, the impact that formed the basin could have created some circular cracks or faults in the outer layer of the planet, and subsequent crustal readjustments could have created these rings.

FIGURE 4.9 Mare Orientale, the youngest basin on the moon. Most of this basin is not visible from the earth, and the discovery of the multiringed "bulls-eye" pattern was not made until orbiting spacecraft photographed the lunar far side. (NASA Photo No. 67-H-934)

The lunar basins have completely inundated the surface with their ejecta blankets, obliterating traces of the early lunar rocks. No moon rock has been found with an age of more than 4 billion years, though a few older soil fragments have been recovered. Lunar scientists have identified most of the lunar near side, the side visible from the earth, with lunar basins or with the ejecta blankets of Mare Imbrium, Orientale, or Nectaris. The far side contains fewer basins, and some older, prebasin material may not have been buried. One of the hopes of the lunar missions was that we might find some "genesis rocks"—rocks dating from the time when the solar system formed. Such rocks might elucidate some of the processes of solar-system formation. Although none have been found on the near side, such rocks may exist on the far side.

The higher surface gravity of Mercury, responsible for the smaller extent of crater ejecta blankets, also limits the influence of basin formation on the

Mercurian surface. About 30 percent of the area mapped by Mariner 10 is covered with intercrater plains, structures that apparently have survived from the era preceding basin formation. A landing mission to Mercury might have a better chance of finding very old material than the Apollo astronauts had.

Thus, impact is the primary agent in sculpting the landscapes of tiny, airless planets such as Mercury and the moon. A meteorite impact excavates a circular crater, which can range in size from the 10-μ impact pits that create the lunar soil to the large 1000-km basins. Ejecta blankets are so extensive on the moon that the ejecta from large basins have completely covered the lunar near side. Can we explain every feature of the lunar surface by impact, or are there some other processes at work?

★ 4.4 VOLCANIC ACTIVITY ★

Look back at Figure 4.4. A rather obvious feature of any lunar photograph is the division of the moon's surface into heavily cratered areas called highlands or terrae, and the darker, smoother mare regions. While the shape of the craters in the mare regions can be explained as a result of their formation by impact, what explains the dark color of their surfaces? What is the origin of the hills and ridges in Figure 4.1? Some process other than impact must be sculpting the landscape of the maria.

A very easy and natural way to produce a smooth, relatively uncratered surface in a lunar basin is to let lava flow into the basin through cracks produced by the impact shock. Gradually, over hundreds of millions of years, the lava would work its way through the cracks and flow downhill toward the center of the basin (Figure 4.10). Many different lava flows would fill the basin. Confirmation of the lava flow sequence is given by crater counts in the Imbrium basin. Different sections of the basin have different numbers of craters in them; the more heavily cratered regions have been exposed to meteorite impact for a longer time and are thus older. One can also reconstruct the basin-filling sequence by determining the ages of rocks, from different parts of the basins, returned by different Apollo missions. The age differences among different sets of lunar rocks show that hundreds of millions of years passed before the basins were completely covered with lava flows.

The types of moon rocks returned by the Apollo astronauts reinforce the theory that the mare basins are filled with lava. The mare rocks are all *basalts*, a type of rock known on the earth and formed by the cooling of molten lava. These lunar basalts are rather similar to terrestrial basalts, though there are some differences in the rocks' chemical make-up. Some lunar basalts are remarkably rich in titanium; the Apollo 11 basalts would be considered high-grade titanium ore if they were found on the earth. Mare basalts have a higher density than highland rocks, and as a result the lunar gravity is slightly stronger over the mare than it is over the lower-density highlands.

Lunar volcanism has done more than simply fill the lunar basins. The Marius hills (Figure 4.1), in the Oceanus Procellarum basin, were formed by the pressure from various lava flows that created the basin floor. The domes

ORIGINAL BASIN

LAVA RESERVOIR

FIGURE 4.10 While meteorite impact was the original event that led to the formation of a basin, the basin floor was formed when lava (color) flowed upward, from reservoirs of molten rock beneath the surface, and filled the bottom of the basin. Several successive lava flows were responsible for the creation of these basin floors.

in the hills may well be the result of hydraulic pressure exerted by the cooling lava. The lunar surface contains a number of *rilles* or depressions, some of which are straight and some of which are winding (Figure 4.11). The straight rilles probably result when material located between two parallel cracks or faults falls downward. The origin of the sinuous rilles is much more uncertain. Are they collapsed tubes of lava that contained a great deal of gas? Most investigators believe that volcanic processes were involved.

Before trying to collect the various impact and volcanic processes into a coherent picture of lunar evolution, we must consider evidence from other types of geological investigations. Analysis of Mercury is limited to examining pictures and trying to figure out what made various surface features. But we have landed on the moon. Rocks returned from the lunar surface and the measurements of seismometers detecting moonquakes provide evidence bearing on the history of our satellite.

★ 4.5 THE MOON ROCKS ★

Some 400 kg of rock and soil were returned to terrestrial laboratories by the Apollo astronauts. Many of these rocks remain in a repository in Houston, awaiting more sophisticated analyses of the future, but the studies that have been made so far add a great deal to our understanding of the lunar surface. The picture presented by the moon rocks is fragmentary, since the rocks have been collected at nine isolated sites on the lunar surface by six Apollo missions and three unmanned Soviet Luna probes. It is not always easy to tie these sites together.

One major rock type, the *mare basalt,* has already been mentioned. Most of the Apollo missions landed in the flat mare regions because of the difficulty of landing the lunar excursion module on a mountainside. These

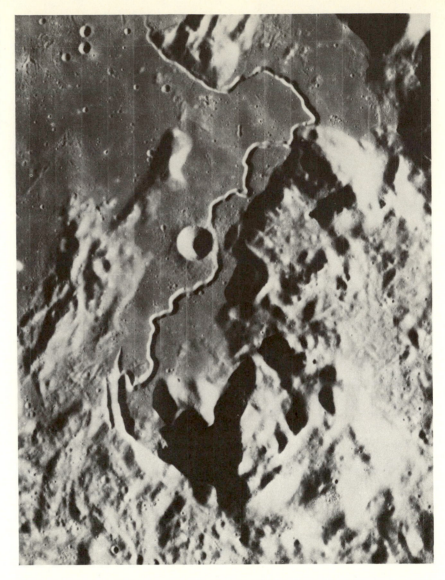

FIGURE 4.11 Sinuous rilles, such as the Hadley rille here, are a feature of the lunar basins. No one yet knows the origin of these features, although most investigators suspect that volcanic processes were involved. (NASA Photo No. 70-H-126)

basalts vary considerably in chemical composition and texture, probably reflecting different conditions in different lava flows. Age measurements indicate that the maria were formed between 3 and 4 billion years ago.

The highland rocks come from the regions that cover four-fifths of the lunar surface. The last four Apollo missions landed on the edges of mare regions in order to sample these rocks. They discovered that in general the highland rocks are more complex than the mare basalts. Most of the highland rocks show evidence of being disturbed by the mare-forming events, since they are generally found as *breccias*. A breccia is a rock composed of many individual rocks welded together by some force. Terrestrial breccias are generally fused together by high pressure, but on the moon it was impact that caused their formation. The shock created by a falling meteorite partially melted some of the rocks, gluing them together. The shock of breccia formation makes it impossible to tell when the highland rocks were formed, but radioactive dating has determined that they were made into breccias about 4 billion years ago.

Chemical analysis of the highland rocks indicates that they fall into two main groups, with considerable variation within each group. One class, rich in potassium (chemical symbol K), *Rare Earth Elements*, and *Phosphorous*, has earned the somewhat ridiculous acronym of KREEP. The other major class of highland rocks are more similar to terrestrial rocks and are called *anorthosites* after the corresponding terrestrial rock type. Anorthosites are more common than KREEP and have been found in almost all highland rocks, while KREEP is confined to a few isolated sites. Presumably, anorthosites are typical highland material, while KREEP comes from a few lava flows that occurred after the highlands were formed. Unfortunately, the shock the lunar surface received when the basins were excavated hides the details of this phase of lunar history.

★ **4.6 MOONQUAKES, EARTHQUAKES, AND** ★
PLATE TECTONICS

The activities of the Apollo astronauts were not confined to wandering around, picking up rocks, and photographing stars from the moon for stellar astronomers. Each Apollo mission left a good deal of equipment and other material behind. Some of what remains is just mission debris: life support systems, the bottom half of the lunar lander, and other trash. There are memorials, placcques attached to the lander, and flags. However, some scientific instruments left on the moon are still functioning. Of immediate interest are the seismometers, left there to monitor moonquakes.

☆ **CONTINENTAL DRIFT** ☆

Moonquakes and earthquakes both involve disturbances of large rock masses. Earthquakes (on the earth) occur generally when continental plates grind against each other. California suffers from earthquakes because the part of the state west of the San Andreas Fault is moving northwestward, in abrupt jolts rather than smoothly. Each sudden motion is accompanied by an earthquake.

In the last 10 years, geologists have uncovered a large amount of evidence showing that all continents drift around the earth in the same way as California splits apart. The earth's crust is composed of about a dozen plates that slowly drift around, pulling apart from each other, lurching past each other,

FIGURE 4.12 A few hundred million years ago, the two American continents, Europe, and Africa were joined together in a global jigsaw puzzle. Since that time, the slow movement of the continental plates has pushed the continents apart into their present positions. Tectonic processes, actions involving large-scale movements of crustal material, have no influence on the lunar or Mercurian landscapes. (Adapted from "The Confirmation of Continental Drift" by P. M. Hurley. Copyright © 1968 by Scientific American, Inc. All rights reserved.)

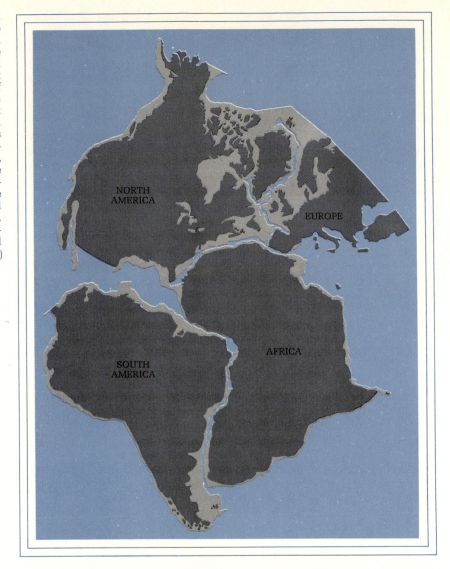

and colliding with each other. The jigsaw-puzzle fit of the American, European, and African coastlines, shown in Figure 4.12, is no coincidence; these continents were joined together a few hundred million years ago. Since then, these continents have been drifting apart, and the space between them has been filled with new crustal material, the Atlantic Ocean floor. The theory that describes the motions of these plates around the earth is the theory of *plate tectonics*. Tectonic processes are geological actions involving the large-scale movement of planetary crust.

An important consequence of the processes of plate tectonics is the continual formation and destruction of large blocks of the earth's crust. Con-

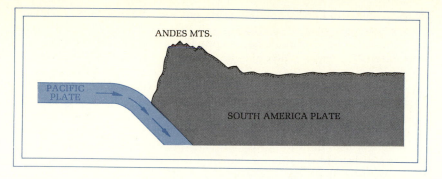

FIGURE 4.13 When the Pacific Ocean floor collides with the west coast of South America, the ocean floor is forced beneath the continental plate and destroyed. The Andes Mountains are pushed up by the collision of the two plates. The motion of the plates against each other results in earthquakes.

sider, for example, what happens when two plates collide (Figure 4.13). The Pacific plate, grinding against the South American plate, is pushed down into the earth's surface, creating an ocean trench. Pieces of the lower plate are scraped off, partially melted, and forced to the surface, forming the Andes Mountains. The crust of the Pacific plate disappears, leaving no trace of the geological processes that formed it billions of years ago. Plate-tectonic processes continually create and destroy crust: the only rocks on the North American continent that have survived from the distant past are rocks in northern Canada. The rocks of this Canadian shield, a billion years old, are among the oldest rocks on the earth, but they seem rather young when compared to moon rocks, with ages between 3 and 4 billion years.

Plate-tectonic processes occur on the earth because of the earth's tremendous heat. Temperatures reach 1000°C about 100 km below the surface; although these high temperatures are not sufficient to melt the rock, its strength becomes quite low. Heat moves continuously through this layer, from the earth's molten core outward. In the plastic layer of the earth, this heat transfer occurs through very, very slow motions of hot rock upward and cold rock downward. The plates of the earth's surface ride on these convection currents, causing the processes that are collectively known as plate tectonics. Earthquakes almost invariably accompany the elephantine movements of the continents past and against each other, and these quakes are the telltale sign of the existence of tectonic processes.

☆ **TECTONICS ON TINY PLANETS** ☆

The earth is tectonically quite active, with major earthquakes occurring every year or so. In contrast, the moon is a very quiet place. All moonquakes are very weak; they would not be perceived by a person on the lunar surface. Apparently large-scale motion of the lunar crust does not occur. The lunar basins, evidence of impact 4 billion years ago, are preserved because of absence of plate tectonics on the moon.

A moonquake, even a small one, can provide much information about the lunar interior. A quake sends pressure waves out in all directions. These pressure waves are analogous to sound waves in air. The speed of these waves as they travel through the lunar interior provides information about its composition. About 60 km below the surface, the crust gives way to a different form of rock that can only exist under high pressures. The nature of this rock layer is still undetermined. Most moonquakes occur about 1000 km below the surface, indicating that any mass motions that do exist on the moon occur deep in the interior, too far down to cause the crust to move around as the earth's continents do.

Tectonic processes may have played a role in the early evolution of Mercury. The planet is covered with long, winding cliffs called lobate scarps. The most reasonable explanation of the origin of these features is the compression of the Mercurian crust, a compression that created high promontories where crust was pushed up and over other Mercurian material. Detailed analyses indicate that if the entire planet had shrunk by a few kilometers, features like the lobate scarps would have been produced. Of course, such analyses do not prove that the scarps must have been created by a large-scale tectonic process like crustal shrinking.

☆ MAGNETISM ☆

More indirect evidence about the nature of the interiors of the moon and Mercury is provided by observations of magnetic fields. The moon has no large-scale magnetic field, but one of the highlights of the Mariner 10 mission was its discovery of the Mercurian magnetic field. This field is only one-hundredth as strong as the earth's field, the magnetic field that makes compasses work. The earth's magnetic field probably comes from the swirling motion of currents in its iron core. The moon lacks such a large-scale magnetic field. The existence of the Mercurian field is a puzzle. While the planet's high density suggests that about two-thirds of its mass is in the form of an iron core, models of Mercury's interior indicate that by now such a core would have solidified. Could the Mercurian field be a fossil of some earlier time when the core was hotter?

Lunar magnetism, absent on a large scale, is present in some of the iron-rich lunar rocks. When a rock rich in iron cools in the presence of a magnetic field, the iron itself becomes magnetized. Medieval mariners often used compasses made of lodestone, an iron-rich mineral that became magnetized when it cooled in the magnetic field of the earth. Because the moon has no planetary field, it is rather surprising that the moon rocks are magnetized. Some kind of a magnetic field must have been present when these rocks cooled, between 4.6 and 3.2 billion years ago. A strong field is not needed; something one-tenth as strong as the earth's present field will do. Where did that magnetic field come from? Did the moon once have an iron core? An iron core is difficult to visualize, since there is no magnetic field now. The mystery remains.

☆ PLANETARY CORES ☆

The earth is generally thought of as a geologically active planet, because events taking place in its interior have visible effects at the surface. The whole process of plate tectonics is a consequence of the transport of heat outward from a hot core. The moon is geologically inactive, since no massive core exists there and there is considerably less global transport of heat from interior to surface. Mercury may be an intermediate case, but our knowledge of Mercury is about comparable to our knowledge of the moon in 1956, before the space age. The temptation to draw too many parallels between Mercury and the moon should be resisted. Still, Mercury's density argues for some kind of a planetary core that is chemically distinct from the Mercurian surface. The formation of this core may account for the presence of lobate scarps and the existence of the Mercurian magnetic field.

★ 4.7 LUNAR AND MERCURIAN EVOLUTION ★

☆ HISTORY OF TINY PLANETS ☆

All the information we have about the moon and Mercury can be put together into a tentative scheme describing the past history of these planets. Any history, such as the one presented here, inevitably contains a certain amount of speculation—especially in regard to Mercury, for which data are sparse. This scheme is not to be treated as religious dogma handed out by the high priests of lunar science, but as a working model of lunar history to be supported or shot down as future facts are unveiled by the continuing investigation of the lunar samples. For the sake of simplicity, I tell the moon's story; Mercury is probably quite similar.

A number of individual stages can be distinguished:

1. *Formation.* Small planets such as Mercury and the moon formed when many small rocks agglomerated together, forming one large mass. The heat liberated when these small fragments smashed into each other probably melted the primeval planets from center to surface.

2. *Crustal differentiation.* The first part of the molten moon to cool was the surface, although some of the deep interior might have been under sufficiently high pressure to solidify quickly. The anorthositic rock characteristic of most highland rocks was the first part of the crust to cool from the molten magma.

3. *KREEP formation.* Since KREEP exists in the highlands, apparently there was a lava flow that produced it after the crust formed. There is no way

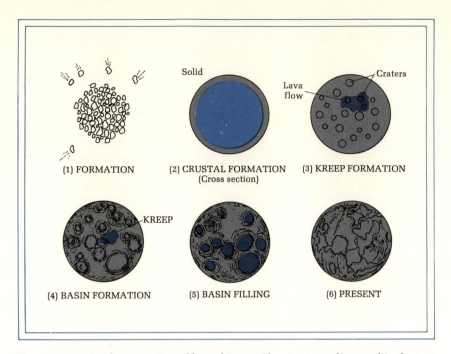

FIGURE 4.14   A schematic view of lunar history. The six stages discussed in the text are depicted very roughly; the geographic locations of such things as the KREEP lava flow are not completely determined, and a certain amount of scientific as well as artistic license has been applied in this drawing. While the overall picture is probably correct, the details have yet to be filled in by lunar scientists.

of telling whether a distinct highland rock type like KREEP exists on Mercury.

4.   *Basin formation.* About 4 billion years ago, just 500 million years after the moon formed, the development of impact craters on the surface climaxed with the collision of a number of large meteorites with the lunar surface. These events excavated huge basins, up to 1200 km in diameter, and covered the lunar near side with the ejecta blankets. On Mercury, because of the higher gravity, the ejecta blankets were less extensive.

5.   *Mare formation.* About 1 billion years after the formation of the lunar basins, lava flowed onto the lunar surface through the cracks left by the impact. These successive lava flows filled the lunar basins with dark material.

6.   *The postmare era.* Very little has happened to the moon in the last 3 billion years, since almost all the present features of the lunar landscape were formed 3 billion years ago. Lunar evolution in this epoch has been minimal: The surface is slowly churned by the impact of tiny micrometeorites, and occasionally a larger crater is dug by a larger particle. But the Apollo astronauts probably did more to change the landscape in the vicinity of their landing sites than all the meteorite impacts since basin formation. The moon just sits passively, helpless and exposed to whatever meteoritic insults are thrown at it.

☆ **COMPARATIVE PLANETOLOGY** ☆

Why are the moon and Mercury so different from the earth? All three planets are made of rocky material. What makes the two small planets desolate rock piles, and the earth a verdant, tropical paradise? An important result of the exploration of different planets is the opportunity to observe the same geological processes in different circumstances.

One cause of the differences between these small planets and the earth is the absence of erosion on Mercury and the moon. Erosion requires the presence of an atmosphere, which the tiny planets cannot retain. The terrestrial landscape is continuously modified, while the lunar surface retains its present form for a long time.

The most critical difference between the earth and the moon is the age of their crusts. The earth is a young planet; the very old rocks of the Canadian shield are less than 2 billion years old. A lunar surface feature younger than 2 billion years is a rarity. Plate tectonics on the earth consumes crust at a great rate, destroying continents and ocean basins when plates collide with each other. Mercury and the moon are not subject to these types of tectonic processes.

The absence of plate-tectonic processes on the smaller planets is a consequence of their smaller masses. Plate tectonics on the earth is the result of a plastic layer 100 km below the surface, a layer on which the continents and ocean basins drift around the earth. The transport of heat from the hot terrestrial interior sets up slow currents in this plastic layer. The plastic layer in the moon is much deeper, and the absence of heat within the moon means that the currents necessary to push continents around just do not exist there. No direct information about Mercury's interior is available, but the indirect evidence provided by the Mariner 10 photographs indicates that Mercury is much like the moon. The smaller masses of these tiny planets mean that whatever initial heat they possessed from the impact of protoplanetary bodies has long since been lost; for the most part these planets are just cold bodies of rock.

SUMMARY

Almost all the features seen in photographs of the moon and Mercury can be explained as results of the impact of meteorites on the surfaces of these small, airless planets. Volcanism filled the large basins and created some smaller surface features, but the large-scale processes of plate tectonics are not found on Mercury and the moon. What we know about these planets—our understanding of the formation of impact craters, data gleaned from the moon rocks, and the results of studies of the lunar interior—can be fitted into a coherent picture of lunar evolution.

This chapter has covered:

The general properites of the lunar and Mercurian surfaces;

The formation of impact craters;

The formation and filling of the lunar basins;

The role of plate tectonics on the earth and its minor importance on the moon and Mercury;

The meaning of the moon rocks;

The state of the lunar interior; and

A working model for the evolution of these tiny, airless planets.

KEY CONCEPTS

Basalt	Ejecta blanket	Plate tectonics
Breccia	Highlands	Rays
Density	Impact crater	Rille
	Mare	

REVIEW QUESTIONS

1. Which planet is hotter in the daytime, the moon or Mercury? Why?

2. Why do lunar craters have larger ejecta blankets than Mercurian craters of the same size?

3. What feature of the crater formation process makes it possible to draw a parallel between bomb craters and lunar craters?

4. What is the difference between density and mass?

5. Why is it more likely that a space mission would find rocks dating from the very early days of the solar system on Mercury rather than on the moon?

6. If the moon were completely solid from surface to center at the time the basins were formed, would it still have dark maria and light highland regions?

7. Name two lunar surface features that are the result of volcanism.

8. A moon rock that had never been brecciated could be (a) a piece of mare basalt, (b) a piece of typical highland rock, or (c) the long-sought "genesis rock" dating from the earliest days. More than one answer may be correct. Explain your answers.

9. Why are the lunar basins, formed 4 billion years ago, still present on the lunar surface? Why would such basins not be seen on the earth's surface?

10. Why are the moon and Mercury so different from Earth?

FURTHER READING

Head, James W., Charles A. Wood, and Thomas A. Mutch. "Geologic Evolution of the Terrestrial Planets." *American Scientist* 65 (January-February 1977): 21–29.

Short, Nicholas M. *Planetary Geology*. Englewood Cliffs, N.J.: Prentice-Hall, 1975.

The Solar System, a reprint of the September 1975 issue of *Scientific American*. San Francisco: Freeman, 1975, chaps 4 and 7.

★ CHAPTER FIVE ★

VENUS AND MARS

Figure 5.1 shows telescopic views of Venus and Mars, the second and fourth planets from the sun. These two planets are roughly similar in size to the third planet, Earth (Color Photo 2). Their distances from the sun are also similar, but their surfaces are very different. Venus is forever shrouded in clouds, but beneath those clouds lies a hot, ovenlike surface. Mars, through the telescope, is a small, enigmatic orange disk, showing distinct surface features. The earth, seen from a distance, is a blue, water-covered globe whose clouds dance through the atmosphere in a symphony of storms.

★ **MAIN IDEAS OF THIS CHAPTER** ★

The surface temperature of a planet depends on three factors: the planet's distance from the sun, the lightness or darkness of its surface, and most important the thickness and composition of its atmosphere.

Venus has an extremely thick atmosphere and a hot, dry surface.

The Martian surface is extremely varied but generally resembles a terrestrial desert, although the temperatures are extremely low.

Strong winds in the Martian atmosphere produce the surface markings that are visible from the earth.

Martian channels, produced by the action of flowing water, indicate that at one time Mars was warmer than it is now.

The search for life on Mars, conducted by the Viking probes in 1976 and 1977, produced ambiguous results.

★ **5.1 GENERAL PROPERTIES** ★

Venus, Earth, and Mars are physically rather similar. Mars is the smallest of the three planets, with a radius of 3400 km; Earth and Venus are almost identical in size, with radii of 6400 and 6100 km, respectively. Their densities, too, are similar, with Venus and Earth having relatively high densities of 5.2 and 5.5, while Mars is a little less dense, only 3.9 times as dense as water. These densities indicate that the planets are composed of rock and iron, like Mercury and the moon. The Martian day is a little longer than the earth's, 24 hours and 37 minutes. Venus does not share the rapid rotation of the other two planets, since it takes 243 days to rotate once on its axis; the direction of its rotation is opposite the directions of the orbital motions of all planets, and the rotations of all but one of the other planets. The appearances of these three planets are more different than their bulk physical properties.

Most of the research interest in Venus has been concerned with its hot, dry atmosphere. Before the space age, astronomers had detected radio emissions

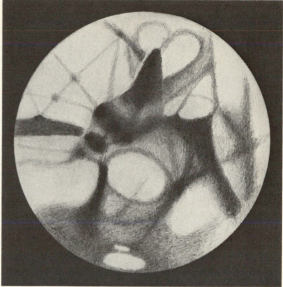

FIGURE 5.1 *Left:* The planet Venus, photographed from Earth with the 5-m (200-inch) telescope of the Hale Observatories. All that is visible is the cloud deck, 50 km above the planet's surface. (Hale Observatories) *Right:* Drawing of the planet Mars, made at the 36-inch refractor of Lick Observatory in 1924. The dark areas, so prominent from Earth, are just darker areas of the Martian deserts. (Lick Observatory photograph)

from a hot layer on Venus, but controversy raged over whether the radio radiation was coming from a hot surface or a high layer of the atmosphere. Early space probes confirmed that the surface temperature was very high, 750 K, making Venus the hottest planet in the solar system. The visible light that reaches the earth from Venus comes not from the surface but from a high cloud layer, 50 km up in the Venerean atmosphere. Venus is no place for our kind of life.

Mars, once thought to have canals on its surface, has turned out to be a geologically fascinating planet. The Mariner 9 mission orbited the planet for almost a year, finally running out of attitude-control gas on October 27, 1972, unable to direct its antenna at NASA's ground stations on the earth. Additional photographs of the surface were provided by two orbiting spacecraft in 1976, but the primary focus of the 1976 Viking mission was the two landers that touched down on the Martian deserts. The photographs from Mariner and Viking form the most complete set of photographs of any planetary surface outside the earth-moon system, and they have revealed a number of geological features that would make any admirer of terrestrial landscapes gasp in wonder: a canyon complex that stretches one-quarter of the way around the planet and is up to 6 km deep, a volcano 24 km high, a series of giant staircases descending from the polar uplands.

A thorough understanding of a planetary environment does not spring from the television sets at Mission Control when they display pictures of the Martian or Venerean surface for the first time. In 1976, several years after the Mariner 9 mission had ended, the analysis of these photographs had progressed to the point where we were beginning to understand the forces that shaped the Martian landscape. Even now the information from the Viking landers has not been fully digested, but in general it confirms the conclusions of the Mariner mission. The exploration of Venus has progressed more slowly, since its surface can only be mapped by radar. Most Venerean research has focused on the atmosphere of this cloud-draped planet.

★ 5.2 PLANETARY TEMPERATURES ★

The most dramatic difference between Venus, Earth, and Mars is the difference in their surface temperatures. Venus at 750 K is an oven; Earth at 298 K is a tropical paradise; and Mars at 213 K (the average temperature of the Viking 1 landing site in late summer) is a frigid desert. Why are these temperatures so different? An answer to this question involves an analysis of the *energy balance* that determines a planet's temperature.

Perhaps almost everyone realizes that the earth is warm because the sun shines on it. This simple idea contains more information about planets than might appear at first glance. Manipulate, analyze, and contemplate the idea, and it becomes possible to calculate what the temperature of any planet should be, as long as any possible atmospheric effects are neglected. Sunlight strikes the planet's surface, and some fraction of this light energy is reflected back to interplanetary space. The ground absorbs the remainder of the incident sunlight and heats up to the point at which it re-emits that energy in the form of infrared radiation. The ground thus maintains an energy balance; the energy absorbed from the incident sunlight equals the energy re-emitted by the ground.

One of the factors that affects a planet's energy budget is its distance from the sun, which determines the amount of energy that strikes the planetary surface. Mercury, very close to the sun, feels the full power of solar heat beating down on it, and it absorbs a good deal of solar energy in the process. It becomes quite hot so that it can reradiate this energy into space and balance its heat budget. To far-off Pluto, on the other hand, the sun is just a faint glimmer in the distance, a not very bright star. Receiving very little solar energy, Pluto need only be a few degrees above absolute zero in order to reradiate this same energy back into space.

Yet distance from the sun is not the whole story. A black object, a dark-colored car for example, becomes very hot when it is left in the hot sun, while a lighter-colored object is somewhat cooler. The solar energy that falls on any object, such as a planetary surface, is not all absorbed. Some fraction of it just bounds off the surface and is reflected back into interplanetary space; this reflected energy cannot heat the planet. The ratio of reflected energy to incident energy is called the planet's *albedo*. Light-colored planets like Venus have a very high albedo, since most of the sunlight that falls on them is reflected. Dark planets like Mercury have a low albedo. All other things being equal, a dark planet with a low albedo will be hotter than a light planet with a high albedo, because the dark planet reflects less and absorbs more of the sunlight that falls on it. A planetary surface, absorbing incoming radiation, heats up to the point at which thermal infrared emission balances the absorbed solar radiation.

The heat budget of a planet can be expressed in quantitative numerical terms, shown in Box 5.1. The equations in the box simply express the ideas of the last two paragraphs in a way that allows us to calculate precise values of theoretical planetary temperatures. The results of such calculations are shown in Table 5.1. It is curious that the temperatures calculated for Earth and Venus are rather similar; the higher albedo of Venus means that it absorbs less of the incident sunlight, compensating for its being closer to the sun.

Is this theoretical scheme any good? Three of the four planets in Table 5.1 have had their temperatures measured directly, by landers that took thermometers to the planetary surface. Mercury's temperature was measured by

---

BOX 5.1 CALCULATION OF SURFACE TEMPERATURES OF PLANETS

Consider a planet with a radius R, orbiting the sun at a distance D. The sun has a luminosity L, emitting L ergs/sec of radiant energy into space.[1] How hot does the planet become if its atmosphere does not affect its surface temperature? Assume, to start with, that the planet has a uniform surface temperature.

Draw a sphere around the sun with a radius equal to D, the planet's distance from the sun. The area of that sphere is $4\pi D^2$, and the planet shades a small area on the sphere, equal to πR^2. Thus the planet intercepts a small fraction, $\pi R^2/4\pi D^2$, of the total energy crossing the sphere. The total energy intercepted by the planet every second is then $L(\pi R^2/4\pi D^2)$.

The planet reflects a fraction A of this incident energy back into space and absorbs the rest. The albedo A is high for a light-colored planet like Venus ($A =$ 0.71) and low for a dark planet like Mercury ($A = 0.058$). The planet thus absorbs a fraction $1 - A$ of the incident energy, and so

Energy absorbed

$$= (1 - A) \times L \times \frac{\pi R^2}{4\pi D^2} \quad \text{(B5.1)}$$

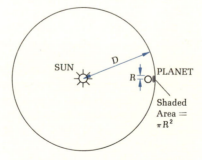

The planet must, in a steady state, re-emit into space all the energy that is absorbed by its surface. The amount of energy emitted by a surface increases as

[1] Recall the discussion of energy units in section 1.2.

cont'd on next page

its temperature increases; every square centimeter of surface emits σT^4 of energy (see Box 1.3). Since the surface area of the planet is $4\pi R^2$,

$$\text{Energy emitted} = 4\pi R^2 \sigma T^4 \quad \text{(B5.2)}$$

Since the energy emitted must equal the energy absorbed, we put the two equations together as:

$$4\pi R^2 \sigma T^4 = (1 - A)$$
$$\times L \times \frac{\pi R^2}{4\pi D^2} \quad \text{(B5.3)}$$

Divide each side by 4σ, take the fourth root of each side, and obtain

Temperature
$$= \left(\frac{L}{16\pi\sigma} \frac{1-A}{D^2} \right)^{1/4} \quad \text{(B5.4)}$$

To apply this to the solar system, let L equal the solar luminosity of 3.90×10^{33} ergs/sec, and express D in astronomical units to obtain, finally,

Temperature (in K)
$$= 281 \left(\frac{1-A}{D^2} \right)^{1/4} \quad \text{(B5.5)}$$

where A is the planet's albedo, and D its distance from the sun in astronomical units. The numerical factor 281 in equation B5.5 comes from plugging the values for L, σ, and $D = 1$ AU into equation B5.4; that is,

$$281 = (3.9 \times 10^{33})$$
$$\div [16 \times 3.14 \times 5.67 \times 10^{-5}$$
$$\times (1.49 \times 10^{13})^2]^{1/4}$$

detecting the thermal emission from its surface. A comparison of theory and observation for all the inner planets is also given in Table 5.1.

Evidently the theory, while producing reasonable agreement with observation in some cases, is quite inadequate in one crucial place. For Mercury the fit is adequate, since the theory assumed that the day and night sides of the planet had equal temperatures. For Earth and Mars the fits are not perfect but not bad. One might be disturbed because the simple theory says that Earth should be colder than it is, and 253 K is below the freezing point of water (273 K). For Venus, however, the disagreement between theory and observation is far too large to be explained as some minor shortcoming of the theory. The theory states that Venus should be a frigid 244 K, below the freezing point of water, but its surface is a torrid 750 K. What's wrong?

Take a closer look at the numbers for Venus. Theory and observation agree quite well if we compare theory with the temperatures at the cloud tops, the layer visible from the earth in Figure 5.1. It is only at the surface, deep below the insulating blanket of the Venerian atmosphere, that the theory falls down. Go back a bit, and you will note that the theory makes a critical assumption. The effects of an atmosphere were ignored. The whole theory need not be thrown out; it can just be modified to include the effects of an atmosphere.

TABLE 5.1 PLANETARY TEMPERATURES

PLANET	AVERAGE DISTANCE FROM SUN (Astronomical units)	ALBEDO	THEORETICAL TEMPERATURE (K)		ACTUAL TEMPERATURE (K)
Mercury	0.387	0.058	442	← Reasonable Agreement →	623 (day) 103 (night)
Venus	0.723	0.71	244	← Poor Agreement →	750 240 (cloud tops)
Earth	1.000	0.33	253	← Reasonable Agreement →	298
Mars	1.524	0.17	216	← Agreement →	220[a]

[a] This is the temperature measured at the surface of the Martian soil at the Viking 1 landing site at latitude 22° north. The temperatures measured by the Viking meteorology experiment, 1.5 m above the Martian surface, are considerably lower.

★ **5.3 THE GREENHOUSE EFFECT** ★

☆ **ENERGY BUDGETS OF PLANETS** ☆
WITH ATMOSPHERES

Figure 5.2 displays the energy budgets of two planets. The left panel depicts the airless planet of the previous section. Sunlight strikes a planet's surface, two-thirds is absorbed by the planet, and the remaining one-third is reflected back into space. The planet's surface balances the energy inflow by reradiating this heat in the infrared part of the spectrum, shown by the two wiggly arrows.

The right panel of Figure 5.2 introduces an atmosphere into the picture. The infrared energy emitted by the ground is used to heat the atmosphere, because the atmosphere absorbs it. The atmosphere becomes hot enough so that it too radiates energy. Some of the radiation emitted by the atmosphere escapes into interplanetary space, and some is directed downward toward the ground. (Here the atmosphere is idealized as a single layer.) The atmosphere has thus trapped radiation between it and the ground.

The heated atmosphere modifies and complicates the energy budget of our planet. Each arrow in Figure 5.2 represents one unit of energy flow. The surface receives four units of energy, two from the sun and two from the atmosphere. It is hot enough to radiate four units of energy in the infrared and keep itself in balance. The atmosphere receives four units of energy from the surface and reradiates all four, two back to the surface and two to interplanetary space. From the point of view of a distant observer in interplanetary space, the planet as a whole is in balance. Three units of incoming solar energy strike it; one is bounced or reflected off the ground, and two are reradiated by the atmosphere. There is an important difference between the planet with an atmosphere and the airless planet, though; the surface of the planet with an atmosphere radiates four units of energy per second, not just two, and must be correspondingly warmer.

115

All that is needed to produce this effect is an atmosphere that can trap infrared radiation. Fortunately most molecules are quite good at absorbing infrared radiation, and two of the best infrared absorbers, carbon dioxide and water, are rather common constituents of planetary atmospheres. The

FIGURE 5.2   The working energy budgets of an airless planet (left) and a planet with a thin atmosphere (right). Each arrow represents one unit of energy transferred per second; straight arrows represent visible light, and wiggly arrows are infrared radiation.

Left: The airless planet's surface receives three units of energy from the sun and reflects one, absorbing two. The surface is heated to the point at which it emits two units of energy per second as infrared radiation. The surface is in balance, absorbing two units of radiation and emitting two.

Right: The working energy budget of a planet with a thin atmosphere demonstrates the need for the surface to become hotter than the airless planet's surface. The atmosphere traps infrared radiation. Check the energy balances: The ground receives four units of energy, two from the sun and two from the atmosphere, and is heated to the point at which it re-emits these four units of energy in the form of infrared radiation. The atmosphere receives four units of energy from the surface and re-emits them, two to interplanetary space and two to the ground. Clearly the surface of this planet must be hotter than the surface of the airless planet; it emits four units of energy while the airless planet's surface emits only two. This figure is a moderately accurate representation of the earth's energy balance.

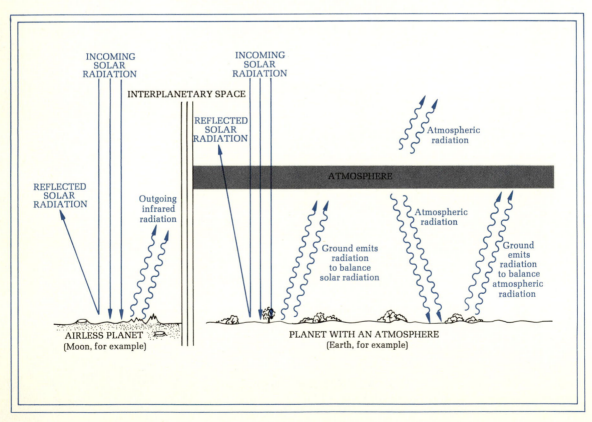

glass in a greenhouse also absorbs infrared radiation and is partly responsible for making the interior of a greenhouse warmer than the cold world outside. Thus, the effect in which an atmosphere warms a planetary surface is called the *greenhouse effect*.

☆ **THE GREENHOUSE EFFECT ON VENUS** ☆

The theoretical model of the greenhouse effect, presented in the previous section, demonstrates that an atmosphere can warm a planetary surface by trapping infrared radiation. But checking whether the greenhouse effect is the real cause of Venus's high surface temperature requires a more precise comparison of theory and observation. A first check is offered by the close agreement of the cloud-top Venerean temperature with the temperature of an airless planet with the same albedo. Figure 5.2 shows that an atmosphere emits the same amount of radiation upward as an airless planet's surface does, so that the upper-atmosphere temperature should (with no other heat sources) be the same as the airless planet's surface temperature. The model checks.

The principal test of the greenhouse effect is a more demanding and more quantitative one. We ask not only whether atmospheres produce warm planetary surfaces but also whether the atmosphere we observe on a planet can create a surface with a temperature in agreement with observation. Figure 5.2 is an approximate representation of the energy balance in the earth's atmosphere, and a quantitative analysis of the energy balance indicates that the greenhouse effect provides a temperature of 300 K, agreeing with the average terrestrial temperature of 298 K. The situation on Venus is considerably more complex since the Venerean atmosphere is very thick. The Russian Venera landers have measured the surface pressure of Venus as 90 times the pressure at the surface of Earth, or 90 atmospheres. This thick layer is 95 percent carbon dioxide, which is a good infrared absorber. Detailed theoretical calculations indicate that the greenhouse effect is probably the cause of the high Venerean surface temperature, but the theoretical models are not completely free of assumptions. It is possible that the Venerean atmosphere may trap infrared radiation only imperfectly, letting a few wavelengths through. Such leaks in the atmospheric blanket would make the greenhouse effect less powerful. This imperfection in the model can be tested rather inexpensively, since laboratory studies of the infrared spectrum of carbon dioxide can determine whether there are any wavelengths that are not trapped by the atmosphere of Venus.

★ **5.4 THE VENEREAN ATMOSPHERE** ★
AND SURFACE

The thick atmosphere of Venus, in addition to heating the surface, makes it extremely difficult to observe the planet. All we can see from the earth is a high, featureless cloud layer, shown in Figure 5.1. This cloud layer can be studied by dispersing the light it reflects into all the wavelengths of the electromagnetic spectrum and using the tools of spectroscopy to analyze its composition. Lower layers of the atmosphere can only be studied by atmospheric entry probes and some "flybys." We have only a fragmentary knowledge of the dynamics of the Venerean atmosphere. Therefore, this section consists mostly of descriptions rather than analyses.

☆   **THE CLOUDS OF VENUS**   ☆

Spectroscopic analysis of the clouds of Venus, the layer we see from the earth, indicates that the principal constituent of the clouds is sulfuric acid, a rather corrosive liquid. The cloud droplets are 75 to 80 percent acid, considerably stronger than that found in an average college chemistry laboratory. The cloud temperatures are near 250 K, and, depending on what else is in the cloud droplets, some of the sulfuric acid may even be solid. Other corrosive acids are also present in the clouds: Hydrochloric acid (HCl) and hydrofluoric acid (HF) have been found in small quantities. Analyses of the water content of the cloud layer vary, but all investigators agree that the cloud layer is quite dry, with only a few parts per million of water. (In contrast, average terrestrial air has a water-vapor content of about 1 percent, or 10,000 parts per million.) Chemical reactions among these various cloud constituents are only partly understood and are currently being investigated by atmospheric chemists. One of the more nasty compounds that could form in the Venerean atmosphere is fluorosulfonic acid (HSO_3F), the strongest acid known.

Studies of the chemistry of the Venerean cloud layer may shed some light on processes taking place in the earth's upper atmosphere or stratosphere. The stratosphere contains a minute quantity of ozone (O_3) which absorbs the sun's ultraviolet radiation and prevents it from reaching the ground and breaking up biologically important molecules. In the 1970s, atmospheric chemists have become concerned that the addition of small amounts of certain chemicals to the stratosphere could modify the chemical reactions taking place there and reduce the amount of ozone. A reduction in the amount of stratospheric ozone would lead to an increase in the skin-cancer rate. Freon (CCl_4) is a common propellant in aerosol spray cans and a refrigerant in air conditioners and refrigerators. In the stratosphere, sunlight can break freon into carbon and chlorine atoms. These chlorine atoms can then destroy the fragile ozone molecule and reduce the thickness of the ozone layer. Is the use of freon a threat to the ozone layer? Unfortunately, we do not know what happens to stratospheric chlorine once it is produced. Does it remain free chlorine and eat ozone molecules, or does it react chemically and become incorporated in another molecule that is not a threat to ozone? The answer is not known at this time, but there is cause for concern because of the widespread use of aerosol cans.

The upper layer of the Venerean atmosphere contains many of the same compounds that are found in the terrestrial stratosphere. Some astronomers, Harvard's M. B. McElroy, the University of Arizona's D. M. Hunten, and their colleagues, have done considerable research on the freon problem.

These investigators had previously worked extensively on trying to understand the chemistry of the Venerean atmosphere. Comparative planetology may play a role in the investigation of the freon threat, even though there are considerable differences between the atmospheres of Earth and Venus.

The Venerean atmosphere, photographed in ultraviolet light by the passing Mariner 10 spacecraft, shows a rather persistent circulation pattern (Figure 5.3). This pattern rotates around the planet once every 4 days. Since the planet itself rotates only once ever 243 days, and backward, this rotation of the cloud pattern indicates that there are 300 to 360 km/h winds whizzing around the planet at cloud level, about 50 km above the surface. It is believed that this circulation is caused by solar heating, since the center of the pattern is located in the center of the sunward side of the planet (on the left side of Figure 5.3), but the detailed dynamics of these powerful winds have not been worked out.

FIGURE 5.3   View of Venus taken by Mariner 10 on February 6, 1974. The Mariner 10 cameras formed images of Venus in ultraviolet light and the Image Processing Laboratory at the Jet Propulsion Laboratory mosaicked and retouched the images to produce this picture. The circulation pattern in the upper Venerean atmosphere is visible. (NASA/JPL Photo No. P–14400)

An astronaut who penetrated the clouds of Venus would leave the corrosive acids above, in the cloud layer, but would encounter a still more hostile place. The lower Venerean atmosphere consists mostly of carbon dioxide, but other gases such as neon and nitrogen may be present in quantities no larger than 5 percent of the total. Temperatures and pressures steadily increase toward the surface. The Russian Venera landers have provided consistent measurements of the surface conditions in a variety of places, with a surface pressure of 90 atmospheres and a surface temperature of 750 K. Venus's slow rotation means that the rapid cycle of day and night would not occur on Venus, and the insulating effect of the atmosphere is so great that daytime and nighttime temperatures are the same. Weak winds, about 4 to 10 km/h, blow on the fearsome oven of the surface.

☆ **THE VENEREAN SURFACE** ☆

Very little is known about the nature of the surface of Venus, because our telescopes cannot penetrate the sulfuric acid cloud layer in the upper atmosphere. Radar waves can penetrate these clouds, and studies of the radar echoes from Venus have begun to yield some preliminary information in the last year or two. Sophisticated procedures are needed in order to determine just what part of the surface is returning a particular segment of the radar echo. As a result, only very large features can be distinguished (Figure 5.4). To date, it is definitely known that there are craters on the surface, and some are hundreds of kilometers across. Such craters are found only in the geologically old regions of the earth, and counts of these craters confirm that the cratered Venerean terrain is probably a few billion years old. However, other radar data have revealed large, irregularly shaped features that are interesting because they are not round. Basins formed by meteorite impact must be round generally, and these Venerean features resemble terrestrial continents in their irregular shapes. These radar results indicate that there may be some plate-tectonic activity on Venus. More radar maps of Venus, to become available in the future, will lead to more definitive interpretations of Venerean geology.

Two Russian spacecraft, Venera 9 and 10, have landed on the surface and returned some pictures (Figure 5.5). The difficulties of operating a spacecraft at a pressure of 90 atmospheres and a temperature of 750 K should not be underestimated, and the picture quality is, therefore, understandably low. However, some large boulders are distinguishable, and it is interesting to note that some light does manage to penetrate the Venerean atmosphere.

In contrast to Venus, the Martian surface is easily seen from Earth. While Venus is a good subject for the atmospheric scientist, geologists have practically no data to work with. Mars, however, can be geologically studied and compared to Earth. It is not geologically inactive like the moon or Mercury, but it is not as geologically active as Earth. Many of the processes that shape the Martian surface are similar to processes that occur on Earth, but these

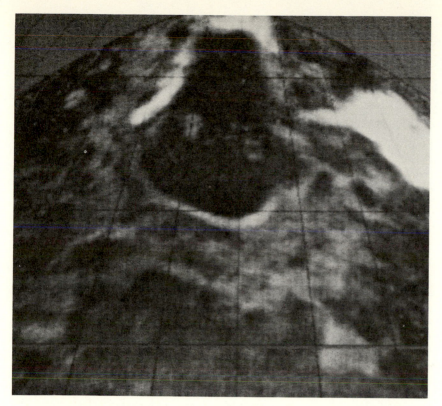

FIGURE 5.4 Radar waves
can penetrate the clouds of
Venus and produce pictures
of the surface like this one.
The large, dark basin is
1500 × 1000 km, about the
size of Hudson Bay. The
bright feature, tentatively
named Maxwell, is about the
size of the state of Okla-
homa. The irregular shapes
of these features suggest
they were not formed by impact processes alone. (From
D. B. Campbell et al., "New
Radar Image of Venus," Science, Vol. 193,
pp. 1123—1124, fig. 1, 17
September 1976. Photograph
courtesy of D. B. Campbell,
Arecibo Observatory.)

FIGURE 5.5 Panorama of the surface of Venus from the Venus spacecraft. The fish-
eye effect is due to the manner in which this photograph was taken and is not real.
The light area at the bottom is part of the spacecraft; the vertical stripes are gaps in the
picture where technical data were transmitted. Rounded boulders are visible on the
surface. (Wide World Photos)

processes function in rather dissimilar environments. Comparative planetology can greatly benefit from such comparisons, as demonstrated by our earlier discussions of cratering on different planets, planetary interiors, and the upper layers of the terrestrial and Venerean atmospheres.

★ 5.5 THE DESERTS OF MARS ★

☆ MARS BEFORE MARINER ☆

Seen through a small telescope in a moderately stable atmosphere, Mars looks like a small, orange blob. Under normal conditions, the average observer can distinguish a polar cap and can sometimes see a dark area or two. Atmospheric turbulence prevents us from seeing fine details because air currents in the earth's atmosphere distort the light coming from the planet. Normal viewing conditions obscure any features smaller than 300 km across. Most terrestrial photographs of Mars are also affected by atmospheric turbulence, and none are shown in this book. The photograph of Mars in Color Photo 5 was taken from the Viking 1 spacecraft.

A few persistent souls, notably Giovanni Schiaparelli, E. M. Antoniadi, and the redoubtable Percival Lowell, spent many long, cold, and frustrating nights at the telescope in search of those few moments of exceptional seeing when the dancing image of Mars becomes steady and the surface features appear clearly. Figure 5.1 is an example of one drawing of Mars made under good conditions. The human eye is superior to the photographic plate when it comes to observing detail on a planet like Mars, because a skilled observer can keep those exceptionally steady views in mind, while the photographic plate averages the good and bad moments together. The picture of Mars that emerged from decades of optical observations was that of a red planet covered with dark markings. The overall pattern of the markings remained fairly constant, and these features were named and mapped.

Drawings have one major disadvantage as a medium of recording observations. The light pattern striking the astronomer's eye has to pass through the mind before it is recorded on paper. Most planetary observers were meticulously careful in their renderings, but on nights of good but not superb observations, some of the dark splotches on the planet tended to be connected into linear dark features. Some of these linear features are shown in Figure 5.1; they were given the unfortunate name of *canali* by Father Pietro Angelo Secchi in 1869. Perhaps he was just following convention. The Martian dark areas, seemingly connected by the canali, were named after seas, following the traditions of lunar nomenclature.

The unfortunate name of canali cursed the study of Mars for a long time. Extensive disputes raged between those who believed the canals were real and those who believed that the canals were optical illusions, caused by the tendency of the mind to connect features at the limits of visibility into straight lines. (We now know the latter view is the correct one; the canals do not exist.) Perhaps it was the romanticism of the late Victorian era that drove Percival Lowell, an otherwise respectable Boston Brahmin, to write the fanciful book *Mars as the Abode of Life*. Lowell seriously proposed that the canali were in fact huge ditches constructed by intelligent beings seeking to transport water from the polar caps to the warmer equatorial regions. Edgar Rice Burroughs, also known as the creator of Tarzan, immortalized the canals of Mars in a series of science fiction novels chronicling the adventures of John Carter, who roamed the surface of this desert planet like the hero of a nineteenth-century romance, saving damsels in distress and battling wicked monsters.

The canal episode gave Mars in particular and planetary astronomy in general a bad reputation in the scientific world. A working scientist could not address the interesting problem of the origin of the light and dark surface markings because the question of the existence of the canals was extremely persistent. An astronomer who denied the importance of canals would be greeted with howls of protest from science fiction fans who, after all, wanted to believe that the desert kingdoms of Edgar Rice Burroughs were real (or at least that they could be real). Astronomical research was often misinterpreted as supporting the existence of canals or dismissed as academic hedging. In such a climate, many astronomers chose to work on stars rather than on planets.

Those few early investigators who did not abandon planetary science in that troubled first half of the twentieth century nevertheless had a frustrating time with Mars. The existence of light and dark areas on the surface, the classical markings of Mars, provoked the question of why the two types of areas differed. Space-age research showed that this question led nowhere, for the light and dark areas are not geologically different.

☆ THE ORIGIN OF THE ☆ CLASSICAL MARKINGS

Figure 5.6 is a more modern map of Mars, made from photographs taken by the orbiting Mariner spacecraft. The pattern of light and dark areas has disappeared, and the whole map is covered with craters, a canyon, four volcanoes, and a variety of other geological features. Closer examination of one

FIGURE 5.6 A map of the most geologically interesting region of Mars, made from the Mariner spacecraft. The four objects labeled "mons" (mountain) are volcanoes. Olympus Mons is the largest known volcano in the solar system. Valles Marineris is a huge canyon complex, and Chryse Planitia is a volcanic plain somewhat similar to the lunar mare. The first Viking lander touched down in Chryse Planitia. (USGS/NASA/JPL)

of the classical dark areas of Figure 5.1 shows that they are made up of a group of streaks (Figure 5.7). These streaks are all very small, at most a few kilometers across. The dark areas of the classical maps are simply regions that have more dark streaks than the lighter colored regions. Both types of terrain are just cratered desert. A close examination of the Mariner 9 photographs revealed no canals or other unusual features in the areas where Lowell and others had drawn them. Science fiction fans must realize that the Mars of Edgar Rice Burroughs is purely fictional.

The revelation of the dark areas as a pattern of streaks guided planetary scientists toward the right approach to understanding the classical markings of Mars. A search for the origin of these dark streaks provides considerable insight into Martian geology and meteorology. Close and extensive study of the Mariner 9 photographs indicates that the streaks are the result of strong winds. When a wind encounters an obstacle in the Martian landscape, the airflow pattern downwind of the obstacle is disrupted. Downwind turbulence causes wind-blown dust particles to be deposited on the surface, creating a dark streak. Analysis of Martian meteorology indicates the existence of the 100- to 200-km/h winds that are needed to carry dust in the thin Martian atmosphere.

FIGURE 5.7 A closeup photograph, from Mariner 9, of Syrtis Major, the prominent dark area in the drawing of Figure 5.1. Syrtis Major is the most prominent of all the classical markings and is visible even under poor observing conditions. The close view of Mariner 9 shows that Syrtis Major contains many dark streaks that are aligned with the prevailing wind direction. (NASA Photo No. 72–H–787)

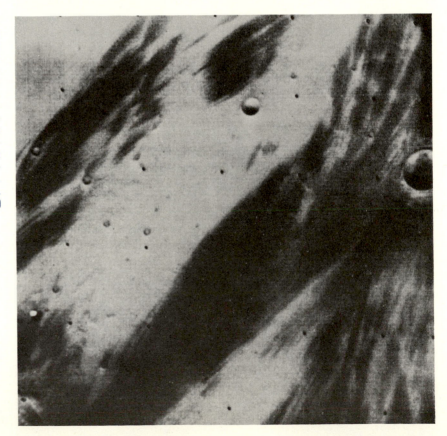

This explanation of the classical markings, while satisfactory, raises some further questions. The overall pattern of dark markings persists from one year to the next, but individual areas show seasonal variations. This phenomenon can be explained as a result of strong winds existing only in certain regions of the Martian surface. Such regions appear dark. Syrtis Major, the largest Martian dark area and one clearly visible in average seeing, lies on a sloping area with strong winds. Current research is trying to clear up the details, to relate the pattern of light and dark areas to the distribution of strong winds over the Martian surface.

★ 5.6 ANOTHER DESERT PERSPECTIVE: ★
THE VIKING LANDER

On July 20, 1976, NASA's Viking 1 lander touched down on the western slopes of the Chryse basin, one of the desert regions of Mars. It took the first recognizable picture of the Martian surface (Figure 5.8) almost immediately after touchdown. A second lander, Viking 2, touched down in a more northern region, nearer the Martian pole, on September 3, 1976; it, too, took a picture of its own footpad and the surrounding terrain (Figure 5.9). Neither craft was the first to land on Mars, for a Russian probe, Mars 3, landed in the middle of a giant dust storm in 1971. It stopped transmitting data after 20 seconds, producing a picture showing nothing recognizable.

FIGURE 5.8   The first intelligible photograph ever returned from the Martian surface. The Viking 1 lander photographed the area surrounding its footpad on July 20, 1976, in Chryse Planitia. (NASA/JPL Photo No. P−17043).

FIGURE 5.9 The Viking 2 lander, like the Viking 1 lander, also photographed its footpad, showing a scene rather similar to that from Viking 1. (NASA/JPL Photo No. P–17681)

Although the 1960s and 1970s have been characterized by continuous exploration of the solar system and expansion of human horizons, a few achievements stand out as technological and philosophical milestones: October 4, 1957—launch of Sputnik 1, the first earth satellite; April 12, 1961—the first manned orbital flight (Yuri Gagarin); January 31, 1966—the first soft landing on the moon (Luna 9); July 20, 1969—the first manned lunar landing (Apollo 11); and July 20, 1976—the Viking 1 soft landing on Mars. Only the preliminary, first-look results are back from the Viking mission at this time, and only some preliminary analyses of the data have been made. More insights from the Viking missions lie in the future.

Both Viking landers touched down in desert regions of Mars. Their data have filled out and confirmed our picture of the Martian plains as frigid deserts containing large quantities of dust. Figure 5.10 shows a panoramic view of the terrain surrounding the Viking 1 lander, and Figure 5.11 shows a comparable panoramic view from Viking 2. Turn back a few pages and examine the lunar surface photograph of Figure 4.2, the lunar picture of Color Photo 1, the Venus photo of Figure 5.5, and the Viking views in Color Photos 3 and 4. The lunar photographs are of volcanic plains; the Viking spacecraft were targeted toward flat, volcanic regions of Mars so that they would not inadvertently land on a steep hillside or on top of a boulder and tip over. The volcanic plains of Mars are quite similar to those on the moon; they were carved out by meteorite impact and have since been filled by lava flows and wind-blown dust.

The classical markings of Mars were explained earlier as being due to the transport of dust particles by wind. The Viking photographs reinforce the idea that wind-driven or aeolian transport of dust is prevalent on the Martian surface. Dust, raised by the retrorocket blasts, is evident in the Viking 1 and Viking 2 footpads (Figures 5.8 and 5.9). A dune field is found in the Viking 1 panorama. It results from the presence of 2- to 3-m boulders, visible at the left

FIGURE 5.10 A panoramic view of the landscape on Mars in the vicinity of the Viking 1 lander. The largest rock visible, at the left, is about 10 m from the space-craft and about 2 m across. (NASA/JPL Photo No. P–17430)

FIGURE 5.11 The Viking 2 panorama, showing far more rocks than the comparable view from Viking 1 (Figure 5.10). The flatness of the landscape in the vicinity of the Viking landers, shown in these two photos, is probably not typical of Mars, since the flat Viking sites were chosen for safety. (NASA/JPL Photo No. P–17689)

of the panorama (Figure 5.10), that partially block the winds and encourage deposition. Closeup photographs of the rocks in the vicinity of Viking 2 indicate aeolian erosion by the action of wind and dust. The vesicular or bubbly texture of the rocks at the Viking 2 site (Figure 5.9) may be the result of their volcanic origin as well, but wind certainly plays a role in shaping these blocks.

Some color photographs of the Martian surface are shown in Color Photos 3 and 4. The vivid and true colors of the American flag provide assurance that the computer has rendered the colors correctly. These photographs illustrate another consequence of the prevalence of wind and dust on Mars: the pink Martian sky. The earth's sky, unlike the Martian sky, is blue, because the very small atoms in the earth's atmosphere scatter sunlight and scatter blue light more effectively than they scatter red light (Figure 5.12). Someone looking at the sky sees scattered sunlight; most of the scattered sunlight in the earth's sky is blue. The pink color of the Martian sky indicates that the particles doing the scattering are not small like those in the earth's atmosphere; rather, the Martian atmosphere is full of dust particles about a micron across. These large particles scatter light according to their color; the pink Martian dust scatters red radiation more effectively than it scatters blue radiation. The sky color is yet another demonstration of the prevalence of dust on Mars.

But dust, by itself, is insufficient to explain the classical markings and the pink sky color. There must be winds strong enough to blow the dust around. The Viking landers carried sensors to measure the weather conditions on Mars, although neither lander is in a very windy location. At the Viking 1 site, the wind blows from the south or southwest most of the time, averaging 4 m/sec and gusting to 7 m/sec [4 and 9 miles per hour (mph), respectively]. The Great Plains of North America have a similar regional slope, and there, too, the winds blow from the south. At the Viking 2 site, the winds are from the southeast and the wind speeds are similar. These low-velocity winds cannot raise dust particles from the surface, but late summer is expected to be a season of light winds because the temperature and air pressure on the Martian surface are relatively uniform. Further, the Viking landing sites were selected, for safety reasons, to be in windfree regions. Mariner 9 provided evidence for stronger winds in winter. The winter winds will be measured at the Viking landing sites.

The Viking landers are admirably situated to monitor the Martian temperature, and measurements of the air temperature 1.5 m above the surface show that Mars is a very cold place by terrestrial standards. Temperatures in the late summer season, shortly after the landings took place, varied between 190 and 240 K at both sites, or roughly between $-80°$ and $-30°C$. It is expected that winter temperatures at the Viking 1 site will be rather similar, since it is located near the equator at Martian latitude $22°$ north. The Viking 2 lander site, at latitude $48°$ north, will probably become considerably colder in the winter season. The climate in late summer at the two lander sites is comparable in temperature to the coldest place on the earth, the Vostok research station in the Antarctic interior, which has temperature extremes of 229 to 250 K in the summer and 190 to 217 K in the winter.

The volcanic plains of Mars, then, are frigid desert areas, where surface rocks abound. The appearance of these desert areas from the earth is domi-

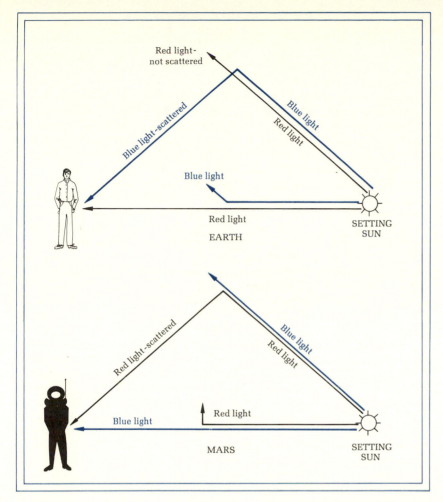

nated by wind-blown dust. Why is the dust pink? The Viking landers contained surface samplers to measure the composition of the lunar soil; they confirmed earlier guesses, based on spectroscopy from terrestrial telescopes, that the principal coloring agent is iron oxide or rust. The red rusty coating is only a thin layer on each dust particle. Chemical analysis of four soil samples shows a uniformly high proportion of iron, between 18 and 20 percent of the total weight of any sample. The remainder of the Martian soil consists of oxides of other common elements like silicon, magnesium, calcium, sulfur, and some aluminum. Such a composition indicates that most of the material in the Martian soil consists of iron-rich clay. On the earth, clays form when there is abundant surface water. Thus the soil analyses indicate that some time in the past water must have been more prevalent on Mars than it is now. No free ice was found near the lander sites, and the water that is there is probably chemically combined with other soil material.

★ 5.7 THE SEARCH FOR LIFE ★
ON MARS

Ever since the canal episode, the idea of life on Mars has intrigued the human mind. The lander evidence that there must have been more water on the surface in the past (to account for the existence of clays) is consistent with the existence of Martian channels, huge riverbeds that at one time contained flowing water (section 5.9). Much of the instrumentation on the Viking landers was connected with one of the chief purposes of the mission, the search for life on Mars. Hostile as the Martian environment is, a number of investigators believed, or hoped, that the climate may have been better in the past and that life would have evolved.

Each of the three life-detection experiments on the landers was an attempt to detect chemical activity in the Martian soil that might be associated with living organisms. Two experiments used a technique adapted from modern medicine, the technique of labeling compounds that are used by living organisms. Carbon compounds were labeled by including a larger than normal porportion of radioactive carbon-14 atoms in the molecules. Soil was then exposed to these carbon compounds in an incubation chamber. If the soil were to react with any of the carbon compounds, it would become radioactive. Instruments measuring the radioactivity of the soil could thus show whether the soil reacted with carbon compounds in a lifelike way.

The experiments differed in their details. One, the "carbon assimilation" (or pyrolytic release) experiment, placed the soil in a chamber and exposed it to labeled carbon dioxide molecules. If there were life in the soil, it would breathe in the labeled carbon dioxide and assimilate it into more complex biological chemicals. After a reasonable incubation time, the chamber was evacuated and the soil burned to see how much radioactive carbon dioxide had been chemically altered by the soil or organisms in it. Another experiment, called the "labeled release" experiment, fed the soil with a mixture of radioactively labeled likely foods and monitored the radioactivity of the air above the soil; living organisms would eat some of the labeled food and breathe out radioactive carbon dioxide, causing an increase in the radioactivity of the air. (The amount of radioactivity used in these experiments was quite low, just enough to allow the experimenters to measure whether the food had been eaten or whether the air had been breathed.) A third experiment, the "gas exchange" experiment, moistened the soil and monitored the chemical composition of the air above it, seeking to detect the byproducts of chemical or biological reactions in the soil.

In the first few weeks of August 1976, the scientific community was startled to learn that the first Viking lander reported some kind of activity in all three biology experiments. The soil did indeed react in the appropriate way; it seemed to breathe the air above it and eat chemicals that were fed it. The Viking investigators then sterilized a soil sample by heating it to 170°C for 3 hours and repeated the experiments. Any life forms would be killed by this treatment, intended to sterilize the soil sample. None of the three experiments reported any activity from the sterilized soil sample, although carbon assimilation seems to be inhibited by the addition of water. The Viking 2 lander biology experiments produced similar results.

The results of the Viking experiments, however, are unclear. The Viking investigators and the scientific community would be more optimistic about the existence of life on Mars were it not for the results of another experiment that measured the abundance of large organic molecules in the Martian soil. All terrestrial life is composed of large organic, carbon-containing molecules. When an organism dies, these molecules remain in the soil. Even the most sterile terrestrial soil samples, from remote cold valleys in Antarctica, contain large organic molecules at the level of 10 to 1000 parts per million. The Viking 1 lander found no large organic molecules in the Martian soil. Such molecules would have been detected if they were present at the level of 1 part per billion of the total weight of the soil sample. Thus large organic molecules are at least 1000 times less abundant in Martian soils than they are in terrestrial soils, even those terrestrial soils that do not support much life. This information indicates that if life on Mars exists, there must be many scavenging bacteria that eat up biological debris very rapidly. While the biology experiments indicated that the soil eats and breathes carbon compounds, the organic compounds that would associate the eating and breathing with life rather than with soil chemistry are just not there.

How are the Viking experiments to be interpreted? Is there life on Mars? No answer to this intriguing question is possible at this time. Experimentation continued until the necessary on-board chemicals ran out in mid-1977. Soil chemists in terrestrial laboratories are attempting to duplicate the activity of the Martian soil shown by the biology experiments, to see if these reactions are caused by chemical compounds not found in terrestrial soils. Two explanations of the results of the biology experiments exist, the biological and the chemical. Neither is totally satisfactory: the biological explanation has to cope with the absence of organic molecules in the Martian soil, and the chemical explanation has no firm candidate for a life-imitating compound that might exist on Mars. Something in the soil acts like a living organism, eating and breathing but not leaving large numbers of carbon compounds in the soil as terrestrial organisms do.

Viking scientists avoid a direct answer to the life question. The team reporting on the biology experiments stated, in October 1976, "The experiments described above give clear evidence of chemical reactions. The essential question is whether they are attributable to a biological system. We are unable at this time to give a clear answer to that question."[1] The team then discussed the soil analysis that showed no organic molecules and, discussing the Viking 1 analysis, noted: "These results, especially if reinforced by analyses at a second site [as they were by the Viking 2 analysis], would tend to make biology on Mars less likely, at least in the terrestrial mode."[2] In December 1976, the director of the Viking program summarized: "No conclusions were reached concerning the existence of life on Mars."[3] Each of those who have followed the situation has a personal interpretation of the results, and mine is that it is unlikely that the Viking spacecraft has found life on Mars. Others are less pessimistic than I.

[1] H. P. Klein, N. H. Horowitz, G. V. Levin, V. I. Oyama, J. Lederberg, A. Rich, J. S. Hubbard, G. L. Hobby, P. A. Straat, B. J. Berdahl, G. C. Carle, F. S. Brown, and R. D. Johnson, "The Viking Biological Investigation: Preliminary Results," *Science* 194 (October 1, 1976): 104.

[2] Ibid.

[3] G. A. Soffen, "Scientific Results of the Viking Mission," *Science* 194 (December 17, 1976): 1276.

The Viking lander has thus provided an ambiguous answer to the life question, but its pictures and soil analyses have provided a good picture of the volcanic plains on Mars. These plains contain craters as well, which are quite similar to their counterparts on Mercury, the moon, and Earth. If the volcanic plains were the only type of Martian terrain, one could say that Mars is quite moonlike in its overall character. But there are some far more exciting Martian landscapes, with too many hills and cliffs to make a Viking landing safe. These regions are almost featureless when seen from the earth. To a geologist, the regions of Mars where volcanism and tectonics have played an active role in shaping the surface are the most fascinating parts of Mars.

★ **5.8 VOLCANISM AND** ★
TECTONIC ACTIVITY

☆ **THE GREAT VOLCANOES** ☆

When Mariner 9 approached Mars in 1971, the anxious investigators waited for the first pictures to come back from the spacecraft. Would the camera system work? Yes, it did work, but the first pictures showed a disappointing featureless disk, less interesting than the worst views seen by terrestrial observers under average conditions. Earthbound astronomers, peering through their telescopes, confirmed that the lack of Martian surface features was not due to poor focus in the camera system: Mars was being ravaged by the worst dust storm seen in years.

The Martian investigators waited; soon the Mariner 9 cameras showed four dark spots peeking through the dust layer. These spots were all in a region that nineteenth-century astronomers had named Tharsis, and one of these spots had itself been named Nix Olympica, Snows of Olympus. These dark spots were all giant mountains, tall enough to protrude through the dust layer that was settling as the mission progressed.

Soon the dust settled to the surface, and the Mariner 9 mission returned an enormous number of photographs of the Martian surface. The four spots turned out to be gigantic volcanic mountains. The largest, renamed *Olympus Mons* because it is a mountain, is shown in Figure 5.13. This volcano is the largest mountain known in the solar system. Its summit is about 20 km above the surrounding plain and 24 km above the Martian surface. Its base, 500 km across, would stretch from Philadelphia to Pittsburgh, from St. Louis to Kansas City, or four-fifths of the way from Los Angeles to San Francisco. A comparison of this enormous feature with its closest rivals on the earth is shown in Figure 5.14, where Olympus Mons and part of the Hawaiian Island chain are drawn to the same scale. The other three volcanoes of the Tharsis ridge are not as gigantic as Olympus Mons, but they still dwarf most terrestrial mountains.

The Martian volcanoes and their statistics make good copy for the *Guinness Book of World Records,* but we want to probe beyond the superficial statement that they are large. Such features are not found on the moon. Their presence indicates that volcanic activity, the ejection of large volumes

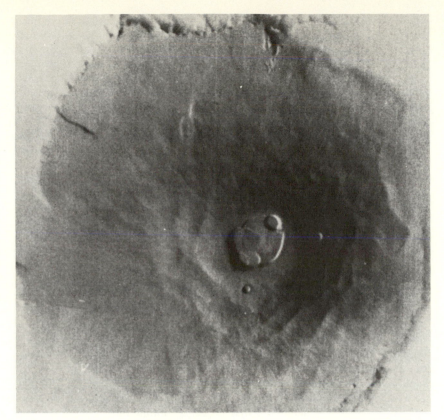

FIGURE 5.13 Olympus Mons, the largest known volcanic mountain in the solar system. (NASA)

FIGURE 5.14 Profiles of two sets of volcanic mountains drawn to the same scale. These mountains are far flatter than shown, because their vertical height has been exaggerated. Olympus Mons, on Mars, is far larger than any of the Hawaiian Islands.

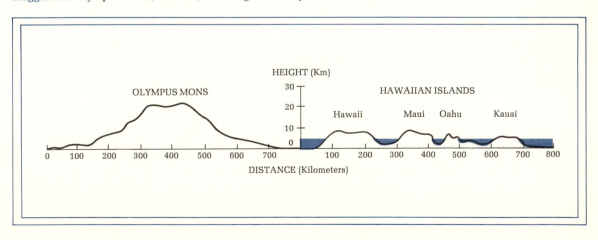

of molten lava from the planetary interior, played a large role in the shaping of the Martian surface. Why, though, are the Martian mountains so much larger than their terrestrial counterparts? Consider how volcanic mountains like Olympus Mons or the islands in the Hawaiian chain are formed.

A volcano forms when lava, or molten rock, is extruded from a planet's interior through a fissure or crack in the crust. Successive eruptions of the volcano build up a mountain of cooled lava, and when the lava flows moderately freely as the Hawaiian Islands' lava does, a large *volcanic shield* with a gradual slope is built up. Some terrestrial volcanoes are built of less freely flowing lava and have steeper slopes, and some are made of cinders and are extremely steep.

There is one crucial difference between Olympus Mons and the Hawaiian Island chain: Olympus is but one volcanic mountain, while there are many islands in the Hawaiian chain. Current geological research suggests that the Hawaiian Islands all originated from the same fissure in the earth's crust, but that continental drift carried the Pacific plate over this hot spot, causing the volcanoes to form at different places. Presumably Olympus Mons has sat over the same hot spot for a long time; we have no evidence of continental drift on Mars. Lava has been able to build up one huge volcano instead of a chain of smaller ones.

Olympus Mons is the centerpiece of the geologically exciting half of Mars centered on the Tharsis region. Half of this enigmatic planet is a sandy desert, pockmarked with craters, but the other half contains many volcanic features. The volcanic half of the planet bulges upward about 11 km above the average height of the Martian surface. Presumably, pressure from hot rock in the interior produced this rise in the level of the terrain. The volcanoes are about a billion years old, younger than the deserts, and so a reasonable conclusion is that a huge mass of molten rock bulged upward toward the Martian surface a billion years ago. This inferred history is a great contrast to those of the lower-mass rocky planets—Mercury and Mars—where the volcanic episodes occurred much earlier.

☆ CANYONS AND TECTONICS ☆

The upward pressure in the Martian volcanic hemisphere created a number of cracks in the crust of the planet, because the crust had to stretch in order to bulge outward. The record-breaking crack is a long canyon complex called *Valles Marineris*, shown in Figure 5.6, stretching one-sixth of the way around the planet, 200 km wide in spots and 6 km from top to bottom. Those who have seen the majesty of the Grand Canyon should be impressed by comparison: Valles Marineris is 27 times as long, 7 times as wide, and 3 times as deep as the Arizona landmark. Yet the Grand Canyon is a different type of geological structure because it was cut by the action of flowing water. The Martian canyon is a huge crack in the crust, produced when pressure from the molten rock beneath the Tharsis uplift pushed pieces of the crust apart. Similar cracks exist on the earth: the Red Sea formed when the Arabian peninsula and Africa drifted away from each other under the inexorable pressures of continental drift. The Red Sea crack actually extends into eastern Africa (look at a map, and note that sea level is an arbitrary way to distinguish high and low terrain levels on the earth's surface). When the Red

Sea—East African Rift complex is considered as a whole, it is longer than Valles Marineris on Mars but only about one-third as deep.

Valles Marineris is just the largest example of a number of cracks in the Martian surface. All these cracks point to a central region near Tharsis and Elysium, in the center of the Martian volcanic hemisphere. One can speculate as follows: Surges of molten rock, pushing toward the Martian surface from reservoirs deep in the interior of Mars, created an intense pressure on a small area of the Martian surface. The surface did the only thing it could do; it bulged upward, creating a domed area that extends over half the planet. The rocky Martian crust is not like rubber or bread dough; it could not stretch to accommodate this pressure. The vast chasms, including Valles Marineris, opened up slowly as the pressure intensified.

Further study of these chasms will show whether or not this speculative scenario is correct and will fill in many details. The cracking process is a *tectonic* process because these chasms were produced by land movement, but it lacks the large-scale drama of plate tectonics or continental drift on the earth.

The volcanic hemisphere on Mars somewhat resembles the earth, but on Mars volcanism does not have to compete with another form of mountain building and canyon carving, plate tectonics. We can better understand the formation of volcanic shields like the Hawaiian Islands by studying their Martian analogs. Rift valleys like Valles Marineris are formed somewhat differently than their terrestrial counterparts, and where there is a difference there is tremendous potential for understanding. A geologist can do an "experiment" by comparing the evolutions of similar land formations under different conditions, and Mars is an ideal laboratory for such studies. Future analysis of Mariner and Viking photographs of the Martian surface will probably yield profound insights for terrestrial geology. Hopefully, there will be government support for researchers who wish to spend years examining the wealth of data that the probes have returned. No one can guarantee that Martian studies will produce new insights into terrestrial geology, but it seems likely.

Mars, as seen today, is a rather inhospitable planet. The Viking pictures confirm the earlier data of Mariner 9—the terrain is forbidding, and the weather extremely cold. Has Mars always been a frigid, rocky desert, where the silence is broken only by the thin Martian winds whipping up the pink sands? Perhaps the most significant finding of the Mariner mission was the discovery of a number of surface features indicating that Mars may have been more temperate in the past. The Viking orbiter further explored these surface features, and the lander's atmospheric analysis indicates that the planet may well be in an ice age. The rocky deserts of Mars are occasionally broken by the channels of streams that once flowed over the Martian surface.

★ 5.9 THE MARTIAN ARROYOS ★

Figure 5.15 is a photograph of a complex of Martian arroyos or channels. An *arroyo*, a common feature of the southwestern United States, is the bed of a stream that flows only occasionally. In the rainy season, the stream can be a raging torrent, but in the dry season only the streambed is left to show that

FIGURE 5.15   A mosaic of photographs from the Viking orbiter, showing an arroyo complex in Lunae Planum, just west of the Viking 1 landing site in the Chryse basin. The terrain slopes from left to right in this photograph. The tributary structure is evidence that these arroyos once carried water into Chryse. (NASA/JPL Photo No. P–17698)

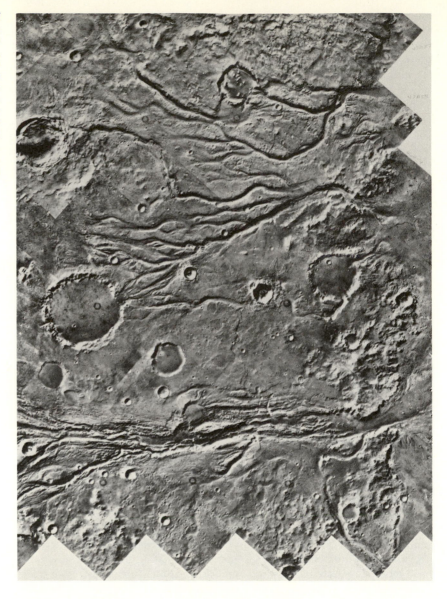

water once flowed there. Figure 5.15 shows that similar structures exist on Mars. The terrain in the figure slopes from left to right, dropping about 3 km over the region in the photograph. Tributary structure, where small streams coalesce to make a larger one, is quite evident, and this structure is a sign of fluid flow. Some investigators call such features on the Martian surface "channels" so that their name carries no implications about their origin. But since the evidence clearly indicates that they were cut by flowing water, the name arroyo is rather appropriate and will be used here.

Why is the presence of arroyos on Mars so shocking? After all, terrestrial deserts contain such features (Figure 5.16). What's wrong with desert features like arroyos in a desert landscape—Mars? Mars has a far colder and thinner atmosphere than Earth has. The average Martian surface pressure is 6.1 millibars, 0.6 percent of the earth's surface pressure of 1013 millibars. Martian temperatures above 0°C, the freezing point of water, are the exception rather than the rule. With such a thin, cold atmosphere, liquid water cannot exist on Mars. Molecules in a liquid are always escaping through the liquid's surface, and there must be enough vapor in the air to replenish the liquid molecules that evaporate. The Martian atmosphere is so thin that such replenishment cannot take place. Water evaporates spontaneously.

The cool temperature of Mars also discourages the possible formation of arroyos in the present environment. The average temperature is 220 K, or −53°C. While the equatorial regions may be warmer at times, data from the Viking landers suggest that it is only at noon on the warmest of summer days that the temperature rises above the freezing point. In the winter, the temperature sinks as low as 150 K (−123°C) in the polar regions, allowing the carbon dioxide in the Martian atmosphere to condense as dry ice.

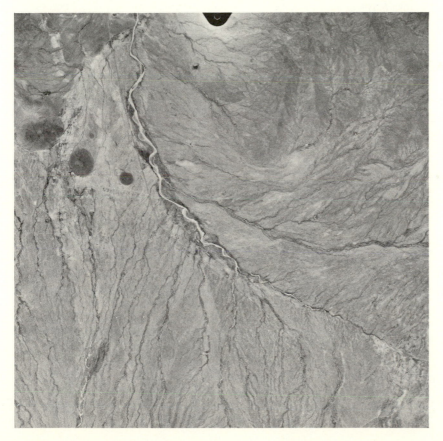

FIGURE 5.16   An arroyo near the city of Tucson, Arizona. These arroyos, streambeds that carry water during the rainy season only, are quite common in the southwestern United States; the arroyos on Mars (Figures 5.15 and 5.17) are quite similar in appearance. (NASA/USGS/Eros Data Center; Photo No. 572000397ROLL, frame 905)

Where did the flowing water come from to make the arroyos? Rain is needed to make a river, and rain is condensed water vapor. If all the water vapor in the present Martian atmosphere were precipitated in the form of rain, the rainfall record would be 0.002 to 0.005 cm, hardly enough to make a slim trickle of water, let alone a 1200-km river. A channel complex west of Tharsis, in the Amazonis region, is 300 km long; the channel complex ending in Chryse, the landing site of Viking 1, is 1200 km long. How can such long channels exist? Under present conditions, any water flow would evaporate before it traveled a few tenths of a kilometer.

The clear impossibility of producing the arroyos under present Martian conditions led some investigators to suggest that the arroyos were not produced by water flow. Lava flows, which may have produced the sinuous rilles on the moon, have been suggested as possible sources of these surface features. However, the Viking photographs have proved that the arroyos are carved by running water. Several characteristics of these arroyos support this argument:

Braided channels often occur in terrestrial arroyos when sandbars are built up in the channels and subsequent water flows cut across the bars. Such braids are commonly found in the Martian arroyos (Figure 5.15, left center; Figure 5.17).

Tributary structure (Figure 5.15) indicates that several flows coalesced at a low point in the terrain to form the main line of water flow.

Multiple flow lines (Figures 5.15 through 5.17) in the channel bottoms argue that the liquid that cut the channels flowed quite rapidly. Viscous, gummy liquids like lava would fill the channel more uniformly.

Meandering patterns indicate liquid flow.

The downhill direction of the channels tends to rule out an explanation of the channels based on wind patterns. All the arroyos are directed from higher regions to lower regions.

What else, other than flowing water, could produce the arroyos? Carbon dioxide does not exist as a liquid unless the pressures are extremely high, 5 times the atmospheric pressure at the earth's surface and 800 times the present Martian surface pressure. Aeolian or wind erosion might carve features with some arroyo characteristics, but it would be difficult for winds to reproduce all the properties listed above. The reasoned yet reluctant conclusion of all the papers that I have read is that the arroyos were cut by flowing water. Photographs taken by the Viking orbiter reinforce this conclusion.

Since present conditions on Mars do not allow the existence of the long rivers needed to carve the arroyos, Mars must have had a different climate in the past. A thicker atmosphere and warmer surface temperatures would have enabled rivers to flow and cut stream channels. The added warmth would permit the existence of liquid water, and the thicker atmosphere would prevent the water from evaporating before it reached the river mouth. The present climate of Mars would then be representative of an ice age— cold and dry; the past climate was presumably warmer and wetter. Counts of

FIGURE 5.17 Braided channels in an arroyo on Mars, associated with Vallis Mangala in Amazonis. The photographed area is about 50 km from top to bottom. (NASA Photo No. 72–H–343; Mariner 9, P–12926)

the number of craters in channel bottoms provide some preliminary indications that the channels were cut hundreds of millions of years ago. Thus the warm era in Martian history is probably not a recent event.

What caused Mars to become so cold and dry? This question takes us from observational fact and reasonably solid deductions based on the Mariner and Viking photographs to educated guesses and speculation. The last section of this chapter outlines a possible evolutionary history for Venus, Earth, and Mars—a history that explains both the great differences among these superficially similar planets and the existence of a warmer climate in the Martian past. Various parts of this history have appeared in scientific journals, but the appearance of an idea in research literature is not a guarantee of truth. Most planetary scientists regard this evolutionary scenario as a good working hypothesis, a useful idea to be tested by future investigations.

★ 5.10 THE CLIMATIC HISTORY OF ★
VENUS, EARTH, AND MARS

In the beginning, each of these planets was a ball of molten rock that condensed out of the gas cloud that produced the solar system. While a giant cloud of hydrogen and helium gas may have enveloped each planet, the light gas atoms were moving too fast to remain fettered to these small planets with their feeble gravitational attractions. Such primeval atmospheres as existed soon dissipated, leaving these planets airless globes.

Soon the planets cooled and formed solid crusts; volcanoes belched forth large quantities of noxious gases—ammonia, methane, and the more benign water vapor. Hydrogen atoms in the methane (CH_4) were stripped away from the molecule by ultraviolet solar radiation, and the carbon atoms picked up some oxygen atoms, forming carbon dioxide.

At this point the story diverges, being different for each planet.

☆ VENUS ☆

Venus was so close to the sun that its atmosphere created a greenhouse effect strong enough to raise its surface temperature above the boiling point of water. Or, perhaps, the clouds did not exist, and its surface was dark enough to be heated to that point. In either case, while volcanoes kept dumping more gas into the atmosphere, the water remained as vapor, contributing more and more to the greenhouse effect that was building up rapidly. More water might have come from the Venerean rocks, as the high temperatures cooked water out of them, further contributing to the dense atmosphere.

We have now succeeded in making a hot Venus, but one problem remains. How does this steambath evolve to the arid oven of today? Planetary scientist James C. G. Walker has made one intelligent guess that demonstrates the possible role of atmospheric chemistry. At the high temperatures characteristic of the Venerean environment, water (H_2O) dissociated into H atoms and OH radicals. The H atoms then escaped from Venus, since its gravity was not strong enough to hold onto these light, fast-moving atoms. The oxygen remained, but soon combined chemically with the surface rocks and disappeared from the atmosphere. Something must have happened to the oxygen, because it is not there now. The sulfuric acid clouds soaked up whatever water remained.

The whole scheme seems to be a reasonable way to make a Venus like the one we see now—a hot, dry planet with an atmosphere of mostly carbon dioxide surrounded by sulfuric acid clouds. Walker rightly points out that the idea is quite speculative; there is little hard evidence to support it, but it seems to be a reasonable working hypothesis. It suggests promising lines for research. A chemical analysis of the Venerean rocks might determine whether they have played the role visualized in the scenario. More work on atmospheric chemistry might confirm, complicate, or even disprove the evolutionary scheme.

☆ EARTH ☆

We understand the past history of the terrestrial climate better, because several good clues provide evidence on the earth's temperature history. The existence of fossil algae in 3.5-billion-year-old rocks indicates that both liquid water and life existed on the earth's surface that long ago. Such evidence places rather firm bounds of zero and 100°C on the average terrestrial temperature. The oceans could neither freeze over nor boil away and still sustain life.

The earth started with the same primeval atmosphere that Venus had: methane, water, ammonia, and carbon dioxide. But the earth was just far enough away from the sun so that water could form a liquid pool—the ocean, the birthplace of life. The atmosphere did not thicken as the Venerean atmosphere did, so that the surface temperature stayed within reasonable bounds. The ocean acted as a stabilizer. A small greenhouse effect remained, so that the water did not freeze. While the chemical composition of the atmosphere eventually changed from the methane-rich primeval atmosphere to the oxygen-rich atmosphere of the present epoch, the global heat balance has not changed drastically.

The earth's climate is not completely stable, however; ice ages have occurred in the past and may occur in the future. Why? An extensive project called CLIMAP has been reconstructing the Ice Age climate in order to understand just what it is that causes them; a working hypothesis has emerged.

Suppose that, next summer for some reason, the winter snow did not evaporate from the plains of northern Canada and the vast steppes of Siberia. Snow is lighter in color than land and has a higher albedo. Because snow reflects more sunlight back into space than land does, a snow-covered Canada would be colder than the present landmass, and the following winter would be colder still. If the trend continued, the land would eventually be covered with a thick ice sheet, and an ice age would occur.

The existence of ice ages is a sign of an *instability* in the earth's climate. The global heat balance can be satisfied in one of two ways: by an earth that is more or less the same as it is now, and by a lighter-colored, ice-encrusted planet. Computer models have verified this statement with detailed calculations. We do not know what causes the earth to shift from one state to another, nor how long the change takes. Clearly human civilization could adapt more easily to an ice age that appeared gradually, over hundreds of years, than to an ice age that appeared in a decade. The instability that causes ice ages is not well understood, but a similar instability may have played an important role in the history of the Martian climate.

☆ MARS ☆

Mars presents a slightly different picture. Once again we start with a planet with a methane-water-ammonia atmosphere and carbon dioxide forms. The water did not form pools of liquid because Mars was too cold, perhaps because of its greater distance from the sun or perhaps because some subtle difference in the atmospheric chemistry prevented a greenhouse effect. Carl Sagan and other planetary scientists guessed that the water was frozen at the polar caps; many of us were very pleased to have this guess confirmed by the Viking orbiter discovery that the permanent polar caps are in fact composed of frozen water. The white polar caps of Mars consist of two substances: frozen water forms the permanent polar cap, while in winter a thin layer of dry ice forms and the visible cap expands. Because most of the Martian atmosphere froze at the surface, the greenhouse effect was of minor importance.

This scenario accounts for the present Martian ice age. But where did the arroyos come from? An instability similar to the earth's may be operating. If the polar caps are thick (a hopeful guess) and if they were to evaporate, they would give Mars an atmosphere substantial enough to produce a greenhouse effect. The water, presently locked up in the polar caps or chemically part of the soil, could be released and cut the channels. The thick atmosphere and the added warmth would allow water to flow in rivers rather than freeze or spontaneously boil.

But what would cause the polar caps to evaporate? We know less about this than we know about the cause of the terrestrial Ice Age, but one can make reasonable guesses based on the terrestrial analogy. Perhaps a gigantic dust storm could cover the caps with a dark dust layer, raising the temperature of the caps and causing them to disappear. A few visionaries have even

FIGURE 5.18 View of the permanent north polar cap from Viking 2. The location of the north pole is shown; the horizon is at the upper right of the picture. The dark bands are spiraling ice-free regions, shown in more detail in Figure 5.19. (NASA/JPL Photo No. P–17679)

MARTIAN NORTH POLE

142

FIGURE 5.19 A close view of one of the ice-free regions near the Martian north polar cap. The laminated or layered structure of the polar deposits is evident. The absence of craters in the polar terrain indicates that the polar ice deposits were formed geologically recently. (NASA/JPL Photo No. 80477)

proposed that the polar caps could be intentionally blackened in order to change the Martian climate.

The speculative scenario of a past Martian spring gets some ambiguous support from observations of the polar caps by the Mariner and Viking orbiters. The polar caps contain some of the most intriguing terrain on the planet, but little is understood about their origin. A Viking orbiter photograph of the north polar cap in late summer (Figure 5.18) shows a number of ice-free areas. The orbit of the Viking probe was changed, and later photographs of these ice-free areas show that they are composed of terraced slopes (Figure 5.19). Measurements of the temperature and water-vapor content of the atmosphere over the polar cap show that the permanent polar cap is made of water ice. The absence of fresh impact craters on the cap indicates that it is far younger, geologically, than the rest of Mars. The laminated terrain at the pole, best shown in Figure 5.19, must have been laid down as a series of layers deposited at regular intervals like tree rings. The freshness of the polar caps indicates that large-scale changes on Mars, called for in the Ice Age scenario, are not inconsistent with the appearance of the polar regions.

SUMMARY

Venus is enveloped by sulfuric acid clouds; Earth is a habitable paradise; Mars is a frigid desert with thin winds howling in the sparse atmosphere. The overall heat balance of each of the three planets determines its surface temperature, and the planetary atmosphere plays a role through the greenhouse effect. Most studies of Venus have concentrated on its atmosphere, since little is known about its surface. The Martian surface has been thoroughly photographed by the cameras of the Mariner and Viking orbiters. The dusky markings visible from the earth are actually myriads of dark streaks, caused by the deposition of dust on the surface. The Viking landers have provided a closer perspective of the surface. The most geologically interesting parts of Mars are the volcanic regions and the accompanying huge canyon named Valles Marineris after the spacecraft that discovered it. The great differences among these three superficially similar planets—Venus, Earth, and Mars—probably result from different evolutions of their atmospheres.

This chapter has covered:

The details of a planet's heat balance;

The role of the atmosphere in heating a planetary surface through the greenhouse effect;

The chemistry of the Venerean atmosphere;

What little we know about the surface of Venus;

The origin of the dark markings on Mars and the nonexistence of Martian canals;

The Martian deserts seen from the Viking lander and the prevalence of dust on Mars;

The discovery of strange soil chemistry on Mars, and the ambiguity of the results of the Viking biology experiments;

The geological processes that produced the Martian volcanoes;

The Martian arroyos; and

A speculative history of the atmospheres of Venus, Earth, and Mars that may explain the great differences among these three planets.

KEY CONCEPTS

Albedo	Climatic instability	Olympus Mons
Arroyo	Energy balance	Valles Marineris
Canali	Greenhouse effect	Volcanic shield

REVIEW QUESTIONS

1. Describe the principal similarities and differences among Venus, Earth, and Mars.

2. The lunar highlands are light in color and the maria are dark. Which is hotter during the daytime? Why?

3. Should someone wishing to keep cool in the summer wear light- or dark-colored clothing? Why?

4. If you could coat the glass of a greenhouse with a substance that would absorb infrared radiation more effectively than glass, would the greenhouse become warmer? Why or why not? (This principle is the basis of some experimental efforts directed toward the harnessing of solar energy.)

5. Summarize the difference between the Venerean environments at the cloud tops and at the surface.

6. Why does the existence of craters on Venus indicate that at least some parts of the surface of Venus are old?

7. What is the difference between the "canals" that Percival Lowell claimed to see and the arroyos discovered by Mariner 9?

8. Would you expect to find dark markings visible from the earth on a Martian desert with no mountains or craters?

9. What discoveries of the Viking lander support the explanation of the surface markings discussed in section 5.5?

10. Summarize the findings of the Viking landers regarding the possible existence of life on Mars.

11. Summarize the similarities and differences between Olympus Mons and the Hawaiian Island chain.

12. What is the difference between Valles Marineris and the Grand Canyon?

13. If Mars were hotter with an atmosphere as thin as it is now, could the arroyos have formed?

14. If carbon dioxide were discovered to be transparent in part of the infrared part of the electromagnetic spectrum, to what extent would the speculative atmospheric histories of Venus, Earth, and Mars have to be changed?

15. Suppose the plains of northern Canada were covered with snow for several summers. What could be done to attempt to prevent the coming of an ice age?

16. Suppose someone dumped tons of carbon-dioxide-eating algae into the atmosphere of Venus. What would happen to the Venerean surface temperature? (Make reasonable and/or optimistic assumptions about the temperatures the algae could survive in.)

FURTHER READING

Hartmann, W. K., and Odell Raper. *The New Mars.* NASA SP–337; obtainable from the Superintendent of Documents, Washington, D.C. 20402, stock No. 3300–00577.

The Viking results have been published in the following issues of *Science*: August 27, 1976; October 1, 1976; November 19 and 26, 1976 (summary); and December 17, 1976.

★ CHAPTER SIX ★

JUPITER AND BEYOND

How do the outer planets of the solar system differ from the inner planets, and which of these differences furnish critical clues to the evolution of planets and the solar system?

The four inner planets in the solar system—Mercury, Venus, Earth, and Mars—are generally similar to each other. We can understand how the landscapes and environments of these planets came to be the way they are by appealing to their terrestrial analogs. The craters of the moon, the volcanoes of Mars, and the searing greenhouse effect of Venus are all phenomena that occur on our own planet to a greater or lesser degree. But the outer solar system presents an entirely different aspect. Gone are the rock balls of the inner solar system; the outer planets are gigantic gas spheres, and most of their satellites are made of ice rather than rock. Planetary scientists are just beginning to progress beyond the stage of simply describing what's out there.

★ **MAIN IDEAS OF THIS CHAPTER** ★

The interiors of the outer planets are mostly fluid.

Photographs of the outer planets show cloud layers in their atmospheres; features visible in these photographs are the results of circulation currents.

The environment near Jupiter is strongly affected by the Jovian magnetic field.

Some satellites of the outer planets are made of ice, in contrast to the rocky satellites of the inner planets.

★ **6.1 OVERVIEW** ★

Our knowledge of the outer solar system is quite limited. Five planets lie beyond Mars: Jupiter, Saturn, Uranus, Neptune, and Pluto. These planets are so far away that ground-based telescopes would have difficulty distinguishing a feature the size of the Gulf of Mexico on Jupiter, the nearest planet. As a result of their immense distances and our limited knowledge, the questions asked about these worlds are much less incisive than the questions asked about the inner planets. The study of the outer solar system is very immature; we are still trying to describe what's there and interpret the descriptions with very simplified models.

Jupiter (Figure 6.1) and Saturn (Figure 1.7) are quite bright in the sky, in spite of their large distances from Earth (590 million km for Jupiter and 1200 million km for Saturn at their closest approach). Uranus, Neptune, and Pluto are smaller, farther away, much fainter, and were only discovered after the invention of the telescope in the seventeenth century. The planets in the outer solar system are surrounded by many satellites, and one of the first discoveries made with the telescope was Galileo's finding the four largest (or

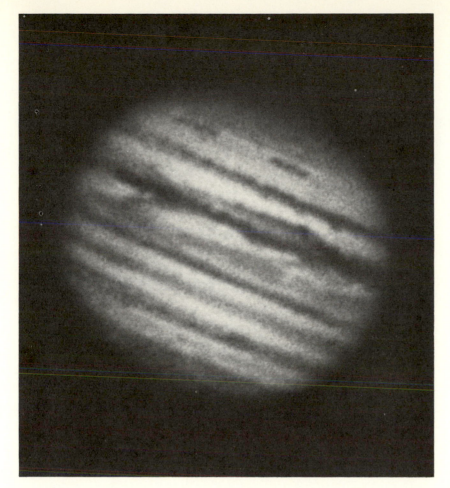

FIGURE 6.1 A telescopic photograph of Jupiter, showing its banded structure in remarkable detail. These bands are not on the surface of the planet, but are different cloud layers in the atmosphere. Jupiter has no solid surface. (Lick Observatory photograph)

Galilean) satellites of Jupiter. Later in the seventeenth century Christiaan Huygens constructed a superior telescope and discovered the rings of Saturn.

In recent years, the renaissance of planetary science has resulted in an explosion of new information about these planets. Space probes have played an important role in this research. Pioneer 10 flew by Jupiter in December 1973 and then moved into a trajectory that will take it out of the solar system. Pioneer 11 passed Jupiter a year later, and will move past Saturn in 1979. In late 1977, two Voyager probes, using a different type of spacecraft, were launched to fly by Jupiter in 1979 and by Saturn in 1981. A variety of other outer solar system missions, such as the Mariner Uranus probe originally scheduled for a 1979 launch, may become victims of tight budgets. The spectacular space probes only provide part of the research story, and a very active, continuing program of ground-based research has provided an enormous amount of information about these planets.

★ 6.2 OUTER-PLANET INTERIORS ★

☆ JUPITER AND SATURN ☆

In some ways, the two largest outer planets more closely resemble stars than they resemble planets. Their sheer size is not easy to appreciate. Eleven Earths would stretch across the visible disk of Jupiter; this gargantuan planet has some 125 times as much surface area as Earth. But for all that size, these planets do not carry as much mass, cubic centimeter for cubic centimeter, as the earth does. The *density* of a planet (recall Box 4.1) measures the concentration of mass per unit volume in a planet, and the densities of the gas giants are much lower than the density of the earth. Saturn is so light that it would float on water, if a tub could be found to hold it (and if one could find some way to keep it from dissolving before the experiment was over!). The physical properties of the outer planets are summarized in Table 6.1, with the earth shown for comparison.

The densities of the outer planets show that they must be fundamentally different in character from the earth. Rocky material, made up of elements such as iron, silicon, oxygen, and aluminum has more or less the same density, wherever it is found. Terrestrial rocks are denser than water, as anyone

TABLE 6.1 PROPERTIES OF THE OUTER PLANETS

PLANET	EQUATORIAL RADIUS (km)	DENSITY (g/cm³)	ROTATION PERIOD	OBLATE-NESS[a]	COMPOSITION
Jupiter	71,600[d]	1.3[f]	9.9 h[a]	0.06	Solar
Saturn	60,000[d]	0.7[d]	10.2 h[a]	0.1	Solar
Uranus	25,400[d]	1.2[b]	10–30 h[c] (retro-grade)	0.06?	Solar; core of ice or rock
Neptune	24,800[d]	1.7[d]	15–17 h[d]	0.02?	Solar; core of ice or rock
Pluto	1,500[e]	1–2[e]	6.4 days[a]	0	Ice[e]
Earth	6,378[a]	5.5[a]	24 h	0.003	Rock

[a] C. Sagan, "The Solar System," *Scientific American* 233 (September 1975): 26.

[b] Recent data (D. Morrison, private communication) indicate that the density of Uranus may be less than 1.

[c] M. J. S. Belton, in L. Goldberg, "Kitt Peak National Observatory Report," *Bulletin of the American Astronomical Society* 8 (1976): 135.

[d] R. L. Newburn, Jr., and S. Gulkis, "A Survey of the Outer Planets Jupiter, Saturn, Uranus, Neptune, Pluto, and their Satellites," *Space Science Reviews* 14 (1973): 179–271.

[e] D. P. Cruikshank, C. B. Pilcher, and D. Morrison, "Pluto: Evidence for Methane Frost," *Science* 194 (1976): 835–837.

[f] H. C. Graboske, J. B. Pollack, A. S. Grossman, and R. J. Olness, "The Structure of Jupiter: the Fluid Contraction Stage," *Astrophysical Journal* 199 (1975): 265–281.

in the mood to throw a rock into a pond can show. The rock sinks. Saturn, however, would not sink; it is not made of rock but rather of hydrogen gas. Jupiter, too, is made substantially of hydrogen gas, but it is denser than Saturn because its tremendous interior pressure compresses the hydrogen to a higher density, forcing more mass into each cubic centimeter. All the outer gas giants have lower densities than the inner planets, and thus they differ fundamentally from the inner planets in composition.

A theoretical calculation can lend some more precision to that last statement. The interior structure of any large astronomical object, be it planet or star, is determined by one major requirement: The object must be able to hold itself up. At any point within the interior of, say, the planet Jupiter, the pressure exerted by the interior must balance the weight of the material lying on top of that point (Figure 6.2). In the absence of such a balance, the planet would contract abruptly. The same balance requirement applies to both the earth and the stars.

A theorist can work from the mass and radius of Jupiter and calculate various models for its interior structure. James B. Pollack of NASA's Ames Research Center has made such a calculation. Pollack needed to know or guess how much pressure Jovian matter exerts as it is compressed. To a limited extent, this knowledge comes from high-pressure laboratory measurements; theoretical extrapolations of the measurements are needed for the inner regions of Jupiter, where the pressures are too great to be reproduced in the lab. Theorists then adjust the chemical composition of the models to fit observations of Jupiter's mass and radius. The composition of Jupiter is almost identical to the composition of the sun: 90 percent hydrogen and 10 percent helium, by numbers of atoms. It would be more difficult to do a similar calculation for a solid inner planet, since the behavior of rock and iron at

FIGURE 6.2   At any point between the interior and the surface of Jupiter, the weight of the overlying surface layers must be balanced by pressure in the interior.

high pressures is not understood as well as the behavior of the gases that make up Jupiter and Saturn.

The resulting theoretical picture of Jupiter is shown in Figure 6.3. From the outside, we only see the top cloud layer, perhaps 1000 km thick. A few thousand kilometers below this cloud layer, the pressure reaches the point at which hydrogen liquefies. The hydrogen is liquid all the way into the planet's core. At a pressure of 3 million atmospheres, one-fifth of the way in, the hydrogen becomes an electrical conductor. If Jupiter contains the same proportion of elements heavier than hydrogen and helium as the sun does, these heavier elements may have sunk to Jupiter's center, forming a small, rocky core. There is no direct evidence regarding the existence of this core, however. Jupiter, as a result, has no solid surface on which an astronaut could land. Someone wishing to explore the Jovian environment would need a well-built submarine, not a vehicle like the lunar rover.

We now know enough about Jupiter to verify our vision of its interior structure to some extent. As Table 6.1 indicates, Jupiter rotates rapidly. This rapid spin causes the planet to bulge outward at the equator in the way that children are flung out toward the edge of a merry-go-round. Jupiter's "midriff bulge" is visible even in small telescopes, and the *oblateness* of Jupiter is a quantitative measurement of the size of this bulge. (The oblateness is the difference between the equatorial and polar radii, divided by the equatorial radius.) The magnitude of the oblateness depends on the mass distribution in the planet. The effect of Jupiter's rotation is largest at the planetary surface; if most of the planet's mass were concentrated at the center, the rotation would act less strongly on the bulk of the planet's mass and the oblateness would be less. Comparing the theoretical and observed oblatenesses, the model fits.

A different measurement of the mass distribution within a planet results from close examination of the path of a space probe when it whips around the planet. Jupiter deflects the path of a Pioneer probe, and the precise distri-

FIGURE 6.3   A cross section of Jupiter. There may be a small rocky core in the interior. Most of Jupiter is liquid hydrogen, and the inner 80 percent (by radius) of this hydrogen, shown in color, is a good electrical conductor.

The outer atmosphere is about 1000 km thick. [This cross section is based on information in H. C. Graboske *et al.*, "The Structure of Jupiter: The Fluid Contraction Stage," *Astrophysical Journal*, 199 (1975), p. 265.]

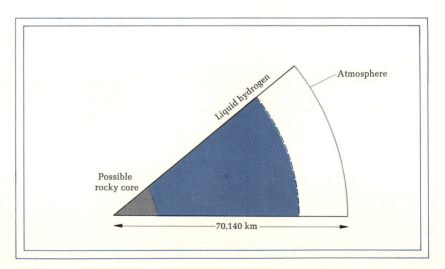

bution of mass within Jupiter affects the trajectory of the craft. A comparison between theory and the actual trajectories of the Pioneer probes is less certain than the oblateness measurement, but the theory still agrees with the observations.

Thus we can explain Jupiter's present interior structure with a model that agrees with observational data in several different ways. How did it come to be that way? A possible clue to its past history is Jupiter's infrared luminosity. Infrared observations, initially made from the earth but confirmed by the Pioneer spacecraft, show that Jupiter radiates something like twice as much energy as it receives from the sun. Jupiter must draw on some internal energy source in order to satisfy its energy-balance requirement. Where does this internal energy come from?

When a planet like Jupiter initially condenses, the compression of its constituent gases causes them to heat up. A similar event occurred when the inner planets formed; the impact of falling rocks melted their surfaces. Thus the Jovian interior was much hotter when the planet was first formed than it is now. In its early days Jupiter was somewhat like a star; its central temperature was 50,000 K, and its luminosity was one-ten-thousandth of the present solar luminosity when the planet was hottest, according to Pollack's calculations. This heat is slowly radiated away in the infrared part of the electromagnetic spectrum while Jupiter slowly cools.

The overall model of Jupiter's interior seems quite convincing. But there are a very small number of observations that bear directly on the internal make-up of this planet. We only know that a model planet of Jupiter's mass must reproduce the observed radius and infrared luminosity; and it must have a mass distribution that is compatible with oblateness measurements and the orbits of space probes. Limited as these observations are, they provide useful information. Although the observations are far more uncertain for Saturn, it, too, has a solar composition of 90 percent hydrogen and 10 percent helium, according to the fits between model interiors and the observed mass and radius.

☆ URANUS AND NEPTUNE ☆

Some thought about the densities listed in Table 6.1 indicates that Uranus and Neptune are rather different in character from their giant brothers, Jupiter and Saturn. If they also were made almost entirely of hydrogen and helium, their densities would be less than the density of Saturn, because their smaller masses would compress the central hydrogen-helium mixture less. A smaller central density would suffice to produce the required pressure. Yet the density of Uranus exceeds Saturn's, and Neptune is more dense than Jupiter. Evidently, these planets must contain fairly large cores of something denser than hydrogen and helium—perhaps rock or some kind of ice. Uranus and Neptune contain more heavy elements than Jupiter and Saturn do, and so they are a different type of body. Neither Uranus nor Neptune is known to emit an excess of infrared radiation.

Uranus has one chief peculiarity: Its rotation axis is almost directly in the plane of the ecliptic, unlike the axes of all the other planets (Figure 6.4).

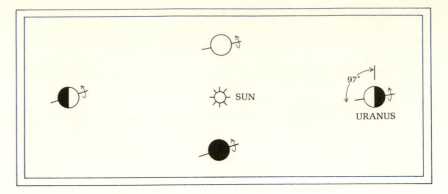

FIGURE 6.4   The rotation axis of the planet Uranus is almost in the plane of its orbit. Consequently, seasons on Uranus are exaggerated; when the pole points almost directly toward the sun, half the planet enjoys midnight sun and the other half is in total darkness.

Consequently, at certain times of the year, the northern hemisphere of Uranus is in perpetual sunlight, the southern hemisphere is in perpetual darkness, and the planet's rotation does not produce a cycle of day and night. At other times its rotation axis is in the direction of its orbital motion, and the day-night cycle occurs. This configuration, unique to Uranus, might produce some rather strange meteorological phenomena, though Uranus's atmosphere might be so sluggish in response to changing solar radiation that the effects would be small.

How did Uranus become tilted at such an outlandish angle? Whatever happened must have occurred in the early days of the solar system, because Uranus has a beautifully regular satellite system that is also tilted at this strange angle. Perhaps some large object nearly collided with the proto-Uranus and tilted it. But how? It's still a puzzle.

While an analysis of the interiors of the outer planets provides a reasonably successful explanation of their sizes, shapes, and masses, such analysis provides no insight into the visible features. These balls of gas have huge atmospheres that produce the features visible in telescopes. The next section examines the atmospheres of the gas giants, drawing heavily from the information we have about the best-known large planet, Jupiter.

★   6.3 METEOROLOGY   ★

☆   APPEARANCE OF THE JOVIAN   ☆
ATMOSPHERE

Perhaps the most compelling feature visible in photographs of Jupiter is the Great Red Spot (Figure 6.5). This blotch, visible in small telescopes and noticed first in the early days of telescopic astronomy in the seventeenth century, presents the most significant problem pertaining to the Jovian atmosphere. Before the 1960s, many speculations (but no firm theories) abounded.

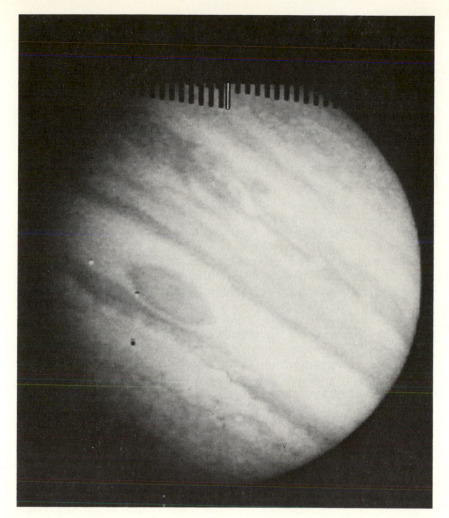

FIGURE 6.5 A photograph of Jupiter from Pioneer 10, showing the Great Red Spot and a number of belts. The comb in the upper part of the photograph is not part of Jupiter, but represents places where the data needed to reconstruct this image were not transmitted from the spacecraft. (NASA Photo No. 74HC—152, 74H—256, A74—9190)

Was it a floating raft of some sort? Was it an atmospheric manifestation of some bump on the Jovian surface?

Photographs from the Pioneer missions (Color Photos 6 and 7) show a number of similar but smaller and less strongly pigmented disturbances in the Jovian atmosphere. The close view from the Pioneer probes resolves Jupiter's atmosphere into a swarm of spots, ripples, wavelets, and plumes; photographs from the Voyager craft should provide still more detail. The broad-banded structure, so evident in small telescopes, determines the overall appearance of the planet, but the boundaries of these light-colored *zones* and dark-colored *belts* are not as regular as they seem to be from the earth. What causes the mottled appearance of Jupiter? Why are the polar regions bubbly rather than banded (Color Photo 7)? Before you read on, stop and look at these photographs for a minute or two. Try to find some regularity in the pattern, something that a theorist could explain with a model.

Our theoretical understanding of Jovian weather patterns is at a tantalizing stage. Jovian meteorologists have made tremendous progress—10 years ago there were only simple descriptions of zones, belts, and spots. Generalized explanations of all these features exist. But our understanding is sufficiently vague, and its connection with the real world so tenuous, that it is quite possible that future observations, both from space and from the ground, will show that the theory as presented here is totally wrong.

The Great Red Spot best exemplifies both our understanding and our ignorance. As long ago as 1951, Seymour Hess and Hans Panofsky recognized that currents in the Jovian atmosphere circled the Spot. While other investigators confirmed this finding, it was not yet clear whether these circulation currents were an atmospheric phenomenon or whether they were connected with the presence of some solid object in the atmosphere or on the surface. The confirmation of the Great Red Spot as an atmospheric phenomenon came in the 1970s, when ground-based astronomers and instruments on board the Pioneer spacecraft mapped the thermal infrared radiation from Jupiter. Hot regions, low in the atmosphere, emit more infrared radiation than cooler regions higher in the atmosphere. The Great Red Spot is one of the coolest regions on the entire planet, and the light-colored zones are cooler than the darker belts. According to this model (Figure 6.6), the observed features in the Jovian atmosphere are caused by high- and low-pressure areas, just as the cloud patterns in the earth's atmosphere (Color Photo 2) are caused by the interplay of fair-weather highs and stormy, cloudy lows. The Great Red Spot is thus interpreted as a particularly strong high-pressure area, high in the Jovian atmosphere.

This model of the Spot has several weaknesses. It is only a qualitative model; no one has yet come up with a mathematical model that predicts temperature differences and wind speeds and that can be compared with observations. Two questions are unanswered. Why has the Spot persisted for centuries? Similar cyclonic features on the earth, storms and high-pressure areas, last for days or at most a week or two. Even when the large mass of the Jovian atmosphere is taken into account (the Great Red Spot is larger than the entire earth!), the expected lifetime of a feature such as the Spot is only extended to years, not to centuries. The Spot's color presents another unanswered question. Why is the Great Red Spot red, while most other spots on Jupiter (Color Photos 6 and 7) are white? A wide variety of compounds have been suggested as possible pigments, but none is compelling. The gases that have been discovered in the Jovian atmosphere—methane, acetylene, ethane, ammonia, and water—are mostly colorless.

The same type of generalized model can explain other large-scale features of the Jovian atmosphere, as Figure 6.6 shows. The light-colored zones and dark belts are the results of rising and falling gas motions. The rapid rotation of Jupiter stretches these high- and low-pressure areas into a band stretching all the way around Jupiter. In the polar regions, rotation is not such an important influence on the atmosphere, and the zone-and-belt pattern gives way to the bubbly appearance shown in Color Photo 7.

These explanations of the Jovian atmospheric features are all quite general. Saturn (Color Photo 8) shows a weaker banded structure. Energy trans-

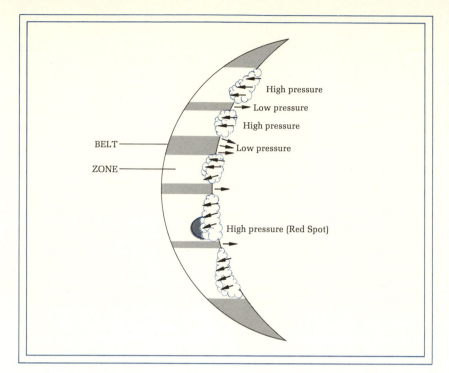

FIGURE 6.6 A rough sketch of the currently accepted model for the Jovian atmosphere, explaining the presence of zones, belts, and the Great Red Spot. The cutaway view shows that these features are caused by high- and low-pressure bands in the Jovian atmosphere.

port in the Jovian atmosphere is by *convection*, large-scale motions of gas, so the existence of circulation currents might be expected. Yet convective features in the terrestrial atmosphere are quite small. The light, puffy cumulus clouds visible on a summer day result from convection, when warm air masses rise and cold air masses fall. We need more data to refine the models. Remember, only two flybys have provided close views of Jupiter. After three similar flybys of Mars (Mariners 4, 6, and 7), Mars was thought to contain nothing but moonlike desert areas.

The atmospheres of the other outer planets might provide some insight into their weather, once again through the analysis of similar processes operating under slightly different conditions. Bands are evident in Saturn's atmosphere, but observations of Uranus and Neptune have not shown conspicuous features. The Stratoscope project, run by Princeton University, sent a telescope up in a balloon to try to overcome the turbulence in the earth's lower atmosphere. This balloon-borne telescope returned some rewarding photographs of the sun, but its photo of Uranus showed no detail at all. Neptune is just a little green ball in the eyepiece of a telescope; at a distance of 30 AU, it is hopeless to try to make out any detail.

The differing colors of the outer planets are caused by the different chemical compounds in their clouds. The major constituent of their atmospheres, hydrogen, is transparent and colorless. The dominant compounds in the Jovian atmosphere, ammonia and methane, combine to produce a white color.

In Saturn, the ammonia layers are further down, perhaps hidden from our eyes, and produce a yellow color. The iridescent green color of Uranus and Neptune, worth seeing through a telescope, is probably caused by strong methane absorption in their atmospheres (recall Chapter 1). Spectroscopic analyses of these planets indicate that their atmospheres are basically solar in composition, consisting mostly of hydrogen and helium.

★   6.4 JUPITER'S MAGNETIC FIELD   ★

Jupiter's core, about half the volume of the planet, is made of hydrogen in an electrically conducting state. Tremendous eddy currents, induced by the planet's rapid rotation, produce electric currents in this conducting core. This dynamo is probably the source of Jupiter's huge magnetic field, 100 times stronger than the earth's field (if the fields are measured at each planet's surface). This magnetic field is also a strong influence on the Jovian environment.

Jupiter's magnetic field chiefly influences the environment near Jupiter by interacting with the *solar wind*. The solar wind is a 400-km/sec blast of electrically charged particles—protons, electrons, and some heavier particles—streaming away from the solar surface. Any planet with a magnetic field can capture these particles and form radiation belts around the planet. The name *radiation belt* is a little confusing because it does not refer to electromagnetic radiation. These belts are regions of space with a high concentration of high-speed, charged particles such as electrons and protons. The earth possesses radiation belts, called the Van Allen belts after their discoverer who conducted an experiment on board the first American satellite, Explorer 1. But Jupiter's belts are far more extensive and far more intense than the earth's.

The inner part of Jupiter's radiation belts is sketched in Figure 6.7. This sketch of the Jovian *magnetosphere,* or the part of space where Jupiter's magnetic field controls the paths of charged particles, is based only on measurements from two flybys, Pioneer 10 and Pioneer 11. The map is extrapolated from measurements made along their two trajectories, and is probably an approximation of the real situation. The axis of the magnetosphere is aligned with Jupiter's magnetic axis and tilted about 10° to the rotation axis of the planet. Because of this tilt, someone in orbit near Jupiter would see the belts wobble up and down while Jupiter rotates.

The high density of charged particles in the Jovian magnetosphere would sterilize any spacecraft passing through them. Analysis of the Pioneer 10 measurements shows that the probe received 1000 times the radiation dose sufficient to kill any humans inside the spacecraft. The Jovian belts are far more intense than the terrestrial ones, and they discourage manned exploration. It would be very difficult to design a craft with adequate shielding against these high-speed particles.

The Jovian magnetosphere extends far beyond the inner radiation belts shown in Figure 6.7. Pioneer 10 measurements hinted at the existence of a

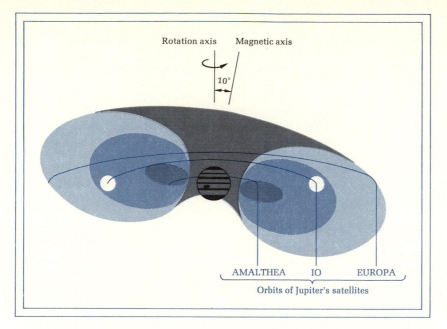

FIGURE 6.7   A sketch of the inner parts of the Jovian magnetosphere, where the number of high-energy electrons is at least 1 percent of the peak value measured by the Pioneer 10 and 11 particle detectors. The belts are tilted with respect to Jupiter's rotation axis by about 10°. Jupiter's three inner satellites orbit within these belts, and one—Io—sweeps up about half the high-energy electrons in its orbit.

huge flat outer belt, 100,000 km thick and extending 7 million km from the surface of the planet. But Pioneer 11 detected concentrations of particles at large distances from Jupiter's magnetic equator, casting considerable doubt on the thin-disk model. The magnetosphere extends a long, long way behind Jupiter, away from the sun; Pioneer 10 passed through the tail of the Jovian magnetosphere in early 1976, when it was halfway out to the orbit of Saturn.

The other outer planets may also have magnetospheres since they, too, rotate rapidly. None of these magnetospheres is as extensive as Jupiter's. Jupiter emits low-frequency radio waves from its magnetosphere, and a similar magnetosphere around one of the other outer planets would emit a detectable amount of radio emission. One potentially interesting magnetospheric phenomenon would be the interaction of Uranus's magnetic field with the solar wind in the 1980s. At that time, Uranus's magnetic pole, presumably not too far from the rotation pole, will point at the sun. It is usually the equatorial region of other planetary magnetospheres that traps the solar wind and contains high-speed particles, but with Uranus in the 1980s the geometry will be quite different. A future mission to Uranus may be able to explore this novel situation.

★ 6.5 OUTER-PLANET SATELLITES ★

One remarkable characteristic of the outer planets is the number of their satellites. While the inner four planets have 3 satellites among them (our moon and Mars's Phobos and Deimos), Jupiter has 13 known satellites, Saturn 11, Uranus 5, and Neptune 2.[1] Saturn also possesses the famous ring system.

Some of the satellites in the outer solar system are quite large, comparable in size to the moon and Mercury. Four of these large satellites, the Galilean satellites, orbit Jupiter. They are a rewarding sight in any small telescope, and some sharp-eyed observers have even seen them with the naked eye. Titan, with a diameter of 5800 km, is probably the largest satellite in the solar system; it is larger than Mercury and only 15 percent smaller than Mars. Neptune also has a large satellite, Triton, but it is so far away that even its diameter is uncertain. Triton may be larger than Titan.

☆ THE GALILEAN SATELLITES ☆

The four Galilean satellites—Io, Europa, Ganymede, and Callisto—are the most thoroughly investigated of the small bodies of the outer solar system. They are nearly as varied as the four inner planets, and a comparison of these satellites with the inner planets may provide information about conditions in the early solar system. The giant outer planets differ from the inner planets both in distance from the sun and in size, whereas the Galilean satellites of Jupiter are similar to the inner planets in size. They thus provide an opportunity for a clearer test of the effects of distance from the sun on planetary formation and evolution.

What are the Galilean satellites made of? The density of an object provides an important clue to its composition. Astronomers have computed the approximate densities of the Galilean satellites from their gravitational effects on each other's orbits; their effect on the paths of the Pioneer probes allowed for a refinement in the density measurements. Io and Europa, the two inner ones, are the densest, with densities of 3.5 and 3.3, respectively; these densities are similar to the moon's density of 3.35. Presumably, therefore, they are composed primarily of rock. Ganymede and Callisto, the two outer ones, are less dense at 1.9 and 1.6; they are made of ice, and their interiors are probably liquid water. The density progression in Jupiter's satellite system parallels the density progression in the solar system, with the innermost objects being densest.

Spectroscopy of the surfaces of these satellites indicates that two of them, Europa and Ganymede, are at least partially coated with water frost. Pho-

---

[1] A fourteenth Jovian satellite, discovered in 1975, is now lost; we cannot confirm the discovery of the eleventh Saturnian satellite, made by Fountain and Larson at the University of Arizona, until 1981 when the rings will be edge-on to us, reducing their glare.

tographs of Ganymede from Pioneer 10 indicate that there are bright and dark areas on the surface of this object. The bright regions are probably water frost, while the dark regions vaguely resemble lunar mare areas. Europa has a very high albedo, so high that the planet must be almost completely covered with frost. Spectroscopy indicates that five of the larger satellites of Saturn are also covered with water ice.

The progression of satellite compositions, from the rocky inner two to the icy outer two, relates to the past history of Jupiter. When the satellites formed, Jupiter was considerably hotter than it is now (recall section 6.1). Model calculations by Pollack and his collaborators indicate that this heat would have melted any ice that might have condensed in Io and Europa, leaving them rocky bodies. This heat did not affect Ganymede and Callisto as much since they are farther away from the hot primeval Jupiter, and the ice remained. Saturn, smaller than Jupiter, had a smaller effect on its satellites, so this model also explains their icy composition.

☆ **IO** ☆

The innermost Galilean satellite, Io, is presently the most peculiar satellite in the solar system. It orbits in the heart of the inner Jovian radiation belt (recall Figure 6.7), and high-speed particles continually pelt its surface. No human being could survive on this lethal satellite. The continual bombardment of Io's surface releases surface gases which become trapped in a volume of space surrounding Io's orbit. Four elements of this gas cloud have been detected spectroscopically—hydrogen, potassium, sodium, and sulfur. Io also tends to concentrate charged particles in a sphere surrounding it. Io's atmosphere is thus the most peculiar in the solar system: A sphere of metallic ionized gas surrounds the satellite, and a doughnut of other gases surrounds its orbit.

The surface composition of Io is also unusual. All the small bodies we have encountered so far have surfaces of either ordinary rock or ice. Io's albedo is unusually high at 0.63, far higher than rock albedos, which run from about 0.1 to 0.2. Water frost is as bright as Io is, but Io's spectrum indicates that water ice is not present on its surface. Most investigators believe that the surface of Io is one giant salt flat, resembling the Bonneville salt flats of Utah. Gas atoms chipped off the salt flat by high-speed particles continually replenish Io's atmosphere.

Io's presence in the heart of the Jovian magnetosphere is responsible for yet another phenomenon unique to Jupiter. In the 1950s, investigators at the University of Florida discovered that Jupiter occasionally emits bursts of radio radiation. Cal Tech scientist Peter Goldreich noticed that these bursts were correlated with the position of Io, and he proposed that huge electrical discharges triggered by Io cause the bursts of radio emission. Such an explanation requires that Io be an electrical conductor; the discovery of the concentration of charged particles of Io's atmosphere by Pioneer 10 explains the conductivity of Io without requiring a metallic satellite.

Titan, the largest satellite of Saturn, is the only satellite other than Io known to possess a large atmosphere. Io's atmosphere is so peculiar that in some senses it should be considered as an extension of the Jovian magnetosphere; Io would have no atmosphere if it were not orbiting in the heart of the Jovian radiation belts. Titan's atmosphere is more normal. Gerard Kuiper, one of the few planetary scientists working in the 1940s, discovered methane in Titan's atmosphere in 1944. Later work indicated that the visible surface of Titan is actually a layer of reddish clouds. Theoretical analysis of the observations shows that methane and hydrogen are present in about equal amounts. This atmosphere is certainly thicker than Mars's, and it may be as thick as Earth's. If the atmosphere is as thick as the observations indicate, the surface of Titan should be covered with a sea of liquid methane and ammonia. Its low density (less than 2.0) indicates that its interior is fluid rather than rocky. Titan will be a high-priority object in the plans of future Saturn probes that will reach the ringed planet in 1979 and 1981.

Saturn possesses a remarkable ring system in addition to its 11 satellites. These rings are a beautiful sight; a small telescope shows the two brightest ones. Large telescopes, used under good observing conditions, show a filmy third ring; a fourth ring that extends almost all the way to the planet's atmosphere was discovered recently. Once every 14 years, Saturn is in a position in its orbit so that the rings are seen edge-on and almost disappear. It is at this time that small satellites, orbiting just outside the ring system, are discovered. The tenth and eleventh satellites were discovered in photographs of the 1966 event. The eleventh one, found only recently when the 1966 plates were inspected, cannot be confirmed until the next time the rings are seen edge-on in 1981. The thinness of the rings during these events indicates that they are less than 10 km thick.

A variety of observations and interpretations show that the rings consist of a horde of small ice particles. Most of these icy fragments are quite small, between 2 and 15 cm in diameter, but larger particles may exist. The particles are about 1 m apart on the average. Their optical properties and physical nature resemble cirrus clouds on the earth. They are so close to the planet that tidal forces from Saturn would disrupt any satellite that happened to form there. The absence of rings around Jupiter is probably a consequence of Jupiter's greater size. Jupiter was so hot when it formed that any ice condensations forming near the planet would have evaporated.

In March 1977 astronomers observing the occultation of a star by the planet Uranus discovered that the solar system contains a second ringed planet. In an effort to reduce the uncertainty in the measurement of Uranus's diameter, several teams of astronomers measured the amount of time it took Uranus to pass in front of, or occult, a faint ninth-magnitude star. Much to everyone's surprise, the light from the star was cut off five times about 45 minutes before the planet passed in front of the star. The current interpretation is that there are five rings around Uranus, with radii between 44,000 and 51,000 km, that diminished the light from the star at those times. It is not yet clear why these rings had never been observed optically. We do not know the properties of the rings of Uranus; we only know of their existence.

Alert readers may have wondered what happened to the ninth planet of the solar system, Pluto. This planet was discovered in 1930 at the Lowell Observatory in Arizona. Percival Lowell, the proponent of Martian canals, predicted the position of a planet X on the basis of irregularities in Uranus's orbit, in the way in which Leverrier and Adams predicted the position of Neptune. Tiny, faint Pluto was not as easily found as Neptune, but the patience and persistence of Clyde Tombaugh, an assistant at Lowell Observatory, paid off. He scrutinized the images of 2 million stars before he noticed one object, in a star field near Gemini, that moved across the photograph. Stars remain in fixed positions, while planets move. The motion of this object indicated that it was a long, long way from the sun, and this denizen of the interstellar deeps was named Pluto after the god of the underworld.

Lowell's initial work indicated that Pluto had a relatively large mass, comparable to the mass of the earth. But measurements of its diameter, crude as they were, indicated that a planet as small as Pluto could not have such a high mass unless it had an impossibly high density. To fit the data, Pluto would have had to be made of platinum. In the 1970s, two investigations indicated that Pluto's effect on the orbits of other planets is unobservably small, and that the mass of Pluto is less than 0.2 earth masses. The platinum planet vanished. Evidently, Lowell's correct prediction of Pluto's position was a lucky guess.

What is Pluto made of? It is hardly more than a pinpoint in the telescope, and so all we know of its diameter is that it is less than 5800 km. A recent spectroscopic measurement discovered methane frost on Pluto's surface. Pluto is so far from the sun that even a volatile gas like methane remains solid. If the interior composition is similar to the surface, Pluto is an iceball like the two outer Galilean satellites and most of the moons of Saturn. Such an object has a density between 1 and 2 g/cm^3, and Pluto's mass is then only a few thousandths of an earth mass.

Are there any planets beyond Pluto? The Lowell Observatory has conducted an intensive, unsuccessful search for such planets. The sensitivity of their telescopes is limited, but a planet as large as Neptune would have been detected if it were closer than 80 AU. Yet another Pluto could well have escaped detection. At Pluto's location, in the depths of the outer solar system, the sun is a barely perceptible disk, providing precious little heat or light. Since planets shine by reflected light, small bodies are extremely difficult to find at such distances.

SUMMARY

The solar system beyond Jupiter presents a rather different picture from the inner regions. Balls of rock, where geological processes similar to those on the earth shape the landscape, are absent. The planets, except for Pluto, are large bodies of gas; Pluto and most of the satellites are small bodies of ice. Our knowledge of these objects is rather limited because only two space probes have flown by one planet, and their great distance makes telescopic observation difficult. Any outer-planet missions that are to be flown in the mid-1980s must be planned now, since it takes the better part of a decade to develop experiments for, build, and launch an outer-planet probe. Too many planetary probes have been killed by tight budgets in recent years, and if some missions are not at least planned now, exploration of the solar system will experience a hiatus in the next decade.

This chapter has covered:

The interiors of the giant planets, particularly Jupiter;

The compositions of Jupiter and Saturn, mostly made of hydrogen gas, and Uranus and Neptune, which have rocky cores;

The clouds in the Jovian atmosphere and their circulation;

The Jovian magnetosphere and radiation belts;

A remarkable family of worlds, the satellites of the outer planets; and

What little is known about Pluto.

KEY CONCEPTS

Belts	Magnetosphere	Radiation belt
Convection	Oblateness	Solar wind
Density		Zones

REVIEW QUESTIONS

1. Summarize the principal differences between the inner and outer planets in a table.

2. Suppose a planet with the same radius as Jupiter had a density of 6 instead of 1.31. How much more massive than Jupiter would it be?

3. What would the hypothetical planet of question 2 be made of?

4. Why are the zones and belts not found in the polar regions of Jupiter?

5. If a planet rotated more slowly than Jupiter, would it be more or less likely to have zones and belts? Explain.

6. What is a radiation belt?

7. Suppose Jupiter's core were made of a substance that is not electrically conducting. Would you expect to find radiation belts around Jupiter? Why or why not?

8. Suppose Io orbited Jupiter at a much larger distance but retained its present composition. Would it have an atmosphere? Why or why not?

9. How do the densities of the Galilean satellites confirm models of the Jovian interior?

10. Is it fair to say that Pluto more closely resembles the satellites of the outer planets than it resembles the planets themselves? Why or why not?

Cruikshank, D. P., and D. Morrison. "The Galilean Satellites of Jupiter." *Scientific American* (May 1976): 108–116.

Fimmel, R. O., W. Swindell, and E. Burgess. *Pioneer Odyssey: Encounter with a Giant.* NASA Special Publication 349.

Ingersoll, A. P. "The Meteorology of Jupiter." *Scientific American* (March 1976): 46–61.

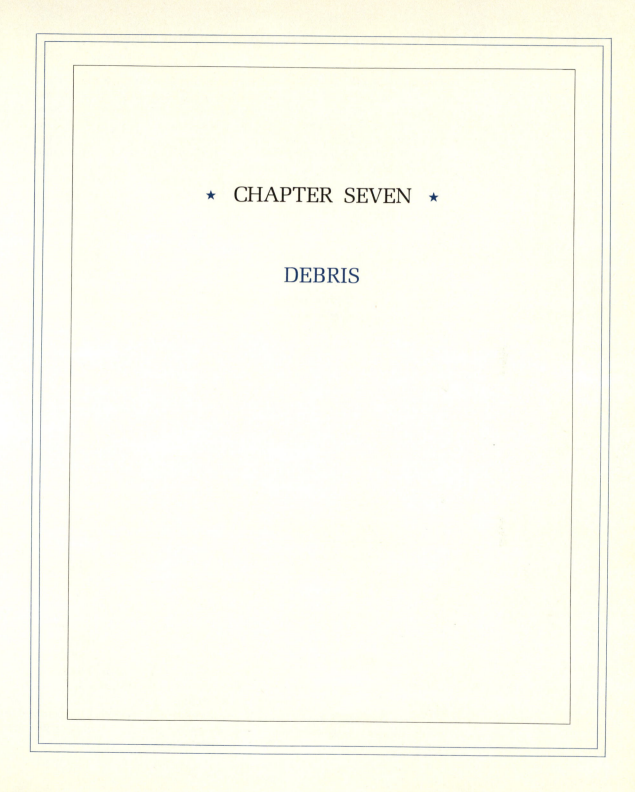

★ CHAPTER SEVEN ★

DEBRIS

Hurtling among the large objects of the solar system—the major planets and their satellites—are a variety of smaller bodies. The largest of these, the *asteroids*, number in the thousands. The spectacular *comets* (Figure 7.1) occasionally produce a sky show when they come close to the sun. A piece of interplanetary rubble becomes a meteor or meteorite when it collides with the earth. These bits of junk, large and small, provide considerable insight into the early history of the solar system. Geological processes occurring on planets have destroyed all the evidence from the early days, and the lunar search for genesis rocks was unsuccessful. Smaller bodies may contain such primitive objects.

★   **MAIN IDEAS OF THIS CHAPTER**   ★

Interplanetary debris comes in a variety of forms: meteorites (objects that strike the earth), asteroids or small planets, comets, and meteors (small objects that leave a vapor trail in the earth's upper atmosphere).

Many asteroids are made of dark, rocky material.

Cometary nuclei are chunks of ice and dust a few kilometers across; the huge visible part of a comet is formed of material that evaporates from the icy nucleus.

★   **7.1 METEORITES**   ★

Meteorites are stones that fall from the sky, bits of solar-system debris that collide with the earth. Until the Apollo landing on the moon, they were the only samples of extraterrestrial material that could be studied in the laboratory; everything else had to be looked at from a great distance. Analysis of these rocks has provided a considerable amount of information about the conditions prevalent when the solar system formed.

Meteorites are broadly divided into two classes, the *iron* and the *stony* meteorites. The iron meteorites are lumps of pure metal (iron and nickel), and they look quite different from the average terrestrial rock. They have a millimeter-thick black fusion crust, formed when they plowed through the earth's atmosphere. If it did not tumble on its way to the ground, a meteorite has shallow grooves, cut into it by the airflow. Iron meteorites are quite dense, and their iron content makes them magnetic. All these properties make iron meteorites fairly easy to recognize. The only similar objects commonly found on the ground are bits of slag from old iron furnaces, which, unlike meteorites, contain bubble-shaped cavities.

An iron meteorite must form in the interior of a large object, or *parent body*, at least 70 km in diameter. Such a large body is liquid when it first forms, and the heavier iron sinks to the center before the large body soli-

FIGURE 7.1 Comet Kohoutek, the infamous "Christmas comet" of the winter of 1973–1974, photographed by the 48-inch telescope of the Hale Observatories. Comets are one form of solar-system debris, appearing irregularly. This particular comet was a disappointment because it was not as bright as some people had predicted. (Hale Observatories)

difies. Collisions then fragment the parent body, leaving the iron meteorite as the remnant of the core.

It is harder to identify stony meteorites on the ground, for they resemble ordinary rocks. However, if you see a meteorite fall, the object you find at the end of the meteorite trail will most likely be a stony meteorite. The trails of falling meteorites, called *fireballs,* are extremely bright, far brighter than the trails of meteors. They resemble the trails of high-altitude aircraft in that they are irregularly shaped and can persist for a minute or so. Sometimes a noise like a sonic boom is heard when a meteorite falls.

Astronomers have classified stony meteorites into a variety of types depending on their chemical contents. The most interesting stony meteorites are the *chondritic* ones, so named because they contain millimeter-sized round mineral spheres called chondrules. Interest in chondrites arises from the abundance of the elements in them, which is very similar to the abundance of nongaseous elements in the sun. This similarity of composition indicates that geological processes have not altered their structure since they condensed from the gas cloud that formed the sun. Analysis of different types of chondrites shows that some are more like the sun than others. The most primitive are the *carbonaceous chondrites* that contain large quantities of carbon.

In order to interpret the evidence about the origin of the solar system that is provided by the carbonaceous chondrites, we must have a better idea of where in the solar system they formed. Are these primitive rocks a sample of the inner solar system, the outer solar system, or someplace in between? Meteorites are but one form of solar-system debris; to appreciate their importance we need to examine some of the other small bodies in the solar system.

★ 7.2 ASTEROIDS ★

☆ DISCOVERY ☆

Astronomers had long noticed a gap in the solar system between Mars and Jupiter. Throughout most of the solar system, each planet is roughly twice as far from the sun as the next one in. A number of eighteenth-century astronomers puzzled over the distribution of planetary orbits and discovered a systematic relationship between the size of a planet's orbit and its order in the solar system. Since Johann Bode discussed this relationship most widely, it has become known as "Bode's law," even though it was discovered by Titius and is not a law in the sense that Kepler's laws are. Table 7.1 shows one formulation of this relationship, illustrating a regularity in the spacing between planets. This same relationship shows up in the regular satellite systems of the major planets, and so it is presumably related to the way in which systems of orbiting bodies form from a gas cloud. Although Neptune and Pluto do not fit this empirical relation, neither of these planets had been discovered in the eighteenth century; Bode's law was regarded as more fundamental by eighteenth-century astronomers than it is now.

Bode's law clearly indicates that there should be another planet between Mars and Jupiter. After the law was published, in the closing years of the eighteenth century, a number of astronomers started looking for the missing planet. These systematic searches were fruitless. The first object found in the gap between Mars and Jupiter was discovered by Giuseppe Piazzi, an Italian astronomer who was making a star catalogue, not looking for a planet. He opened the nineteenth century by noticing, on January 1, 1801, that one of the stars he was measuring for his star catalogue was not a star at all, because it moved relative to the rest of the star field. He observed this wandering object for several nights, and then computed its orbit from the most accurate of his observations. Sadly, his orbit was wrong. The object did not appear at the predicted location a year later; the missing planet had been found and then lost.

This first small planet was recovered, not by an astronomer spending long, cold nights peering through a telescope in search of the lost object, but by a mathematician who devised a technique that used all the information in Piazzi's observations. Karl Friedrich Gauss, discoverer of a large number of mathematical results, was able to calculate an orbit that was the best fit to all of Piazzi's observations, not just the perfect fit to a few. The success of Gauss's method was dramatically proved when he applied it to the measurements of the positions of Piazzi's lost planet. He predicted where the new planet should be found, observers looked, and there it was. This new planet was rescued from the limbo of lost asteroids and was named Ceres.

TABLE 7.1 BODE'S LAW

PLANET	TITIUS'S FORMULA		ACTUAL DISTANCE (AU)
Mercury	0.4 + 0	= 0.4	0.4
Venus	0.4 + (0.3 × 1)	= 0.7	0.7
Earth	0.4 + (0.3 × 2)	= 1.0	1.0
Mars	0.4 + (0.3 × 4)	= 1.6	1.5
———	0.4 + (0.3 × 8)	= 2.8	——— (Asteroids)
Jupiter	0.4 + (0.3 × 16)	= 5.2	5.2
Saturn	0.4 + (0.3 × 32)	= 10.0	9.5
Uranus	0.4 + (0.3 × 64)	= 19.6	19.2
Neptune	0.4 + (0.3 × 128)	= 38.8	30.1
Pluto			39.4

Ceres is one of many *asteroids,* or *minor planets,* that have orbits that fill the Bode's-law gap between Mars and Jupiter. Three other bright ones— Pallas, Juno, and Vesta—were discovered in the first 10 years of the nineteenth century. Once photography was applied to astronomy, hundreds of thousands of asteroids were found, for an asteroid shows up as a short streak on a single photograph of the sky. About 2000 asteroids are now well enough known to have acquired a good orbit, a number, and a name. Large or interesting objects are named after mythological figures: 1 Ceres, 4 Vesta, 433 Eros, 624 Hektor, and 1566 Icarus. Mythological names are too few in number to serve for all 2000 asteroids, and so the less significant ones bear whatever name the discoverer prefers.

☆   **ASTEROIDAL PROPERTIES**   ☆

Ceres, the first asteroid discovered, is also the largest, with a radius of 500 km. Most asteroids are too small to be seen as disks; their sizes must be measured indirectly, utilizing a technique developed by David Morrison and his co-workers of the University of Hawaii. Morrison measures the infrared radiation from an asteroid to determine the total amount emitted as thermal radiation from the surface of the object. One can easily measure the amount of radiation reflected from the asteroid by measuring its brightness in the visual part of the electromagnetic spectrum. The sum of the two types of radiation, the infrared radiation emitted from the hot surface and the visual radiation reflected from the surface, is equal to the total amount of sunlight that the asteroid intercepts. Since bigger asteroids intercept more light, the diameter of the asteroid can be calculated.

Asteroidal diameters measured in this way range from the large diameter of Ceres downward. The very small asteroids are too small to be discovered. There is, then, a whole collection of rocky objects in the solar system— ranging from Ceres down through the smallest visible asteroids to the meteorites—invisible in space and only discovered when they impact the earth.

The smaller asteroids are not spherical, since their gravity is not strong enough to compress the asteroidal rock into a ball. An asteroid that varies the amount of light it reflects is showing its irregular shape, for it dims and brightens as the amount of asteroid facing us changes. The two satellites of Mars, Phobos and Deimos, probably resemble asteroids. The Viking mission obtained a remarkably detailed photograph of Phobos on September 23, 1976 (Figure 7.2); the irregular shape of this object is quite evident in the picture.

The heavily cratered surface of Phobos illustrates the major role that impact has played in its history. There are some rather curious features on this tiny Martian moon, though. The line of craters running across the photograph looks like a line of secondary impact craters, formed when a large crater is excavated and bits of crater ejecta make more craters. The striations on Phobos are not understood; it is possible that some of the surface features were formed when Phobos was part of a larger body.

The measurement of the diameter of an asteroid also provides a measurement of its albedo, the fraction of sunlight that it reflects. Asteroids are surprisingly dark objects; more than three quarters have albedos below 0.07. Asphalt pavement, a dark terrestrial material, has an albedo of 0.15; Mercury, the darkest planet, has an albedo of 0.11. Only coal or black cloth is as dark as the darkest asteroids, which have albedos of 0.02.

What are these dark asteroids made of? The radiation reflected from their surfaces indicate that there are no features at all in their spectra, providing few clues to their composition. Similar types of objects, which are also dark and have no spectral features, are the carbonaceous chondrite class of meteorites. Other, higher-albedo asteroids have reflection spectra that match the spectra of stony meteorites. Further research will probably result in a more extensive matching of asteroid spectra with meteorite types.

This finding that a substantial fraction of the asteroids, the dark ones, may be carbonaceous has prompted a resurgence of interest in these flying rocks. No longer are they merely incidental objects, bits of rubble that someone must keep track of. The level of interest in a particular type of object in the solar system depends on the amount of insight that this type of object can provide into larger questions. These dark, possibly primitive, asteroids may provide some clues to the origin of the solar system.

☆ ORBITS ☆

Most asteroids follow regular, almost circular orbits in the *asteroid belt* between Mars and Jupiter. About 70,000 asteroids are bright enough to be seen with a telescope, but only about 2000 have been discovered and catalogued. These discoverable asteroids are at least half a kilometer across and are about 10^7 km away from each other in the belt. The asteroid belt is not a forest of rocks, but a relatively empty place.

Not all asteroids are found in the belt. Jupiter's gravity enables particles at points 60° ahead of or behind Jupiter in its orbit to remain in stable posi-

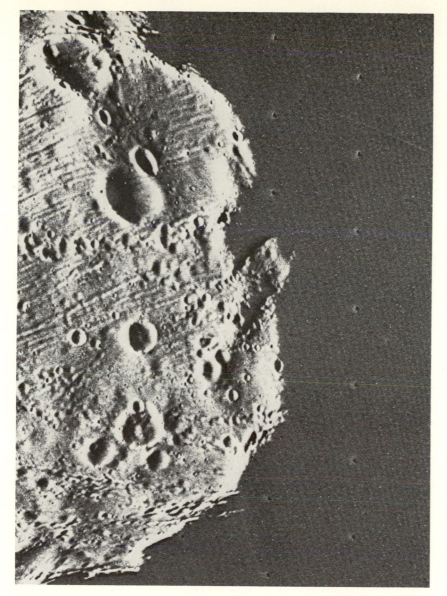

FIGURE 7.2 A detailed photograph of Phobos, Mars's inner satellite, taken in September 1976 by Viking orbiter 2. At the right is the edge of the sunlit portion of Phobos; the stripes and dots in the rest of the photograph are artifacts of the imaging system and are not real. The large number of craters illustrates the role of impact in forming the surface of this object. (NASA Photo No. P-18022)

tions. An asteroid in such a position (Figure 7.3) is in a state of *stable equilibrium;* if it is moved out of position by some perturbing force, the gravitational forces of Jupiter and the sun tend to restore it to its original position. These two points are called the *stable Lagrangian points,* or L_4 and L_5, after their discoverer. The Lagrangian points in Jupiter's orbit contain a group of asteroids known as the *Trojans* and named after heroes of the Trojan War.

Lagrangian points are associated with all orbiting bodies. In particular, the earth-moon system contains Lagrangian points, and Gerard O'Neill of

FIGURE 7.3 L_4 and L_5 are
the two stable Lagrangian
points in a planet's orbit. A
particle placed at one of
these points will tend to remain there; if it is pushed
out of position, gravitational
forces will act to restore it to
its former position at L_4 or L_5.

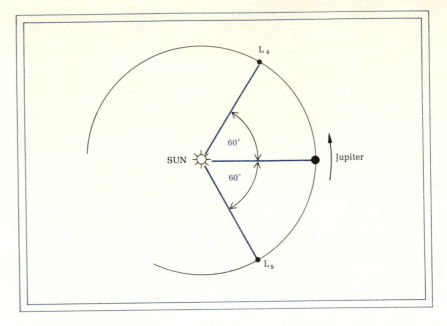

Princeton has proposed that one of these points (L_5) be used as the site for a huge space colony. The advantage of placing a space colony at a Lagrangian point is one of ease of supply. Raw material for this structure could be mined on the moon and catapulted toward L_5. If the cargo load were not aimed quite accurately gravitational forces would tend to take it to L_5, the site of the space station. Presently this idea is just a dream, but such a structure might exist in the future.

Perhaps the most intriguing group of asteroids is the *Apollo*, or Earth-crossing, group. The orbits of these asteroids extend into the inner solar system, inside the earth's orbit. Some of these objects come quite close to the earth. On October 20, 1976, a newly-discovered Apollo asteroid designated 1976 UA came within 0.01 AU, or $1\frac{1}{2}$ million km, of the earth. A systematic search for Apollo asteroids has been conducted by Eleanor Helin of Cal Tech, and at this time about 30 have been found. Helin estimates that between 400 and 1200 Apollo asteroids bigger than 1 km exist.

☆ COLLISIONS BETWEEN APOLLO ☆
ASTEROIDS AND THE EARTH

The asteroid 1976 UA did not collide with the earth in 1976. But this Apollo asteroid, and all others like it, will eventually be drawn into a near collision with Venus or Earth. About 80 percent of these encounters will not result in collisions, for most Apollo asteriods will approach the planet, whip around it in a tight orbit, and be tossed out into the outer solar system (beyond the orbit of Jupiter), never to be seen again. But one encounter in five will put an asteroid into an orbit that will eventually bring it onto a collision course. About 10^8 years will pass before any given Apollo asteroid will encounter Earth or Venus. The known number of Apollo asteroids indicates that one of them will strike the earth about every million years.

While a small meteorite can hit the earth and do relatively little damage, the impact of a 1-km Apollo asteroid would leave a fairly large crater. Laboratory tests and analyses of bomb craters indicate that small objects, less than 20 m across, produce craters that are 25 times the diameter of the impacting object. Provided that these results hold true for larger objects (no one wants to drop an asteroidal-sized object on the earth just to see what happens), a collision between an Apollo asteroid and the earth should leave a crater larger than 25 km in diameter. Smaller objects, not discoverable as asteroids, would leave smaller craters. Such a collision would be quite disastrous if it occurred near a populated area.

The crater left by an asteroid or meteorite colliding with the earth can last for a long time, in the right geological setting. The most famous terrestrial impact crater, to Americans, is the 1.2-km-diameter Meteor crater in northwestern Arizona (Figure 7.4). Larger structures have been discovered in northern Canada, a part of the earth that has remained intact for several billion years. These ancient impact structures on the earth are known as *astroblemes* (star wounds).

A rather unusual group of rocks appears to be related to the formation of astroblemes. The *tektites* are teardrop-shaped bits of fused glass, found in a few isolated regions of the world. Most investigators believe that they formed when large Apollo asteroids or other bodies impacted on the earth.

FIGURE 7.4 Meteor crater in Arizona, formed by the impact of a meteorite on the Arizona desert. This crater is 1.2 km across. (Yerkes Observatory photograph)

★ 7.3 COMETS ★

☆ GENERAL PROPERTIES ☆

Comets are another form of solar-system debris that can be observed in the sky. Occasionally, a wispy comet emerges from the distant reaches of the solar system to spend its few moments near the sun. About once every 5 years or so, a comet becomes bright enough to be seen with the unaided eye, providing a spectacular nighttime sky show for a few weeks (Figure 7.1, Color Photo 9). Like asteroids, comets have been relatively unaffected by geological processes that disturb planetary surfaces. Comet researchers are growing in numbers, for their work can illuminate other aspects of solar-system science. No longer are they confined to cataloguing these occasional visitors and answering questions from the public and the press on where to see the latest sky show.

Comets are named for their discoverers. About 10 comets are discovered each year, most of them very faint and of interest only to comet cataloguers. The three brightest comets of the 1970s have been Comet Bennett (1970), Comet Kohoutek (1973), and Comet West (1976). Bennett and West were very rewarding ones to look at. Kohoutek, heralded as the "comet of the century" when it was first discovered far from the sun, was a disappointment. Halley's comet, named for the astronomer who first recognized that it returns every 76 years, is a bright comet with a fairly short period; it will return in 1986.

The appearance of comets might suggest that they streak across the sky, but they remain nearly fixed in one spot. The *tail* is a large stream of gas and dust pushed away from the comet's *head* by the action of the sun. Since the tail points away from the sun in general, it is usually higher in the sky than the comet's head. Comets are brightest when nearest the sun, and are generally best seen in the west after sunset or in the east before sunrise, in the same general location as the inner planets Mercury and Venus.

Figure 7.5 illustrates the different parts of a comet. The comet's head is composed of a fuzzy *coma* or atmosphere and a small, bright *nucleus*. The coma is an atmosphere produced when the sun's heat evaporates material

FIGURE 7.5 A schematic diagram showing various parts of a comet.

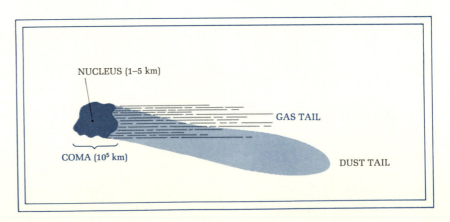

NUCLEUS (1–5 km)

GAS TAIL

COMA (10^5 km)

DUST TAIL

from the nucleus, a small chunk of ice and dust. If the comet comes within 1 AU or so of the sun, a tail is produced. Some comets are surrounded by huge clouds of hydrogen gas, emitting ultraviolet radiation that telescopes in space have detected.

☆   COMA   ☆

Any attempt to identify the chemical composition of comets must begin with the coma; the nucleus is too small to be analyzed. Some light from the coma is emitted by gas molecules, producing radiation when they fall from a higher energy state to a lower one. The wavelengths of the light emitted by these molecules are determined by the energy differences between molecular energy states; as a result, a particular pattern of wavelengths of emitted photons corresponds to a particular molecule. Consequently, analysis of the coma spectrum allows us to determine its composition (recall the discussion of spectroscopy in Chapter 1).

The appearance of several bright comets in the 1970s made it possible for astronomers to find a number of new molecules in comets. Radio astronomers had discovered that many of the transitions between different molecular energy states produced photons of radio radiation, and they turned their telescopes toward these bright comets. In particular, the coming of Comet Kohoutek in 1973 was predicted well enough in advance so that time on the Kitt Peak 36-foot radio telescope could be reserved for searches for cometary molecules. A long list of molecules has been found in the atmospheres of various comets: H_2O, CH, CN, HCN, CH_3CN, NH, NH_2, OH, CH^+, CO^+, and CO_2^+. The elements hydrogen, oxygen, sodium, carbon, calcium, chromium, cobalt, manganese, iron, nickel, and copper have also been detected.

The list of molecules and elements is so long that the discovery of more molecules would produce little new insight. The important question now is the chemical composition of the nucleus itself. Theorists try to model the chemical reactions that take place in the coma after the parent molecules evaporate from the nucleus. Positively identified nuclear constituents include water ice, frozen carbon dioxide (dry ice), and probably the ices of more complex compounds like CH_3CN and HCN. Comets are icy objects, like the satellites of planets in the outer solar system.

☆   TAILS   ☆

Comet tails are the most conspicuous parts of comets. Two types of tail exist: a straight, wispy *gas tail* and a broader *dust tail*. The gas tail can be thought of as an extension of the coma; it is formed when charged particles in the solar wind push coma material away from the sun. Radiation from the gas tail comes from gas molecules dropping from high to low energy states, and most of it is emitted at wavelengths the eye cannot detect. The visually impressive tail is the dust tail, formed when the pressure of the sun's radiation pushes dust particles away from the nucleus. The distinction between gas and dust tails is quite clear in the color photograph of Comet West (Color Photo 9). The gas tail is blue and the dust tail is yellow.

The dust in comet tails has an albedo of 0.2, and so it re-emits 80 percent of the incident sunlight as infrared radiation. Several groups have put considerable effort into measuring the infrared intensity of the bright comets of the 1970s to determine the composition of the dust particles. Their work indicates that these particles are about 1 μ in size; the presence of very strong infrared emission at a wavelength of 10 μ indicates that many of the dust particles contain silicate compounds. Most terrestrial and lunar rocks are silicate compounds, minerals containing metallic elements combined with silicon and oxygen.

The tails of comets, which can stretch halfway across the sky, are surprisingly insubstantial. While it is not possible to make extravagant claims like "you can stuff a comet tail into a suitcase," the total amount of material in them is still small. Comet Arend-Roland, a bright comet of 1957, had a large tail that contained only 10^{13} to 10^{14} grams of dust. While 10^{14} grams is a billion tons of dust, and would make a 300-m cubical pile, the material in a comet tail is spread so thinly that the dust particles are 1 m apart.

One feature displayed by Comet Arend-Roland was a sunward-pointing tail, or *antitail*, that looked like a spike (Figure 7.6). While comet tails generally point away from the sun, a comet passing very close to the sun can develop some strange tail shapes near perihelion, its closest approach to the sun. Figure 7.7 illustrates how these shapes can evolve as the result of our perspective on the earth. Antitails and curved tails can develop at this time. Such an explanation of antitails is quite reasonable, and it was impressively confirmed when the originator of the theory, Smithsonian Observatory's

FIGURE 7.6 Comet Arend-Roland, a bright comet of 1957, showing the spike-shaped antitail. (Hale Observatories)

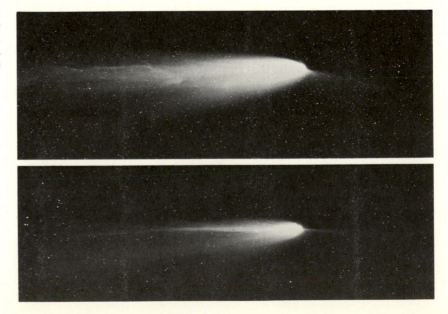

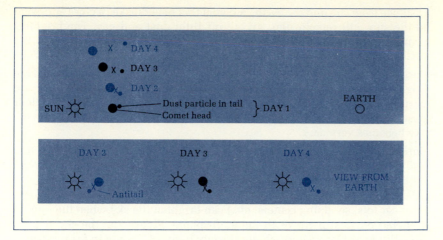

FIGURE 7.7 Near perihelion, the closest approach to the sun, comets can develop strangely shaped dust tails. On day 1, the comet head (large circle) ejects a dust particle, which forms part of the dust tail (dot). On day 2, shown in color, the comet ejects another dust particle (cross). However, at this time, the particle ejected on day 1 (dot) has lagged behind the comet in its orbit because it is farther from the sun. From the viewpoint of the earth, shown in display panels, the dust particle is situated between the comet and the sun, forming an antitail. By days 3 and 4, these dust particles, traveling in their own orbits around the sun, have passed the comet, but their lag behind the comet may cause the dust tail to be curved.

Zdenek Sekanina, predicted that Comet Kohoutek would have an antitail. Earthbound astronomers were blanketed by clouds, but Skylab astronaut Edward Gibson observed this comet shortly after it rounded the sun. He did, as Sekanina had predicted, see an antitail on Comet Kohoutek. Such a prediction is always an impressive confirmation of a theory.

☆ **VISIBILITY OF COMETS** ☆

The photographs in this book probably present a misleading picture of how spectacular comets can be. A telescopic photograph can capture the marvelous detail visible in a comet tail, while the unaided eye often loses the faint tail against a light-struck urban sky. Bright comets, visible to the unaided eye, appear once every 4 or 5 years, often unpredictably. A number of factors determine how visually impressive a particular comet will be:

Size and dust content. How much stuff is in the icy nucleus? How much raw material is there to produce the visible parts of the comet, the coma and tail? Dusty comets are visually brighter than gaseous comets because gas molecules emit most of their radiation at wavelengths the eye does not detect well.

Distance from sun and the earth. A comet that never makes a close approach to the sun will remain dim, giving off little dust and gas. The brightest comets are those that travel close to the sun while remaining close to the earth.

Position in the sky. Comets, like the inner planets Mercury and Venus, are brightest when they are close to the sun. Unfortunately, when they are closest to the sun they can be lost in the glow of twilight, especially if their position in the sky puts them close to the horizon so that they set before the sky becomes very dark.

Darkness of the sky. While a comet may emit a large amount of light, this light is spread over a large coma and tail. Its contrast with the sky can be quite poor, especially when it is viewed from a light-polluted urban or sub-urban sky (Figure 7.8). Light from human activities (lights for ballparks, used-car lots, shopping centers, etc.) can brighten the sky when the light is scattered off dust particles in the atmosphere. Figure 7.9, the sky as seen from Kitt Peak Observatory, shows the city lights of Tucson in the northeast. *Light pollution* makes it difficult to see or photograph faint objects, and presents a serious problem to observatories located near urban centers. The 100-inch telescope at Mount Wilson, once the largest in the world, is now used only for observations of relatively bright objects because the city lights of the nearby Los Angeles metropolitan area swamp the faint objects that lie at the outer limits of the observable universe.

Unpredictability. Comets are inherently unpredictable, and some fall far short of expectations while others are surprisingly bright. We can only guess how a comet will behave when it approaches the sun, since the chemical reactions and dust production in the coma proceed quite differently in dif-

FIGURE 7.8 The darkness of the sky can greatly affect the visual appearance of a faint, extended object such as a comet. In 1974, Comet Kohoutek was an impressive sight from Kitt Peak Observatory, 40 miles west of Tucson, but barely visible from the city. It is hoped that Kitt Peak Observatory's dark sky will continue to remain relatively free of light pollution.

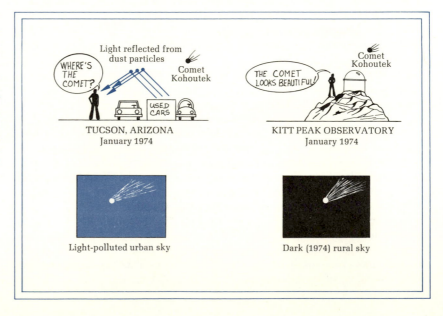

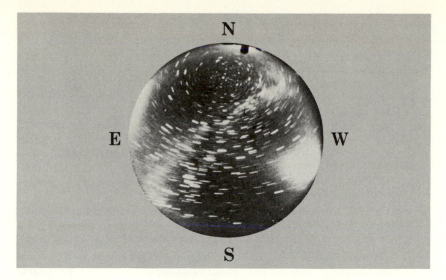

FIGURE 7.9 A photograph showing the Kitt Peak sky. The swath of light in the northeast is scattered light from the city of Tucson, 40 miles away. West of Kitt Peak there are no cities. The band of light stretching upward from the western horizon is the zodiacal light produced when sunlight is reflected by small meteoroids in the plane of the earth's orbit. (© by the Association of Universities for Research in Astronomy, Inc. The Kitt Peak National Observatory.)

ferent comets. Comet Kohoutek, discovered when it was far away from the sun, was a disappointment when it turned out to contain little dust and was much dimmer than expected. When the next bright comet appeared, in the spring of 1976, a number of astronomers (including myself) hesitated to be optimistic and make any promises about good-looking objects in the sky. Comet West was about four times as bright as expected, rewarding those who (despite our gloomy predictions) woke before dawn to glimpse the brightest comet (so far) of the 1970s. Comet watching is like life: Nothing ventured, nothing gained.

Halley's comet was spectacular in 1910, but it will be far less conspicuous on its next visit to the inner solar system in 1986, for many of the reasons listed above. Since it is the same comet that came around in 1910, its size, dust content, and distance from the sun will be the same and will not affect its brightness. But in 1910 Halley's comet came very close to the earth. The earth passed through the flimsy tail of this famous comet, confounding the prophets of doom who forewarned that such an occurrence would mean the end of the world. Next time, in 1986, the closest approach of the comet to the earth will occur on April 10, and at that time it will be 0.42 AU away. At that time, 2 months after it passes closest to the sun, it will be best situated for Northern Hemisphere observers. Even so, it will be poorly placed, low in the sky, difficult to find in the twilight glow, and setting soon after the sun. The most conspicuous and tragic difference between the performances of 1910 and 1986 is related to the problem of light pollution. In 1910, Halley's comet was seen by most people against a dark, primitive, rural sky. This next time, it may not even be visible in a light-struck urban or suburban sky, except through binoculars.

Halley's comet travels around the sun in a long, skinny elliptical orbit that brings it back into the vicinity of the sun every 76.1 years (Figure 7.10). Short-period comets like Halley's can be observed every time they return to the inner solar system, and most of them have been observed on several occasions. Halley's is the brightest of the short-period comets; most remain quite faint and are only visible through telescopes. The spectacular comets that appear occasionally are generally long-period comets, comets that travel on orbits so extended that it takes them tens of thousands of years to go around once. The existence of one of these comets is unsuspected until it emerges from the depths of the solar system, zips around the sun, and then returns to the outer part of its orbit. Cometary orbits are generally very elliptical, unlike almost circular planetary and asteroidal orbits.

The evolution of cometary orbits is not completely understood, but Jan Oort proposed a scheme in the 1950s that still seems to fit the observations. He noted that the longest-period comets are 50,000 AU from the sun at the farthest, and he proposed that most comets exist as large chunks of ice and dust at that distance from the sun, about one-fifth of the way to the nearest star. Weak gravitational perturbations from passing stars then send the comets into the inner solar system, where the sun evaporates the outer ice layers and produces the coma and tail that we observe. Some of these new comets are captured into shorter-period orbits by encounters with planets. Depending on its internal structure, a short-period comet may last for a long time or be broken up after a few encounters. If Oort's hypothesis is correct, comets are icy objects like the satellites of the outer planets; they formed far, far away from the sun when the gas cloud that was to become our solar

FIGURE 7.10   The orbit of Halley's comet. The orbits of several planets are shown for size comparison, but the comet's orbit is inclined relative to the plane of the planetary orbits. Black dots show the position of Halley's comet on January 1 of the indicated years, unless some other date is shown. The positions of both the comet and the earth are shown on February 9, 1986, when the comet is closest to the sun, and on April 10, 1986, when it is closest to the earth.

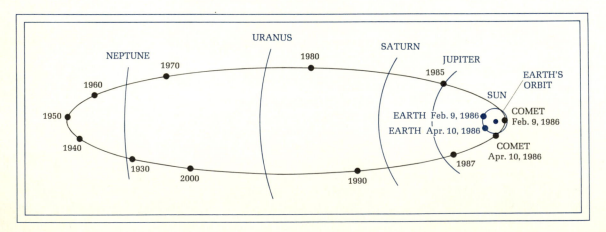

system collapsed. A space probe sent to one of these primordial icebergs might be quite productive.

☆   COMETS IN HISTORY   ☆

The appearance of a bright comet was regarded as an evil omen throughout most of human history. Before the Copernican revolution, when people thought that the planets were little lights in the sky, the appearance of an unusual object in the heavens was interpreted as a sign of divine wrath, and people became alarmed. Ambroise Pare, court physician to Henri II, Francois II, Charles IX, and Henri III, sixteenth-century kings of France, described the comet of 1528 in these vivid terms:

> This comet was so horrible and so frightful and it produced such great terror in the vulgar that some died of fear and others fell sick. It appeared to be of excessive length and was the color of blood. At the summit of it was seen the figure of a bent arm, holding in its hand a great sword, as if about to strike. On both sides of the rays of this comet were seen a great number of axes, knives, and blood colored swords among which were a great number of hideous human faces with beards and bristling hair.[1]

Figure 7.11 is a drawing from Pare's commentary.

Now, being civilized and knowledgeable, we realize that comets are harmless celestial bodies, posing no threat to the human race. These scare stories, a humorous addition to planetarium shows, are of course a relic of the past. Or are they? Two leaflets widely distributed in the United States described the coming appearance of Comet Kohoutek in rather vivid terms, warning of approaching doom: "*Who knows what a climactic effect this may have upon* the earth herself, and what earth-shaking *events she will bear* from the fruits of this *heavenly rape* of her earthly body?"[2] "Governments are notorious for playing down the possibilities of impending disasters and threats of doom. . . ."[3]

★   7.4 METEORS   ★

Go outside on any night during the second week in August, lie back on the grass, look up at the sky, and soon you will see a small pinpoint of light

---

[1] Ambroise Pare, *Monstres Celestes,* quoted in Peter L. Brown, *Comets, Meteorites, and Men* (New York: Taplinger, 1974), p. 17.

[2] Moses David, "More on Kohoutek," The Children of God, 1973. Emphasis in the original.

[3] Moses David, "What Will the 'Christmas Monster' Bring?" The Children of God, 1973. These leaflets were, to my knowledge, distributed in St. Louis, Iowa, and Hawaii; I suspect that their distribution was nationwide.

streaking across the heavens. That streak is a *meteor*, the trail of a small bit of dust encountering the earth's upper atmosphere. Each one of these small bits is called a *meteoroid* before it collides with the earth, and the solar system is full of them. Many meteoroids come in swarms, and a *meteor shower* occurs when such a swarm encounters the earth and produces a great many meteors in a short span of time. The most reliable shower is the Perseids, the shower that occurs in the second week in August. In the early morning, when the earth's orbital motion carries it toward the swarm of meteoroids, the Perseids can produce one meteor per minute. Table 7.2 lists some of the most conspicuous showers.

Some meteor showers are rather erratic, varying in intensity from year to year, because the meteoroid particles are distributed rather unevenly along

TABLE 7.2 MAJOR METEOR SHOWERS

NAME	DATE	EXPECTED HOURLY RATE
Quadrantids[a]	January 3	50
Eta Aquarids	May 4	20
Ophiuchids	June 20	20
Capricornids	July 25	20
Delta Aquarids	July 29	20
Perseids	August 12	50
Orionids	October 21	20
Leonids	November 17	Variable; see text
Phoenicids	December 5	50
Geminids	December 13	50

[a] The Quadrant is an obsolete constellation; the center of this shower is located within the boundaries of Lyra but it preserves its old name.

the shower orbit. A good display only occurs when the greatest concentration of particles happens to be there when the earth crosses the particles' orbit. The Leonids are an extreme example; they are extremely intense once every 33 years, if conditions are right; in intervening years they are sometimes strong, sometimes weak. In these peak years this shower can produce a tremendous rain of meteors, several hundred per minute. In the nineteenth century, the years 1833 and 1866 produced spectacular showers. In 1899 and 1933, the showers were disappointingly meager. In 1966, eastern observers awaited the shower but, when the sun rose on November 18, nothing had happened; it looked like another bad show. But observers in the western United States who waited a few hours until their sunrise saw what one person later described as "the greatest meteor display ever witnessed by man since observing records began and certainly since the Earth began to encounter the stream in historical times . . . much more spectacular than the great displays of the nineteenth century."[4] The next scheduled peak for this erratic shower will occur in 1999. Will the Leonids equal the magnificent spectacle of 1966? No one knows, but I'll be watching.

Meteors appear because a small meteoroid leaves a vapor trail when it burns in the earth's upper atmosphere. Photographs of meteors show that they slow down when the friction of the atmosphere impedes their motion, and measurements of the deceleration rate seem to indicate that the densities of meteors are very low, in some cases less than 0.1 g/cm^3. Such a low density, along with the association of certain meteor showers with the orbits of extinct comets, indicates that the shower meteors are icy particles.

Shower meteors probably come from comets that break up. The physical characteristics of the particles in showers resemble the physical characteristics of comets: icy, wispy particles putting on a brief show. The principal link between shower meteors and comets is the similarity of their orbits. The Andromedids, for example, follow the orbit that Comet Biela followed before it broke up in 1846.

[4] Peter L. Brown, *Comets, Meteorites, and Men* (New York: Taplinger, 1974), p. 213.

★ 7.5 THE ZODIACAL LIGHT AND ★
INTERPLANETARY DUST

While the easiest way to detect a meteoroid is to observe the consequences of its collision with the earth, interplanetary meteoroids can also be detected by the light they reflect. A faint glow extending along the ecliptic can be seen in the western sky after twilight ends, if other sources of sky brightness (light pollution) do not obscure it. The photograph of the Kitt Peak sky (Figure 7.9) shows the *zodiacal light* quite clearly in the western sky. Interplanetary dust particles, small meteoroids, reflect sunlight and produce this effect. Measurements of the intensity and color of the zodiacal light indicate that the particles concentrated in the plane of the ecliptic and producing this effect are about 1 μ across, about the same size as the dust particles in comet tails. This dust is spread more or less uniformly across the inner solar system.

The interplanetary meteoroids that produce the zodiacal light also collide with the earth; some are not vaporized by the upper atmosphere and remain in the stratosphere, where high-altitude balloon samplers can collect them. Figure 7.12 is a photograph of one of these large particles, showing its fluffy structure. Most of the upper-atmosphere particles are more concentrated than the one shown in Figure 7.12. These particles are more or less similar to the shower meteors, but they are a little denser. Are these particles really shower meteors, or are the measurements of the density of the shower meteors wrong? No one knows; this is a question for the future.

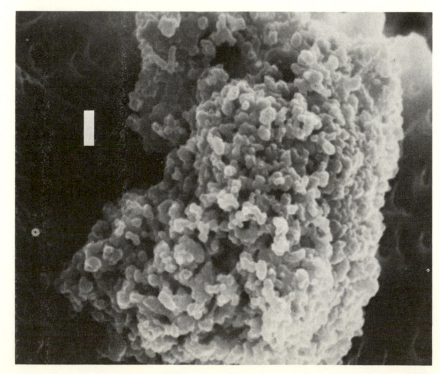

FIGURE 7.12   A small inter-planetary particle found in the earth's upper atmosphere by a balloon-borne particle collector, given the catalogue number U2–5A (21). This is the fluffiest particle ever found in the upper atmosphere, and it may resemble some of the denser meteors. The white scale bar is 1 micron (μ) long. (From D. Brownlee, F. Horz, D. Tomandl, and P. Hodge, "Physical Properties of Inter-planetary Grains," in B. Donn, M. Mumma, W. Jackson, M. A'Hearn, and R. Harrington, eds., *The Study of Comets*, National Aeronautics and Space Administration SP–393. Photograph courtesy of D. Brownlee.)

★ 7.6 EVOLUTION OF ★
SOLAR-SYSTEM DEBRIS

Small objects in the solar system come in a variety of forms, with the icy comets and cometary meteors in a class somewhat distinct from that of rocky asteroids and meteorites. An open question is where in the solar system the various types of debris originated. These objects are so small that their orbits can easily be altered by encounters with the much larger planets. Theorists have constructed a tentative evolutionary scheme that traces each of the forms of debris back to their origin when the solar system formed; it is outlined in Figure 7.13.

In the beginning there were two distinct types of objects, and possibly a third. The parent bodies, rocky objects at least 70 km in diameter, probably circled the sun in the vicinity of the asteroid belt. Cometary, icy objects existed then in much the same form as they exist today, but no one is sure whether they were formed somewhere near the orbit of Uranus or Neptune

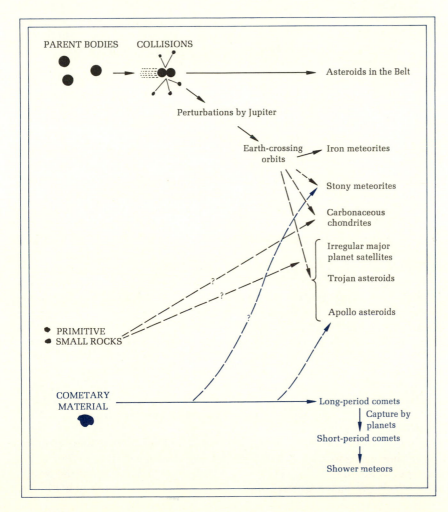

FIGURE 7.13   A tentative evolutionary scheme for the history of solar-system debris. Solid arrows show those evolutionary paths that are fairly certain, and broken arrows show those paths that are uncertain.

or further out in the system. It is possible that there were some rocks smaller than the parent bodies and not dense enough to form iron cores. The distinction between rocky and icy debris parallels the distinction between the rocky inner planets and the satellites of the outer planets, similar in size but different in composition.

The parent bodies collided and fragmented soon after they formed. Calculations show that a collection of objects with a size distribution similar to that of the asteroids would result from repeated collisions. Most of these objects remained in the asteroid belt, forming the asteroids visible today. A Jovian gravitational perturbation could force an asteroidal particle into an earth-crossing orbit, and eventually some of these objects would strike the earth as meteorites. The iron meteorites are definitely connected to the parent bodies, since their composition indicates that they were formed in the core of some large body. The stony meteorites, carbonaceous chondrites, and at least some of the Apollo asteroids probably have the same evolutionary history, but the evidence is not as clear in this case. The identification of the dominant class of dark asteroids as having carbonaceous surfaces supports the connection between meteorites and asteroids. However, some investigators believe that the carbonaceous chondrites never formed part of the larger parent bodies and represent unaltered primitive objects.

Comets are another source of solar-system debris. They presently are in an entirely different part of the solar system, the comet cloud on the outer fringes. Material from this cloud reaches the inner solar system when a long-period comet falls into the inner solar system. New long-period comets are continually created; Comet Kohoutek, the disappointing Christmas comet of 1973, was a new comet on its first passage near the sun. Long-period comets can sometimes be captured by the planetary system, becoming short-period comets; the short-period comets are eventually disrupted by the repeated blasts of sunlight and become shower meteors. Some investigators contend that comets that lose their ability to form tails become Apollo asteroids, although the majority view is that Apollo asteroids originally come from the belt. Asteroid 433 Eros was carefully observed during its close approach to the earth in 1975, and it, at least, is a rocky object.

The scheme outlined in Figure 7.13 seems quite neat and easy. The relatively certain evolutionary paths are shown with solid arrows, but the many dotted arrows and question marks attest to the uncertainties. The Apollo asteroids and stony meteorites probably come from the parent bodies and are thus asteroidal in origin, and it is almost certain that some of these objects come from the belt. Yet the orbit of a comet captured by Jupiter is not so very different from the orbit of an asteroid perturbed by Jupiter into an earth-crossing orbit, and some investigators see comets as the source of some Apollo asteroids and other large meteoroids. The Trojan asteroids are quite different from the others, and their origin is very uncertain. The dotted lines in Figure 7.13 are working hypotheses, to be verified or destroyed by further research.

The importance of unraveling the evolution of the stony meteorites comes from their value as genesis rocks. The study of planets tells us more about geological and meteorological processes on them and less about the origin of the solar system. Planetary study can provide a fascinating new dimension for earth scientists, but it may be the less spectacular members of the solar system that will tell more about its origin. We need more than the present collection of a few random observations of meteors and comets to understand the full picture of what they are and where they come from.

SUMMARY

Key evidence regarding the origin of the solar system may well come from an examination of its smaller members. The asteroids or minor planets are small bodies, ranging from the 1000-km-diameter Ceres to small kilometer-sized bits of rock. The comets, irregular visitors from the outer edges of the solar system, give off gas and light up when they pass the sun; most then return to the distant reaches from whence they came. Some comets disintegrate into meteor swarms that produce showers when their orbits cross the earth's path. When a piece of interplanetary rubble survives the hazardous journey through the earth's atmosphere and collides with the earth, it becomes a meteorite.

This chapter has covered:

The observational appearances of various types of solar-system debris: comets, cometary meteors, meteorites, and other meteoroids;

The composition of these different types of objects, indicating a distinction between icy cometary objects and rocky asteroidal objects;

The events that take place when a comet nears the sun;

The encounters between interplanetary debris and the earth, resulting in the appearance of a meteor or a meteorite fall; and

A tentative scheme for the evolution of solar-system debris.

KEY CONCEPTS

Antitail	Dust tail	Minor planets
Asteroid belt	Fireballs	Nucleus
Asteroids	Gas tail	Parent bodies
Astroblemes	Iron meteorites	Stable Lagrangian points
Carbonaceous chondrites	Light pollution	Stony meteorites
Coma	Meteor	Tektites
Comet	Meteoroid	Zodiacal light
	Meteor shower	

REVIEW QUESTIONS

1. Why are iron meteorites easier to find than stony ones?
2. Which meteorites—the iron ones or the stony ones—are more valuable

to someone trying to understand the physical conditions under which the solar system formed?

3. Is Bode's law an empirical description of planetary orbits or is it a fundamental law of physics like Newton's law of gravitation?

4. Describe the three types of orbital paths that asteroids tend to follow.

5. Why are there fewer astroblemes, impact craters, on Earth than there are on the moon or Mars?

6. Why is the dust tail of a comet more visible to the unaided eye than the gas tail?

7. Would you expect to see a bright comet at midnight? Why or why not?

8. Why are the Leonids so irregular and the Perseids so dependable in producing meteor showers? What is the difference between these two meteor streams?

9. Sketch the evolutionary connection between asteroids and meteorites, and between comets and meteors.

FURTHER READING

Brown, Peter L. *Comets, Meteorites, and Men*. New York: Taplinger, 1973.

Marsden, B. "Comets." *Annual Review of Astronomy and Astrophysics* 12 (1974): 1–22.

Morrison, D. "Asteroids." *Astronomy* (June 1976): 6–18.

Whipple, F. L. "The Nature of Comets." *Scientific American* (February 1974): 48–57.

Part One (Basics) dealt with solid astronomical fact, much of which has been in the research literature for hundreds of years and is well confirmed by observations. In contrast, Part Two concerns results that have been unveiled primarily in the last 10 years. An active field such as planetary science is a controversial one, so this summary table segregates solid fact, secure interpretation, working models (theoretical ideas that will be reinforced or shot down as new data come in), and speculation.

SOLID FACT

Lunar and Mercurian craters are formed by the impact of meteorites.

Volcanoes exist on Mars.

The atmospheres of Mars and Venus are mostly carbon dioxide, and Venus is sizzling hot while Mars is cold.

Martian arroyos exist.

SECURE INTERPRETATION

The evolution of the moon was largely completed several billion years ago.

Water eroded the Martian arroyos (while the interpretation is secure, the consequences are profound!).

Martian volcanoes are shield volcanoes.

Comets are dirty iceballs.

Shower meteors are dead comets.

WORKING MODEL

Venus, Earth, and Mars differ primarily in the degree to which the greenhouse effect determines their surface temperatures.

Meteorites come from asteroids; meteors come from comets.

The distinction between the icy outer Galilean satellites of Jupiter and the inner rocky ones reflects the high luminosity of Jupiter when its satellite system was formed.

SPECULATION

Mars was warmer in the past.

PART THREE

STARS

The picture of the solar system developed in Part Two left out its most important member, the sun. Solar astronomy, describing the physical processes taking place on the solar surface and in the solar interior, differs greatly from planetary science because the questions of primary interest cannot be related to phenomena on our own planet and to earth sciences such as geology and meteorology. Where does the sun obtain its energy? How long will the energy supply last? What will happen when it runs out?

It is difficult to answer such questions by analyzing only our sun. Human beings have observed the sun for millenia and, while scientific study of the sun began almost 400 years ago with Galileo's discovery of sunspots, this is only a very short moment in the life cycle of our star. Stellar evolution is even slower than geological evolution; the sun's present evolutionary stage has lasted 5 billion years and will continue for another 5 billion years. We would have to wait a long time to see the sun evolve. But by studying other stars we can follow the stellar life cycle.

A close examination of the sky reveals stars in all stages of evolution. Giant clouds of gas contain protostars, stars just now forming. Most of the stars are in the same evolutionary stage as the sun, providing us with a comprehensive view of stars in their vigorous middle age. The hugest, reddest stars (red giants) are dying stars—those whose energy sources are running out, whose bloated envelopes would engulf the inner solar system. The stellar graveyard is more difficult to discover, as dead stars are extremely faint. Yet patience, persistence, and a certain amount of good luck have uncovered three types of dead stars: white dwarfs, neutron stars, and the enigmatic black holes.

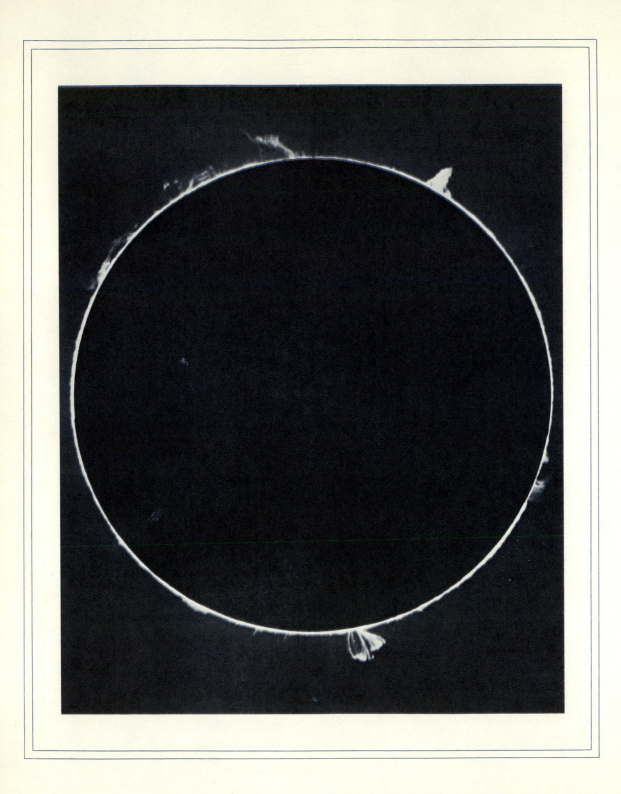

★ CHAPTER EIGHT ★

THE SUN

The fact that primitive algae existed on the earth's surface some $3\frac{1}{2}$ billion years ago is mute testimony to the existence of liquid water on the earth's surface in the distant past. The sun kept the earth warm, just as it does now. What is the nature of this energy source that we depend on? Careful observations of its surface show that it is a gaseous ball, composed mostly of hydrogen. Photographs of the upper layers of the solar atmosphere show a wealth of detail, unobservable on any other star. The sun is both an interesting object in its own right and the nearest example of a star, the one we can analyze most thoroughly.

★ MAIN IDEAS OF THIS CHAPTER ★

The visible surface of the sun is a layer of hot hydrogen and helium gas.

Photographs of the sun's upper atmosphere show many details of structure.

Many of the structures in the sun's upper atmosphere are associated with explosive events on the sun.

The solar interior, like the surface, is gaseous but much hotter; nuclear reactions at the center provide the energy to keep the sun shining.

★ 8.1 THE VISIBLE SURFACE ★

Chapter 1 posed three questions that an astronomer asks about a particular object under study: Where is it? What is it? How does it evolve? The sun's location was determined hundreds of years ago by astronomers who scrutinized the motions of the planets and realized that it was the center of the solar system, some 150 million km from the earth. The second question, "What is it?" was not attacked until the twentieth century, when the powerful tools of spectroscopy allowed us to analyze the composition and nature of the solar surface. The third question, pertaining to the evolution of the sun, can only be answered by examining other stars like the sun that are at different stages of their evolution.

Seen from the earth (Figure 8.1), the sun looks like a glowing ball. A knowledge of its distance and a measurement of the angle it covers in the sky provide a value of 695,000 km for its radius. Its surface emits 3.9×10^{33} ergs of energy every second, indicating that the solar furnace has a power of 3.9×10^{26} watts, equivalent to about 4×10^{17} large electric power plants. The earth intercepts a small fraction of this power, converts it to heat, and provides the human race with a livable environment. The power in the sun is so much greater than anything that human technology has been able to produce that it is impossible to visualize its magnitude.

FIGURE 8.1 The sun, shown here rising near Nairobi, Kenya, is the source of heat, light, and energy for the earth. What makes the sun shine? This chapter focuses on the central object of the solar system. (Nancy Lockspeiser/Stock, Boston)

We seek to do more than just gaze at the sun in awe, but a photograph of the sun shows little detail and provides no information on its physical make-up. The spectrograph is the essential instrument used in an analysis of conditions on its surface. In a spectrograph, light from the sun (or any celestial body) is focused on a slit, and the image of that slit in various wavelengths is then photographed. The dark lines crossing the solar spectrum (Figure 8.2) indicate that at particular wavelengths the sun emits less radiation; something is absorbing this radiation, creating *absorption lines*. The

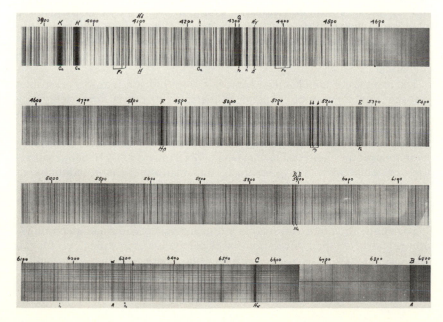

FIGURE 8.2 The solar spectrum. A vertical slit isolates a small portion of the solar surface, which is photographed in various wavelengths to produce a spectrum; the dark vertical lines mark wavelengths at which the sun emits less light. The numbers above the spectrum are wavelengths in angstroms; the symbols below the spectrum identify the elements responsible for various spectrum lines. The way in which these dark lines are formed is explained in the text and illustrated in Figure 8.3. (Hale Observatories)

pattern of absorption lines in the solar spectrum is extremely complex, and it seems that the spectrograph has only changed this featureless disk into an indecipherable mess.

The solar spectrum of Figure 8.2 may be complex, but it is not written in an unknown language. Thanks to spectroscopy, each of the absorption lines in Figure 8.2 can be related to some atom or molecule in the sun's outer envelope. Absorption lines are produced when atoms absorb photons of particular energies from a beam of sunlight passing through the solar atmosphere.

Consider a beam of sunlight traveling from the solar surface to a waiting astronomer on the earth (Figure 8.3). It must pass through an obstacle course of atoms in the sun's outer layers. Various atoms are waiting anxiously to steal photons from this beam of sunlight, but these thief atoms are quite selective in what they can absorb. A photon with just the right energy,[1] just the right wavelength, can collide with an atom and kick it into a higher energy state. Photons with an incorrect amount of energy, an incorrect wavelength, pass right by the atom, untouched. Atoms in the solar atmosphere only remove particular photons from the beam of sunlight traveling toward the waiting astronomer. Atoms in planetary atmospheres act in a somewhat similar way in creating absorption lines.

The wavelengths or energies of the absorption lines in the solar spectrum are a direct indication of the atoms and molecules present in the gas layer that we see. Each element or molecule has its own pattern of energy levels, and thus its own pattern of wavelengths that it can absorb. Scientists have spent long years poring over the solar spectrum of Figure 8.2 in an effort to match each absorption line with the element or molecule that produces it. Some elements have a complex set of energy levels and can absorb a wide variety of photons; most of the absorption lines in the visible solar spectrum come from iron. Some atoms are sufficiently rare and have a sufficiently sparse structure of energy levels so that few of their absorption lines appear in the solar spectrum. Beryllium, for example, shows only one distinct line in the solar spectrum. Scientists have not yet assigned about half the lines in the solar spectrum to a particular element or molecule, probably because the laboratory studies of just what energies particular atoms can absorb are incomplete.

The analysis of the chemical composition of the sun requires more than just the identification of elements in the solar spectrum. The ability of a chemical element to produce absorption lines depends on physical conditions in the solar atmosphere. It might appear from Figure 8.2 that iron is the most common element in the solar atmosphere, and astronomers of the early twentieth century held to this mistaken belief. However, M. N. Saha realized that the temperature and pressure in the solar atmosphere determine an atom's ability to produce an absorption line; for example, only 1 hydrogen atom in 100 million can absorb a photon of visible sunlight, whereas iron atoms absorb photons far more readily.

The application of this principle to the analysis of the solar spectrum requires a fairly detailed calculation. Large computers follow the flow of radiation through various layers of the solar atmosphere and calculate what the

[1] Recall the discussion of energy and photons in section 1.3 and the discussion of absorption lines in planetary atmospheres in section 1.5.

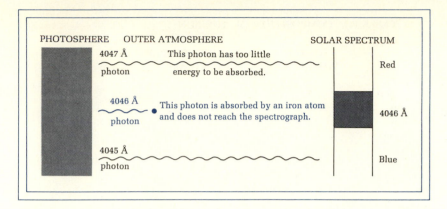

FIGURE 8.3 The formation of an absorption line at 4046 Å in the solar spectrum. Three photons leave the solar photosphere, the visible surface. Two pass unscathed through the outer solar atmosphere, but the 4046-Å photon (color) is absorbed by an iron atom and does not reach the spectrograph. The shortage of 4046-Å photons in the solar spectrum is visible as an absorption line. Since only iron atoms can absorb 4046-Å photons, the existence of an absorption line at this wavelength (or energy) indicates the presence of iron in the sun's outer atmosphere.

spectrum should look like. The temperature and pressure of these layers and the abundance of various chemical elements in them are then varied until the calculated spectrum matches the observed one. In this way, we have determined that the solar *photosphere*, the layer we see when we look at the sun in visible light, has a temperature of 6000 K and a pressure of 0.1 times the pressure in the earth's atmosphere. A recent survey listed the abundance of 66 elements in the solar atmosphere; Table 8.1 lists some of these elements. We may not be able to bring a piece of the sun into the chemical laboratory for analysis, but spectroscopy allows us to determine its composition quite accurately.

As Table 8.1 shows, the sun is 90 percent hydrogen and 10 percent helium. All the other elements are far less common, with the quantities generally decreasing as the elements become heavier. The relative abundances of elements in the sun are more or less consistent with the abundances of elements on the earth, except that the sun contains large quantities of hydrogen and helium gas, whereas the earth is mostly rock. Carbon, nitrogen, and oxygen, for example, are quite common on our planet while the precious metals silver and gold are rare; the same pattern is seen on the sun. Only the giant planets of the solar system, Jupiter and Saturn, possess hydrogen and helium in solar quantities.

Spectroscopic analysis of sunlight provides a good deal of information about the temperature, pressure, and composition of the solar surface. But is the sun just a quietly glowing ball of gas? Other stars are too far away for us to make a detailed examination of their atmospheres and surfaces, but the sun is close enough so that details of the structure of the solar atmosphere can be observed.

TABLE 8.1 COMPOSITION OF THE SUN

ELEMENT	ATOMIC WEIGHT	RELATIVE NUMBER OF ATOMS
Hydrogen	1	1.00
Helium	4	0.10
Carbon	12	4.2×10^{-4}
Nitrogen	14	8.7×10^{-5}
Oxygen	16	6.9×10^{-4}
Silicon	28	4.4×10^{-5}
Calcium	40	2.2×10^{-6}
Iron	56	3.2×10^{-5}
Silver	108	7.1×10^{-12}
Gold	197	5.6×10^{-12}

Source: John E. Ross and Lawrence H. Aller, "The Chemical Composition of the Sun," *Science* 191 (1976): 1223–1229.

★ 8.2 SUNSPOTS ★

A close look at the sun reveals that it is not just a featureless ball of gas; *sunspots* are visible on its surface, as Figure 8.4 shows. Even a small telescope will show these features that Galileo discovered. (*Warning: Do not look directly at the sun through a telescope. Eye damage will result.* Telescope manufacturers have developed filters that reduce the intensity of sunlight at the end of a telescope from eyeball-searing to pleasant. The sun's image can also be projected on a screen and viewed safely.) These spots may look small, but the huge size of the sun conceals the true dimensions of sunspots. A good-sized spot could swallow the earth.

The detailed view of a sunspot group shown in Figure 8.4 demonstrates the rich structure that is often associated with sunspots. They are not simple black dots on the solar surface. Sunspots are generally associated with *active regions,* areas on the solar surface that display a wide variety of phenomena involving energy emission different from the quiescent emission of light from the photosphere. *Solar activity* involves such features as sunspots, the bright spots called *faculae* often found near them, and in some cases the more delicate features found in outer layers of the solar atmosphere. Occasionally solar activity involves *flares,* explosive outbursts of light and emissions of high-speed particles on the solar surface. The heart of all solar activity is the tangled magnetic field of the sun, which, in active regions, reaches strengths 10,000 times that of the earth's surface field.

People have been observing and counting sunspots since their discovery by Galileo in the seventeenth century. Two centuries later, Heinrich Schwabe and Rudolf Wolf analyzed their own observations and the historical records and found that the number of sunspots varied in an 11-year *cycle.* In the peak years of the cycle, as many as 100 spots can be seen on the solar surface; many active regions show well-developed spot groups like those in

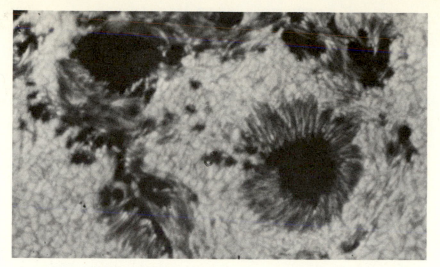

FIGURE 8.4 A sunspot group, in one of the most detailed photographs of the sun's surface ever made. These sunspots show a great deal of structure. This photograph was taken by a telescope that was carried far above the earth's turbulent lower atmosphere by a balloon. (Project Stratoscope, Princeton University)

Figure 8.4. At the low point of the solar cycle, fewer than 10 spots occur, and many of these can be missed by observers with small telescopes or poor observing conditions. Variations in the subsurface currents generating the solar magnetic fields are the probable cause of the solar cycle, although the precise nature of the dynamo mechanism that might drive these currents is not well understood.

The solar cycle is not perfectly regular, for the number of spots visible at solar maximum varies from one cycle to another. The cycle seemed to stop for a century or so during the reign of Louis XIV of France, an individual coincidentally nicknamed *Le Roi Soleil*, the Sun King. Between 1645 and 1700, no spots at all were observed in many years, and the most spots ever seen was 11 (in 1684), a number typical of the minimum years in the solar cycle as we know it today. This pause in the solar cycle is named the "Maunder minimum" in honor of the nineteenth-century astronomer who first called attention to it. A recent analysis of the historical record by John Eddy indicates that the Maunder minimum did in fact represent a real solar phenomenon, since a sufficiently large number of astronomers were observing the sun at that time and, had they been present, sunspots would certainly have been recorded. The cause of this peculiar absence of sunspots and cessation of the cycle is not understood, and it may provide us with an important check on current theories of the cause of the solar cycle.

Sunspots are dark because they are cooler than the surrounding photosphere, with a temperature of about 4000 K (versus the photospheric temperature of about 6000 K). An isolated sunspot would seem to glow, but sunspots look black compared to the brighter sun. No one understands what causes sunspots to form, but once they form, the magnetic field associated with active regions generates a pressure that maintains the spot as a cool region within the surrounding hotter gas of the photosphere. A spot remains cooler because the magnetic pressure plus the reduced gas pressure balances the higher gas pressure of the surrounding photosphere.

While the sunspots may seem like small features on an otherwise smooth solar photosphere, the amount of structure observable in solar photographs increases dramatically when one examines layers higher in the solar atmosphere. These complex patterns of ionized gas are shaped by the interaction of swirling, electrically conductive gases and the magnetic fields of the sun. One of the principal goals of solar research in the last two decades has been the untangling of the complex web of structures visible in the outer layers of the sun (Figure 8.5). Since solar activity is caused ultimately by the changing patterns of solar magnetic fields, the analysis of these magnetic tracers will hopefully lead to a better understanding of the root causes of solar activity.

The same fundamental physics that governs the appearance of ionized gases in the sun governs the operation of a possible source of electric power on the earth, the hydrogen fusion reactor. In this device, ionized hydrogen gas is confined by magnetic fields and heated to a temperature high enough so that hydrogen atoms can collide, fuse, and produce energy. No one has yet designed a reactor that can confine hot ionized hydrogen, because no one has yet developed a magnetic field shape that does not collapse before the fusion reactions begin. It is perhaps too much to claim that the sun will teach us how to tame fusion power, but it is true that the same physical principles are involved. With an understanding of the basic physics comes the insight that leads to new breakthroughs.

☆ **THE CHROMOSPHERE** ☆

The first layer above the visible solar surface is called the *chromosphere*, so named because its light is dominated by the red glow of emission from hydrogen atoms. Most of the radiation from the chromosphere is emitted when atoms drop from higher to lower energy levels. The emitted photons have energies exactly equal to the energy difference between the two levels,

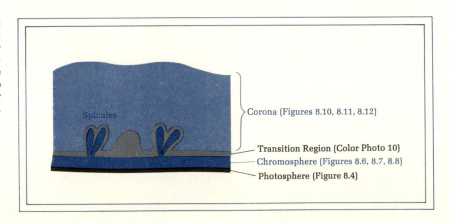

FIGURE 8.5   A schematic representation of the different layers of the outer solar atmosphere. Photographs of these layers are shown in Figures 8.4, 8.6 to 8.8, 8.10 to 8.12, and Color Photo 10.

Spicules

Corona (Figures 8.10, 8.11, 8.12)

Transition Region (Color Photo 10)
Chromosphere (Figures 8.6, 8.7, 8.8)
Photosphere (Figure 8.4)

FIGURE 8.6 Photograph of the chromosphere, taken through a filter transmitting only light emitted by hydrogen atoms at a wavelength of 6562 Å. The bushy structures are spicules. (Sacramento Peak Observatory, Association of Universities for Research in Astronomy, Inc.)

and thus these photons have specific wavelengths or colors. The chromosphere was first discovered by observation of a colored layer of luminous gas that was seen when the moon covered the photosphere at solar eclipses.

It is possible to photograph the chromosphere by photographing the sun through a filter that transmits only the light that the chromosphere emits. Figure 8.6, one such photograph, shows the chromosphere in its quiet state, in a region far removed from the influence of active regions. The vertical, bushy structures are *spicules*, vertical jets of gas that contain much of the chromospheric gas. These spicules protrude far above the photosphere, about 10,000 km above the normally visible surface. Spicule temperatures are about 15,000 K, far above the temperature of the photosphere. The rest of the chromosphere, lower in the solar atmosphere, has a temperature that varies between 4000 and 20,000 K. The high temperature of the chromosphere, also characteristic of the other upper layers of the sun, is maintained by shock waves coursing upward from the photosphere. These shock waves make noise, which carries the energy that heats the outer solar atmosphere. If we could listen to the chromosphere, we would hear the rumble of sound waves providing heat for the outer solar atmosphere.

The most beautiful and intriguing chromospheric structures are found immediately above active regions. Two photographs of the active chromosphere are shown in Figures 8.7 and 8.8. These photographs illustrate the value of chromospheric structures as tracers of magnetic field lines. Just as iron filings trace out the magnetic field of a bar magnet (Figure 8.9), so do the chromospheric fibrils trace out the magnetic fields leading from one spot group to another within an active region. The overall magnetic field within an active region is dominated by the two spot groups, one delineating a south magnetic pole and one delineating a north pole. But the field geometry is more complex than the simple model predicts, as the photographs show. The bright regions in Figures 8.7 and 8.8 are flares, which will be described in more detail later.

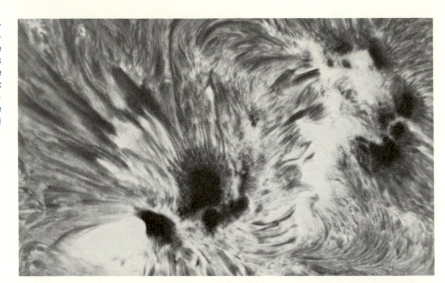

FIGURE 8.7 This photograph of the solar chromosphere above an active region shows the usefulness of chromospheric structure as a tracer of magnetic fields. Compare it with Figures 8.8 and 8.9. (Hale Observatories)

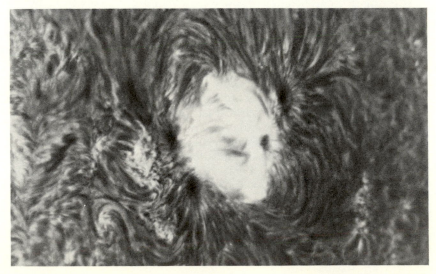

FIGURE 8.8 The chromosphere above an active region that is less complex magnetically than the one in Figure 8.7. (Hale Observatories)

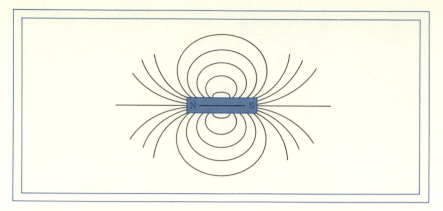

FIGURE 8.9 A bar magnet, with north and south poles, has a pattern of magnetic field lines like that shown. This pattern can be observed by placing a bar magnet under a sheet of paper and sprinkling iron filings on top of the sheet.

☆ THE CORONA ☆

The *transition region*, a thin sheet of gas, separates the 10,000-degree chromosphere from the tenuous million-degree corona, the outer layer of the solar atmosphere. This transition region is wrapped around the chromospheric structures, and its temperature is several hundred thousand degrees. At such high temperatures, atoms lose many of their outer electrons. Oxygen, for example, is stripped of five of its eight electrons at a temperature of about 300,000 K. It is possible to photograph the fascinating structures in the transition region by photographing the sun in wavelengths that are emitted by these highly ionized oxygen atoms. However, such photographs must be made from space, since these emissions, with wavelengths of 1032 Å, cannot penetrate the earth's atmosphere. Color Photo 10 is a photograph of the transition region made from Skylab, using blue to denote the most intensely emitting regions. The loop structure again shows the importance of magnetic fields in shaping this part of the solar atmosphere.

Above the transition regions lies the envelope of million-degree gas known as the corona. The *corona* is the gaseous halo that is visible during a solar eclipse (Figure 3.6). Since the hot corona emits x rays very intensely, and the cool photosphere emits very little x radiation, it is also possible to photograph the corona by photographing the x-ray sun (Figure 8.10).

The x-ray photograph of the solar corona dramatically shows the irregular distribution of the coronal gas. The coronal emission reaches its maximum intensity in a few bright points. These bright regions last only for a few hours, and they are generally associated with sunspot groups. More puzzling still are the dark areas in the photograph, where no coronal gas appears at all. These *coronal holes* are perhaps the most pivotal discovery of the last decade of solar research; recent work has shown that the solar wind probably flows out of these coronal holes. (Figure 8.11 shows another coronal hole.) The looped structures in the x-ray picture indicate the importance of magnetic fields in determining the distribution of coronal gas; further analysis of these complex structures may provide some information about the overall magnetic structure of the sun.

205

FIGURE 8.10 The solar corona, seen in soft x-ray images from Skylab. Note the region of little or no coronal x-ray emission, the *coronal hole* (at the upper left), the looped structure connecting different active regions, and the bright points of x-ray emission. (From G. S. Vaiana, J. M. Davis, R. Giacconi, A. S. Krieger, J. K. Silk, A. F. Timothy, and M. Zombeck, *Astrophysical Journal (Letters)*, Vol. 185, L47–L51, published by University of Chicago Press. Copyright © 1973 by the American Astronomical Society. All rights reserved. Photograph supplied by Solar Physics Group, American Science and Engineering, Inc., Cambridge, Massachusetts 02139.)

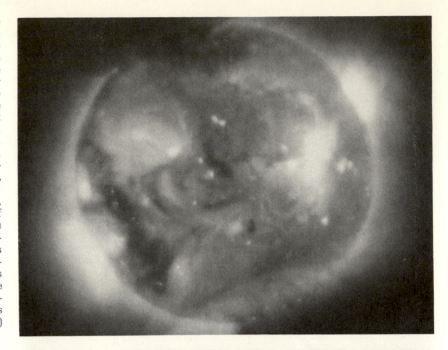

FIGURE 8.11 A coronal hole near the pole of the sun. This image of the sun at a wavelength of 625 Å shows coronal radiation, emitted by magnesium atoms with nine electrons removed. (Photograph courtesy of Harvard College Observatory.)

Chromospheres and coronas have also been observed on other stars. These outer layers of the solar atmosphere emit strongly in wavelengths in which the photosphere emits little radiation. The photosphere, for example, scarcely emits at a wavelength of 1216 Å, while the hotter chromosphere and corona produce copious quantities of such radiation when hydrogen

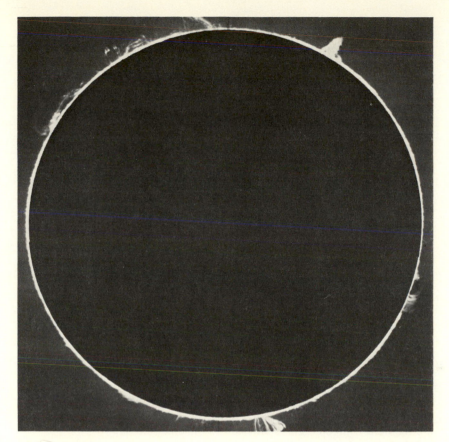

atoms drop from the second to the first energy level. A number of the brighter stars, with temperatures comparable to the solar temperature, also emit at this wavelength. The discovery of these features on other stars will allow us to interpret the differences between these stars and the sun and gain more insight into their nature. The bright star Capella, for example, has a temperature similar to the temperature of the sun, but Capella is considerably larger. An analysis of the differences between the outer envelopes of these two stars provides some insights into the mechanisms responsible for maintaining the high coronal temperatures on these stars.

The most spectacular coronal features are the huge *prominences*, concentrations of remarkably cool coronal gas (Figure 8.12). Prominence temperatures are only 10,000 K, in contrast to the million-degree corona. They come in a wide variety of shapes. Their association with coronal magnetic fields is not obvious in some cases, and these enormous condensations of coronal gas, large enough to engulf the earth, remain one of the coronal curiosities.

Photographs of the solar corona show only the part nearest the surface of the sun. In fact, the solar corona extends throughout the solar system and is observed near the earth's surface as the *solar wind*. One astronomical unit from the sun, the solar wind has a density of about 10 particles/cm³, and it streams by the earth with a velocity of 400 km/sec. This outward motion of the corona is the result of its extremely high temperature, which creates a relatively high coronal pressure far, far out into the solar system. Since the coronal pressure is greater than the pressure of the interstellar gas outside the solar system, the corona flows outward. Recent evidence indicates that much of this outward flow is continually replenished by matter flowing through the coronal holes at the solar surface. It is the solar wind that creates and shapes comet tails and that provides the particles that become part of the radiation belts around magnetically active planets such as Earth and Jupiter.

★ **8.4 EXPLOSIVE EVENTS** ★
ON THE SUN

☆ **SOLAR FLARES** ☆

The complex, fascinating character of the solar chromosphere and corona is rendered still more interesting by the sudden events that cause dramatic changes in these regions within a matter of minutes. Many of the structures of the chromosphere and corona are evanescent, with individual spicules, for example, remaining in one piece for about 10 or 20 minutes. On the sun, the most spectacular changes associated with active regions are solar *flares*.

A flare is observed as a sudden brightening of a small part of the surface area of the sun. Chromospheric or coronal photographs show flares to their best advantage; the bright areas in Figures 8.7 and 8.8 are solar flares. Flares are the most characteristic manifestation of solar activity.

It is believed that flares are produced by the strong, tangled magnetic fields associated with active regions. The intense magnetic fields seem to become unstable and release up to 10^{32} ergs of energy in a matter of minutes. Energy released in this matter may be the source of heat for a parcel of coronal gas, which then expands explosively and brightens. The whole show typically lasts for about an hour. The flare model is quite plausible, although the details are not yet understood.

Most solar flares produce a burst of high-speed charged particles that spiral around magnetic field lines. Those that travel at speeds approaching the speed of light produce a burst of radio radiation that can be detected on the earth. Collisions between fast particles and the coronal gas produce bursts of x rays as well. The detection of x-ray and radio bursts provides another way of studying the activity that accompanies solar flares.

☆ **SOLAR FLARES AND THE EARTH** ☆

We on the earth are not just distant spectators of the drama of solar activity. Some of the high-speed particles produced by a solar flare reach the earth and produce measurable effects on our upper atmosphere and on radio communications.

About an hour after the photons emitted by a solar flare arrive and show us the flare event itself, along with any accompanying radio or x-ray bursts, a stream of high-energy particles or *cosmic rays* emitted by the flare reaches the earth's orbit. Cosmic rays from outside the solar system continuously bombard the earth, but a solar flare increases the quantity of lower-energy cosmic rays. The interaction of these particles and the high-energy photons emitted by a solar flare with the earth's upper atmosphere creates atmospheric disturbances. One reason for the interest in solar flares is that this burst of particles could kill an unprotected astronaut who was exposed to it for a few hours or so. Since most of these particles are absorbed by the upper atmosphere, they present no danger at the earth's surface.

About 2 or 3 days after the solar flare, a cloud of electrons reaches the earth. This cloud of charged particles is deflected by the earth's magnetic field and interacts with it. Sensitive compass needles jitter as the earth's magnetic field responds to the electron influx. Since long-distance radio communication relies on the ionosphere (the earth's upper atmosphere) as a reflector of radio signals, a disturbance of the ionosphere resulting from the solar flare can affect radio communications.[2] Perhaps the most spectacular result of the arrival of these electrons is the *aurora*, produced when an electrical discharge is directed through the earth's upper atmosphere. The geometry of the earth's magnetic field focuses this discharge near the magnetic poles, so that auroras are most clearly visible from the polar regions. An observer in the United States generally has to be north of latitude 40° to have a good chance of seeing an auroral display, although occasionally more southern observers can detect one.

☆ FLARE STARS ☆

Events similar to solar flares have also been observed on other stars. A distant observer could, in principle, detect a solar flare by observing a sudden increase in the brightness of the sun, but the sun is so luminous that this increase in brightness would not be detected with present-day equipment. On a less luminous star, however, a flare would show up quite readily. Many faint, red stars do show sudden increases in brightness, becoming up to ten times as bright in a matter of minutes. Such stars are called *flare stars* because this activity is probably quite analogous to a solar flare.

Observations of flare stars show that, aside from flares, their brightness changes periodically. A number of investigators have interpreted these variations as evidence of huge spots on these stars. When the star rotates, it becomes dim if most of the surface that we see is covered with a huge spot. The association between sunspots and solar flares supports this explanation of the changing brightness of these stars. Again, observations of similar phenomena in different circumstances offer the promise of insight, and it may

---

[2] Such disturbances tend to be more frequent near sunspot maximum, when solar activity is most frequent. Users of CB (citizen's band) radios may be more troubled by "skip" around 1980. Skip, in CB language, is the reception of signals from distant transmitters as these signals bounce off electrically charged particles high in the atmosphere.

be that intensive analysis of the flare-star phenomenon will lead to a better understanding of both solar and stellar flares.

★ 8.5 THE SOLAR INTERIOR ★

The sun is a gigantic energy source, emitting 3.9×10^{33} ergs/sec of energy in the form of sunlight, solar flares, high-speed particles, and radio and x-ray bursts. Where does all this energy come from? The hot surface must transfer energy to the cold interplanetary space just because it is hot; but if there were no internal energy source, the hot solar surface would soon cool off and we would all freeze. What keeps the sun shining? The answer to this question must come from an analysis of the solar interior.

We cannot observe the solar interior directly. Since we cannot drill into it and insert probes to determine its temperature, chemical composition, and physical state, the tools of theoretical inference are the ones that play a primary role in the analysis. We want to guess the state of the interior of a ball of 1.989×10^{33} g, made mostly of hydrogen gas, and test our guess by comparing our theoretical predictions of surface properties with observations. We ask, theoretically, "What holds the sun up? What keeps it from collapsing? How is energy transferred from the inside to the outside? Where does the energy come from in the first place?" We hope that our answers to these questions will produce a solar model that is consistent with what we see. For the most part, the present model agrees with observation.

☆ PRESSURE AND GRAVITY ☆

A schematic picture of the inside of the sun is shown in Figure 8.13. The visible surface, the photosphere, and the chromosphere are shown in the picture; the corona extends above the chromosphere. But we wish to probe the interior. Divide the interior into a hot core and a cooler envelope; it does not matter where the dividing line is. The envelope is heavy. At any point, the core must support the weight of a column of gas extending from the bottom of the envelope all the way to the photosphere. Why doesn't the core give in and collapse under this weight? The core, hotter than the envelope, has a greater gas pressure. It is this excess pressure that supports the weight of the envelope. This balance between excess pressure and weight must hold at any point in the solar interior.

The sun's need to hold itself up produces, in the center of the sun, a tremendous pressure of 2×10^{11} atmospheres, or 2×10^{11} times the atmospheric pressure at the earth's surface. This pressure comes from the rapid motion and collision of hydrogen and helium atoms in a 15-million-degree gas. It is this high temperature that supplies the pressure needed to hold the sun up.

But the heat in the interior leaks out of the sun when the sun shines. The sun loses 3.9×10^{33} ergs of energy to the outside world in the form of sunshine every second. Without some internal energy source, the sun would

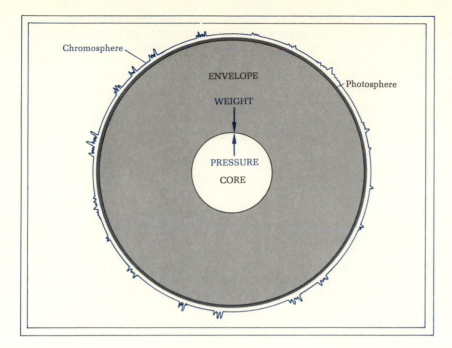

FIGURE 8.13 Pressure in the core balances the weight of the solar envelope and keeps the sun from collapsing. In the sun this pressure is provided by the 15-million-degree temperatures in the solar core.

cool off, the excess central pressure would disappear, and the sun would collapse catastrophically in 15 million years. However, the stability of the earth's climate indicates that the sun has been in the same evolutionary state for 3.5 billion years, over 200 times the length of time it could live off of its storehouse of central heat. What keeps the sun going year after year?

☆   THE SOLAR ENERGY SOURCE   ☆

The sun's center is so hot that in some ways it resembles the center of an exploding hydrogen bomb. The energy source of an H-bomb is nuclear fusion, which is also the energy source that powers the sun. In the sun's center, atomic nuclei fly around at tremendous speeds; the average nucleus collides with another some 10^7 times each second. Most of these colliding nuclei are hydrogen nuclei—simple protons. If two of these protons hit each other hard enough, the nuclear forces that tend to bind nuclear particles like protons together can overcome the electrical forces that tend to push two positively charged protons apart. Such a close collision results in the production of a deuterium nucleus. That is, one of the protons is transformed into a neutron, and this neutron and one proton form a nucleus of deuterium or heavy hydrogen. (The excess positive charge is transferred to a positron, a positively charged electron, which moves away from the new nucleus.) The important result of this nuclear reaction is the production of a heavier nucleus and the liberation of energy.

The nuclear-reaction chain does not stop with the production of deuterium. While reaction paths may differ, more and more protons are added to the growing nucleus until it contains four nuclear particles—a helium nucleus with two protons and two neutrons. (Box 8.1 sketches the details of possible reaction chains.) The fusion of four protons into one helium nucleus by these powerful nuclear forces produces 25 to 26 million electron volts (MeV) of accessible energy. This 25 MeV is not a great deal of energy; it is only about 10^{-5} ergs, or 10^{-14} of the energy a 100-watt light bulb puts out every second. Fusion, however, is a remarkably efficient energy source, for

---

BOX 8.1   NUCLEAR-REACTION PATHWAYS

Stars like the sun use the proton-proton reaction to generate most of their nuclear energy. This chain begins when two reactions fuse three hydrogen nuclei into a nucleus of helium-3, containing one neutron and two protons:

no. of positive charges   no. of nuclear particles

$$_1H^1 + {_1}H^1 \rightarrow {_1}H^2 + {_1}e^+ + \nu$$
$$_1H^2 + {_1}H^1 \rightarrow {_2}He^3 + \gamma$$

where e^+ is a positively charged electron, γ a gamma ray, and ν a neutrino.

Most of the time, two He^3 nuclei from two of the above reactions make a He^4 nucleus and two protons, so that six protons make a He^4 nucleus and two protons:

$$_2He^3 + {_2}He^3 \rightarrow {_2}He^4 + {_1}H^1 + {_1}H^1$$

Occasionally, alternative chains can occur:

$$_2He^3 + {_2}He^4 \rightarrow {_4}Be^7 + \gamma$$

Most of the time:

$$_4Be^7 + {_{-1}}e^- \rightarrow {_3}Li^7 + \nu$$
$$_3Li^7 + {_1}H^1 \rightarrow {_2}He^4 + {_2}He^4$$

A fraction of 1 percent of the time:

$$_4Be^7 + {_1}H^1 \rightarrow {_5}B^8 + \gamma$$
$$_5B^8 \rightarrow {_4}Be^8 + {_1}e^+ + \nu^*$$
$$_4Be^8 \rightarrow {_2}He^4 + {_2}He^4$$

Most of the neutrinos observed by the Davis experiment (pages 215–216) come from the rare B^8 chain, the neutrinos marked with an asterisk.

Another cycle, with carbon, nitrogen, and oxygen nuclei as intermediaries, is used relatively little in the sun but is quite important in hotter stars. It is sometimes called the *CNO Bi-Cycle* because two distinct chains can enter into the cycle. For convenience, the hydrogen atoms that go into the reaction and the helium atoms that come out are printed in a different color from the CNO assembly line that just goes around and around in two circles:

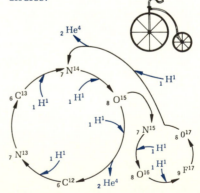

the chemical combustion or burning of four hydrogen atoms and two oxygen atoms would produce 10^{-11} ergs, or only one-millionth of the energy produced by fusion. The sun fuses 600 million tons of hydrogen every second to produce enough energy to keep it shining; the amount of hydrogen that exists in the sun is enough to keep this energy source working for billions of years. When we eventually turn to solar energy to power human civilization, it will be nuclear reactions that are the ultimate power source.

☆ **ENERGY FLOW IN THE SUN** ☆

The presence of a nuclear energy source in the sun's central region does not complete the model. This energy-generating core, occupying less than 5 percent of the solar volume, is so hot that it emits hordes of high-energy x-ray and gamma-ray photons. If we were exposed to that core directly, the blast of lethal gamma rays would kill life on the earth. Further, the core would lose its energy so fast that it would soon exhaust its fuel. The massive envelope of the sun acts as a blanket to hold the heat in, so that energy leaks out of the sun comparatively slowly.

Throughout most of the sun, energy flows by radiation through a tortuous path from center to surface. At the center, the gamma rays emitted by the core can only travel 0.005 cm, or 50 μ before being absorbed in collisions with charged particles. Other photons are soon re-emitted, to travel their own short journeys before being absorbed once again. Eventually many, many of these tiny journeys add up to a net flow of energy from the inside to the outside, as Figure 8.14 schematically depicts. It takes about 2 million years for a single quantum of energy to work its way outward through the solar envelope. Close to the surface, conditions are such that mass motions of gas or convection can carry the energy. Once the energy reaches the surface, it is free to escape into the surrounding space as sunshine.

☆ **MODEL BUILDING** ☆

The job of the theoretical astrophysicist wishing to understand the solar interior is not limited to listing the ingredients that make the sun work. The physical processes discussed in the last few sections have to be described more quantitatively, using equations and numbers. We need to know exactly how many hydrogen atoms fuse every second in a cubic centimeter of a gas of a given composition, pressure, and temperature. This knowledge comes from many painstaking measurements in nuclear-physics laboratories of the rates of the various reactions in Box 8.1. A precise description of the way in which pressure increases toward the solar center is part of the model. Perhaps the most difficult and uncertain component of the calculations is the prescription of energy flow from the solar center to the surface. Just what kinds of obstacles to photon flow are there in solar matter? How tortuous a journey must the photons follow in carrying energy through the envelope?

FIGURE 8.14 Energy, generated by fusion reactions in the solar center, flows to the surface by a tortuous path. Photons travel short distances and are absorbed by atoms; the atoms emit more photons that also travel short distances. In the extreme outer layers, energy flows by convection, the movement of large masses of heated gas.

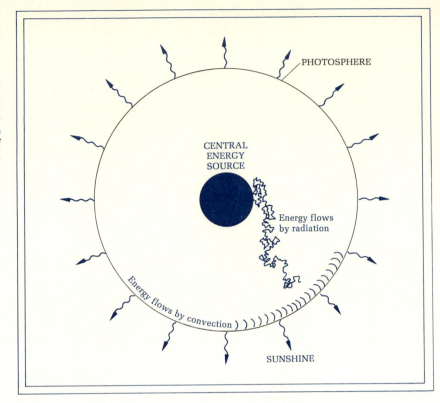

Once the physical processes taking place in the solar interior are described by a sufficiently accurate (and complex) set of equations and tables, a large computer can be used advantageously. The physical laws, nuclear reaction rates, opacities, and so forth are woven into a large computer program. The computer grinds away and produces a large pile of numbers stating how dense, how hot, and how reactive each layer of the sun is. Figure 8.15 is a distillation of these numbers into a short graph illustrating a recent solar model. But the critical question remains. Is the model correct? Is Figure 8.15 just a meaningless graph, or is it a reasonably accurate representation of conditions in the sun?

★ 8.6 TESTING SOLAR MODELS ★

We know that the sun is a ball of gas about 4.5 billion years old. The physical laws that govern the reactions in stellar interiors, embodied in a computer program, provide the recipe allowing us to calculate how this ball of gas will act. The theorist then moves to the computer, dumps 2×10^{33} of gas composed of 9 parts hydrogen, 1 part helium, and a pinch of other stuff into

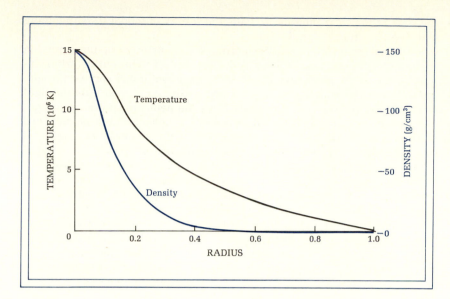

FIGURE 8.15 A recent solar model, showing the run of temperature and density with radius in the solar interior. [Data from J. Bahcall, N. Bahcall, and R. Ulrich, *Astrophysical Journal,* 156 (1969), p. 565 for the inner regions; and from D. D. Clayton, *Principles of Stellar Evolution and Nucleosynthesis* (New York: McGraw-Hill, 1968) for the outer regions.]

the program, and lets the computer cook it for 4.7 billion years. (It will not take the computer billions of years to perform the calculations; but if these calculations were to be done by hand, it might be impossible for the theorist to keep up with the star.) The computer results should match the real sun.

The match between calculation and observation can be tested in several ways. The radius of the sun can be measured as 695,000 km; the computer should and does reproduce the observed radius. The luminosity is 3.9×10^{33} ergs/sec, and the computer can reproduce this figure. Furthermore, geological evidence indicates that the earth's climate has been reasonably stable for $3\frac{1}{2}$ billion years, and the calculations should indicate that the solar luminosity was in the correct range to allow water to exist on the earth's surface for that length of time. The calculations show that in fact the solar luminosity has only increased slightly in the 4.5 billion years since the sun was formed; in its present evolutionary state, it is a very stable star.

☆ **THE SOLAR NEUTRINO EXPERIMENT** ☆

Is that all? Do astronomers now relax, confident that they understand the solar interior? A little thought about these observational tests of our theoretical picture of the solar interior demonstrates how indirect and skimpy they are. The model describes conditions in the interior, and the tests refer to the surface. We cannot drill a hole into the sun, insert a thermometer, and test whether the central temperature is indeed equal to the model temperature of 15 million degrees. However, an elusive type of particle, the *neutrino*, can penetrate the blanket of the envelope and give us some direct evidence on the state of the solar interior. To date, the results of model tests involving neutrino detection are equivocal.

Neutrinos are extremely elusive particles. A neutrino resembles a photon in that it has no mass and travels with the speed of light, but neutrinos interact with matter very infrequently. Photons do not travel very far through ordinary matter, but a neutrino could travel through several hundred light years of lead without being stopped. Thus the neutrinos produced by nuclear reactions in the center of the sun can zip right through the envelope and travel to the outside world. A very small fraction of the energy produced in nuclear reactions comes off in the form of neutrinos; most of the energy is given off as photons (gamma rays) and fast-moving nuclear particles.

It would thus seem that a neutrino telescope could easily confirm our picture of the sun's interior. A theorist can follow the nuclear reactions in the solar core, summarized in Box 8.1, and calculate just how many neutrinos are emitted every second. It turns out that about 6×10^{10} neutrinos cross a square centimeter of the earth's surface every second; 3.5×10^{17} photons would cross a similar area in that short time. But the neutrinos are very difficult to detect; anything that can pass through hundreds of light years of lead is difficult to trap in a detector. Catching and measuring even a small fraction of these neutrinos is quite difficult.

The neutrino telescope that has been used for this test of the solar model is a large vat of cleaning fluid, located a mile below the earth's surface in a gold mine in Lead, South Dakota. This huge tank contains some 10^{31} chlorine atoms. The model predicts that about once a day, one of these chlorine atoms will intercept a solar neutrino, capture it, and be transformed into an atom of radioactive argon. Since argon is inert chemically, it will escape from the cleaning-fluid molecule and sit in the tank as an isolated gas atom. Counting tens of argon atoms among the 10^{31} chlorine atoms in the tank is an experimental challenge.

Dr. Raymond Davis, builder of the neutrino telescope, has developed and tested an elaborate scheme for counting this small number of argon atoms. He lets the tank accumulate argon atoms for a month or so, and then bubbles helium atoms through the tank to extract the radioactive argon atoms. The argon atoms are counted individually when they decay back to chlorine. Like all good experimenters, Davis has checked his procedure by placing 500 radioactive argon atoms in the tank. Tests show that he can recover the argon atoms with 95 percent efficiency.

The solar neutrino counting experiment began in the late 1960s, in an effort to verify the model of the solar interior. The results have been frustrating, since Davis has not yet detected any neutrinos that definitely come from the sun. A small amount of argon is produced by the interaction of cosmic rays with the tank, in spite of its protected location a mile underground. An average of all the runs made prior to 1976 indicates that fewer than 0.3 argon atoms are being produced each day. Counts from the Davis experiment are often expressed in solar neutrino units (SNU); 1 SNU = 1 capture per second per 10^{36} target atoms. A counting rate of 1 SNU corresponds to 1 argon atom produced every 5 days; thus the observed rate is less than 1.5 SNU.

When the Davis solar neutrino experiment began, theorists expected a counting rate of 30 SNU. The failure to detect 30 SNU was, at first, only a partial failure of the model. Most of the neutrinos picked up by the Davis telescope come from a little-traveled branch of the reaction pathway in the sun's core, involving a boron-8 (B^8) nucleus. Small changes in the solar model can greatly change the rate of this rare reaction. In the 1970s, theorists made a considerable effort to bring theory and experiment into agreement by refining some of the quantities used in the calculations. By 1973, the theoretical (calculated) capture rate of the Davis experiment was reduced to 6 ± 2 SNU. The uncertainty in the estimate comes from uncertainties in the theoretical calculations and in the reaction rates. Such a rate still conflicts with the experimental result, in which the counting rate from the sun is less than the 1.5 SNU produced by the cosmic-ray background.

Does the disagreement between theory and experiment indicate that the theoretical model for the solar interior is wrong? One can make major modifications to the model and still retain the basic picture by, for example, assuming that the heavy-element abundance in the solar center is one-tenth of the abundance at the surface, and pushing the calculated rate down to 1.5 SNU. The Davis experiment has produced some rather puzzling results; occasionally, a flushing of the tank produces a large number of argon atoms, indicating a counting rate of 4 to 5 SNU for that particular run. These high runs have become somewhat more frequent in 1976. A number of people have proposed other methods for detecting solar neutrinos, which may be tried in the future.

Thus our attempt to verify our picture of the solar interior is at best equivocally successful. While theoretical models can reproduce the sun's present size and luminosity, the most direct test of the model has not produced agreement with theory. The persistence of astronomers in believing that this theory of the solar interior is correct comes from its success in describing the properties of other stars, a success that will be described in the next chapter.

SUMMARY

Careful, thoughtful analysis of the light that reaches us from the solar exterior provides a good deal of information about the various layers of gas that we can see. Most visible light comes from the *photosphere*, a layer of 6000° gas that contains some sunspots. The outer solar atmosphere, consisting of the chromosphere and corona, is the site of solar activity. Our determination of the physical state of the interior rests on inference, on the computation of a theoretical model, and its confirmation by observations. A thorough test of this theory must come from observations of other stars.

This chapter has covered:

The state and chemical composition of the solar photosphere;

The nature of sunspots and the sunspot cycle;

The state of the sun's outer atmosphere—the chromosphere and corona;

The wide diversity of solar activity;

The physical laws that are used to model the solar interior; and

The puzzling results of the solar neutrino experiment.

KEY CONCEPTS

Absorption line
Aurora
Chromosphere
Corona
Coronal hole

Cosmic ray
Flare star
Neutrino
Photosphere
Prominence
Solar cycle

Solar flare
Solar wind
Spicule
Sunspot
Transition region

REVIEW QUESTIONS

1. What happens in the solar photosphere to cause the formation of absorption lines in the solar spectrum?

2. What feature of atoms is responsible for the existence of absorption lines?

3. Suppose you could isolate a sunspot from the surrounding photosphere. Would it still be black? Explain.

4. What is it about the way in which the photographs of Figures 8.4 and 8.6 were taken that causes their different appearances?

5. Summarize the different ways in which solar activity can be observed.

6. Are auroras more common near sunspot maximum or near sunspot minimum? Explain.

7. What holds the sun up?

8. If the sun's core rotated rapidly, producing a centrifugal force that would help hold the envelope up, would the core be hotter or cooler? Explain.

9. What is the significance of the Davis solar neutrino experiment?

FURTHER READING

Bahcall, J. N., and R. Davis. "Solar Neutrinos: A Scientific Puzzle." *Science* 191 (1976): 264–267.

Friedman, H. "Our Star the Sun." *The Amazing Universe,* chap. 3. Washington, D.C.: National Geographic Society, 1975.

Noyes, R. W. "New Developments in Solar Research." In *Frontiers of Astrophysics,* edited by E. H. Avrett. Cambridge, Mass.: Harvard University Press, 1976.

Vaiana, G., and W. H. Tucker. "Solar X-Ray Emission." In *X-Ray Astronomy,* edited by R. Giacconi and H. Gursky. Dordrecht, Holland: Reidel, 1974.

★ CHAPTER NINE ★

STARS LIKE THE SUN

The last chapter presented a theoretical picture of the interior of our particular star, the sun. Other stars are so much farther away that we must use more indirect techniques to measure their sizes and luminosities. Such measurements indicate that most of the stars are similar to the sun in their overall structure. These stars can thus be compared to the sun in order to confirm our theoretical picture of the solar interior.

★ **MAIN IDEAS OF THIS CHAPTER** ★

A star's brightness depends on both its luminosity and its distance.

A star's luminosity depends on its size and temperature.

Measurements of a star's color or an examination of its spectrum can determine its temperature.

The Hertzsprung-Russell diagram can classify stars and sort out those that are similar to the sun in size.

Stellar masses can be measured when stars orbit each other in a binary system.

Comparison of the theoretical picture of stellar interiors with the measurement of stellar properties confirms the theory.

★ **9.1 BRIGHTNESS, LUMINOSITY,** ★
AND DISTANCE

☆ **BRIGHTNESS** ☆

A rather important and easily measurable property of stars is the intensity of their light. Figure 9.1, a photograph of the star cluster called h and chi Persei, illustrates the extreme variability of stellar brightness. But a simple distinction between bright and dim stars provides no quantitative information that can be used to measure a stellar radius or test a model of its interior. We seek to quantify brightness. What, exactly, is it about a star that makes it appear brilliant or barely visible?

The lens in the human eye focuses starlight on the retina, and it is the amount of energy reaching the retina every second that determines whether a star appears to be bright or dim. A telescope works in the same way in focusing light energy onto a point (Figure 9.2), but the larger collecting area of a telescope can concentrate a greater amount of energy than the small area of the human eye. Thus, the energy reaching the focus of a telescope (or the retina) depends both on the collecting area of the telescope and the brightness of the star. A reasonable quantitative measure of the brightness of the star is then the amount of energy crossing each square centimeter of the lens of the

FIGURE 9.1 The twin star clusters h and chi Persei, showing the great variety of stellar brightnesses visible in a long-exposure photograph. (Lick Observatory photograph)

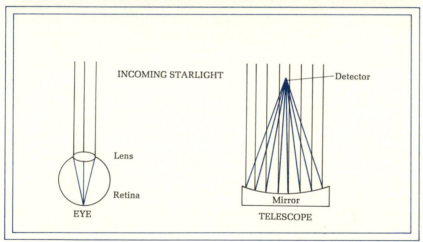

INCOMING STARLIGHT

Lens

Retina

EYE

Detector

Mirror

TELESCOPE

FIGURE 9.2 Both the eye and a telescope concentrate incoming light, with the eye focusing radiation on the retina and the telescope concentrating radiation on a detector. The amount of energy focused on the retina determines the brightness of the star; a bright star is one that sends a large quantity of radiation across the lens or mirror.

eye or the mirror of the telescope. This quantity, the amount of light energy striking each unit of area, is called the *flux* of radiation emitted from the star. The flux can be measured in various regions of the electromagnetic spectrum. It is thus possible to measure and analyze the x-ray flux, visual flux, blue flux, infrared flux, or radio flux, for example.

A telescope can measure the flux of radiation from a star in a very straightforward manner. Some sort of a *detector* is placed at the focus of a telescope to measure the rate at which energy strikes it. The flux is then the amount of energy detected every second, divided by the area of the telescope, with allowances for inefficiencies in the detector, light absorbed by the telescope mirror, and light absorbed by the earth's atmosphere.

☆ PHOTOMETRY ☆

A variety of detectors have been used by astronomers to measure the flux from stars. The most primitive (and inexpensive) detector is the human eye. An eyepiece placed at the telescope focus concentrates the light striking the mirror at one point on the retina; the observer can then look at the star field and estimate the relative brightnesses of the various stars in the field. The eyeball method of estimating stellar brightness is used by thousands of amateur astronomers, members of the American Association of Variable Star Observers (AAVSO), which keeps track of numerous stars that vary the amount of light they emit. The variations of a star can be recorded by comparing the variable star to other stars in the field that emit constant amounts of light.

A similar, but slightly more accurate, method of estimating brightness is photographic photometry, in which a photographic plate is placed at the telescope focus. The sizes of the images of the stars allow one to estimate relative brightnesses.

Neither visual nor photographic photometry directly measures the flux from a star; the brightnesses of at least a few stars in the field must be known so that the rest can be measured. A detector that can directly measure the energy concentrated at the focus is a *photometer*. The heart of the photometer is the detector, a device that emits an electric current when light strikes it. The light meter in a camera is one type of detector; others are described in Box 9.1. The strength of the electric current emitted by a detector is a direct measure of the amount of light energy striking it every second. Accurate photometry, possible when the skies are completely clear with no clouds in the way, can measure the fluxes from stars with uncertainties of about 1 percent.

☆ LUMINOSITY ☆

The flux from a star provides some indication of how powerful the star is. All other things being equal, the bright stars that produce a large flux at the earth are the powerful, high-luminosity stars. But all things are not equal, and a star can be bright either because it is near to us or because it is intrinsically powerful. If we know a star's distance, a measurement of the flux at all wavelengths can determine its total *luminosity*, the rate at which it radiates energy in all directions over the entire electromagnetic spectrum. Like flux, the luminosity can be defined as the energy emitted in any particular part of the electromagnetic spectrum. Thus a star can have an x-ray luminosity, an ultraviolet luminosity, a visual luminosity, an infrared luminosity, and a total luminosity.

The relationship between luminosity, flux, and distance can be made more quantitative with a simple geometric construction (Figure 9.3). Draw a

Color Photo 1   The lunar module and rover from the Apollo 16 mission. Note the stark grayness of the lunar surface. (NASA/JPL)

COLOR PHOTO 2   The most habitable planet in the solar system—Earth. (NASA)

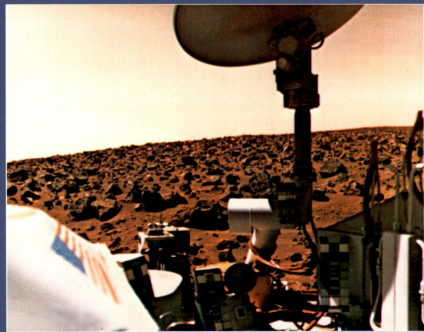

COLOR PHOTO 3   The Martian surface in early afternoon, Mars time, photographed by the camera on the Viking 2 lander. The horizon is tilted because one of the lander legs is on a rock; the landscape at the Viking 2 site is quite flat. The American flag is on the wind shield of the thermoelectric generator that powers the craft, and the dish-shaped antenna at the right is used to transmit data and photographs to the earth. (NASA)

COLOR PHOTO 4   The view southeast of the Viking 1 lander. The orange cable in the foreground is part of the spacecraft. (NASA Photo No. P—17165)

COLOR PHOTO 5   A color photograph of Mars taken by the Viking spacecraft on June 18, 1976. Valles Marineris is visible near the upper or northern part of the photograph. Chryse Planitia, the landing site for Viking 1, is on the nighttime side of the planet, just to the left of the dark patch in the center. (NASA)

COLOR PHOTO 6   A Pioneer 10 photograph of Jupiter and its satellite, Io. The dark spot on Jupiter is Io's shadow. (NASA)

COLOR PHOTO 7   Photograph of the polar regions of Jupiter from Pioneer 11. The banded structure at the equator gives way to a bubbly appearance at the pole because Jupiter's rotation is no longer influential at the poles. (NASA)

COLOR PHOTO 8　The planet Saturn. Saturn, too, has zones and belts like Jupiter. The rings are swarms of small ice particles. (University of Arizona, Lunar and Planetary Laboratory)

COLOR PHOTO 9　The head and inner tail of Comet West on March 9, about 4:45 A.M. eastern standard time, enlarged from a 2¾-minute exposure with an 8-inch f/1.5 Celestron Schmidt, on Fujichrome film, force processed to ASA 400. Photographed by Dennis de Cicco at Duxbury Beach, Massachusetts. Note the straight, narrow gas tail, blue in color, as distinguished from the broad, curved dust tail, of a pearly hue, leaning to the north. (Dennis de Cicco of *Sky and Telescope* magazine)

COLOR PHOTO 10   A false-color photograph of the transition region in the solar atmosphere between the chromosphere and the corona. In this computer-reconstructed photograph, based on data from Skylab, the blue regions are those where emission at 1032 Å is most intense, and the red regions are those where emission, though present, is least intense. The looped structures in the transition region indicate the importance of magnetic fields in the structure of the chromosphere and corona. (Photograph courtesy of Harvard College Observatory)

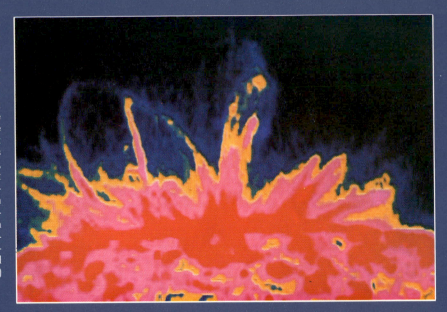

COLOR PHOTO 11   The vivid red colors of the Orion nebula, a cloud of ionized hydrogen gas (an H II region), are produced by hydrogen atoms emitting photons when they drop from the second to the first excited state. Four stars in the overexposed central region, the Trapezium, provide the power for this object. (© 1975 by the Association of Universities for Research in Astronomy, Inc. The Kitt Peak National Observatory.)

COLOR PHOTO 12 The Trifid nebula in Sagittarius, a gaseous nebula or H II region, like Orion. (Copyright by the California Institute of Technology and Carnegie Institution of Washington. Reproduced by permission of the Hale Observatories.)

COLOR PHOTO 13 The open cluster, the Pleiades. The blue nebulosities around the brighter stars are reflections of starlight by dust. Chapter 13 explains why these nebulosities are blue. (Copyright by the California Institute of Technology and Carnegie Institution of Washington. Reproduced by permission of the Hale Observatories.)

COLOR PHOTO 14 Three planetary nebulae. These objects are the remnants of red-giant stars. *Top:* NGC 7293, the Helix nebula in Aquarius. *Center:* M 57, the Ring nebula. *Bottom:* NGC 6781, a planetary nebula in Aquila. (Copyright by the California Institute of Technology and Carnegie Institution of Washington. Reproduced by permission of the Hale Observatories.)

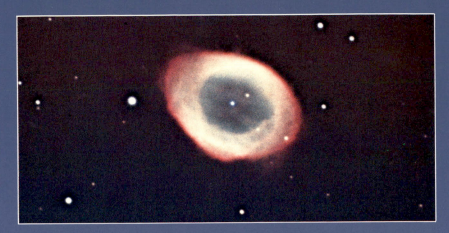

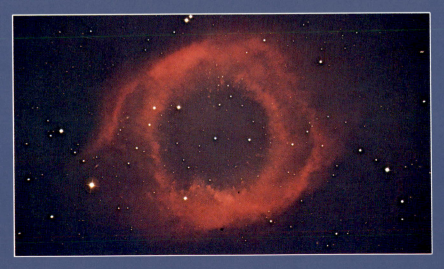

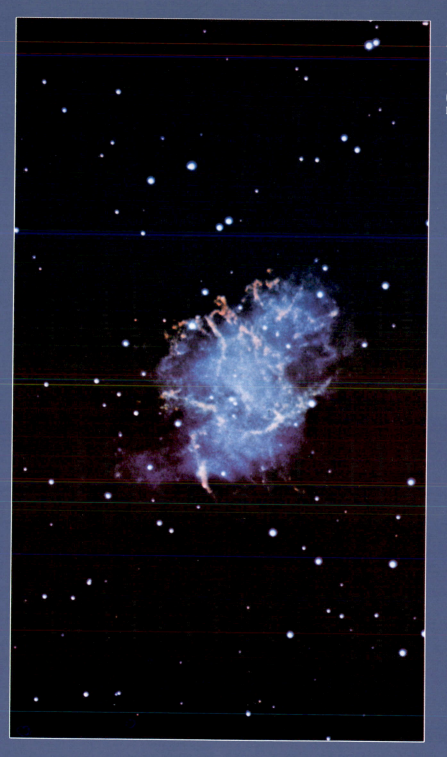

COLOR PHOTO 15 The Crab nebula, Messier 1. This object is the remnant of a supernova explosion observed by the Chinese in 1054 A.D. (Chapter 12). (Lick Observatory photograph)

COLOR PHOTO 16 The Large Magellanic Cloud, one of the galaxies nearest us (Chapter 14). (© by the Association of Universities for Research in Astronomy, Inc. The Cerro Tololo Inter-American Observatory.)

COLOR PHOTO 17 The Andromeda galaxy, M 31. The blue color of the spiral arms shows that young main-sequence stars exist there and that star formation continues in that part of the galaxy. (Copyright by the California Institute of Technology and Carnegie Institution of Washington. Reproduced by permission of the Hale Observatories.)

Box 9.1 Astronomical Detectors

Light concentrated by a telescope at its focus must be detected, and its intensity must be measured, in order to do quantitative observational astronomy. The ideal detector with an efficiency of 100 percent would do something—ring a bell, emit an electron, or change an atom of silver halide to metallic silver—every time a photon struck it. Furthermore, it would record the position of the incident photon, so that the flux from all stars in the star field could be measured at the same time. Such a detector would also be useful in a spectrograph, where photons of different wavelengths strike the detector at different positions; it could measure the intensity of starlight in all wavelengths simultaneously. Real detectors, of course, are not ideal, and astronomers have built a variety of devices in order to make detectors as ideal as possible for particular applications.

The *photographic plate*, a piece of glass coated with an emulsion of silver halide grains, is one of the oldest detectors, first used for astronomy in 1850. The photographic process preserves the location of light-struck areas extremely well and is thus very useful for measuring the position of a star in the field. But the efficiency of this process is very low, for only 1 percent of the photons striking the emulsion contribute to changing a silver halide grain to metallic silver and producing the photographic image. Consequently we have to record light, consuming precious telescope time, for 100 times as long as it would take to accumulate the data with an ideal detector.

Photoelectric devices are pieces of metal or metallic sulfides that emit electrons when photons strike them. We then count these emitted electrons to measure the light intensity. Photoelectric detectors have efficiencies of 20 to 30 percent and can thus accumulate information far more rapidly than photography can. Furthermore, photography is useless in the infrared part of the electromagnetic spectrum; only photoelectric detectors are sensitive to photons with wavelengths longer than 1.1 μ. Unfortunately, a simple photoelectric device does not record the position of the incoming photon, and so only one star or one wavelength can be measured at one time.

In the last 5 years, astronomers have developed a number of devices that have the efficiency of photoelectric detectors but still preserve some information on the position of the incoming photon. An *image intensifier*, for example, magnetically focuses electrons emitted from a light-sensitive surface onto a phosphorescent screen that glows where electrons strike it. Such a device has ten times the overall efficiency of photography; as a result, someone with the small 1.5-m telescope at Palomar can accumulate information as fast as someone on the much larger and costlier 5-m (200-inch) telescope can with photography. The techniques used in television cameras also offer some promise.

sphere around the star, with the radius of the sphere equal to the distance between the star and the earth. As long as the space between us and the star is empty, all the energy emitted by the star must cross that sphere. The flux, the amount of energy crossing each square centimeter of the sphere every

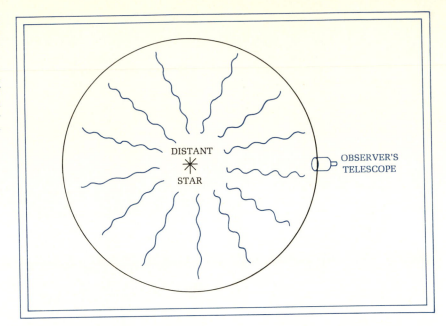

second in all wavelengths, is then just the luminosity divided by the area of the sphere, which is mathematically

$$\text{Flux} = \frac{\text{luminosity}}{4\pi(\text{distance})^2} \qquad (9.1)$$

Thus a measurement of the flux in all wavelengths allows us to measure the total luminosity if we know the star's distance from us. Further, a flux measurement in one wavelength band determines the luminosity in that wavelength band.

☆ THE MAGNITUDE SCALE ☆

For historical reasons, optical astronomers have burdened themselves (and astronomy students) with a somewhat more awkward way of describing stellar brightness and its relation to luminosity and distance. While current research papers are gradually shifting away from this system, it is sufficiently entrenched in the literature to be worth knowing. Hipparchus, the Greek astronomer who made the first star catalogue in the second century B.C., divided all visible stars into six *magnitude* classes, running from the brightest first-magnitude stars down to the faintest sixth-magnitude stars. Virtually all star catalogues ever since have used the magnitude scale to measure stellar brightnesses, refining and defining it more precisely (Box 9.2). Apparent magnitude and flux are simply two different ways of describing the same stellar property. Perhaps the most confusing aspect of the magnitude scale is that magnitudes run backward; large, positive numbers mean

BOX 9.2 MAGNITUDES AND DISTANCES

The apparent magnitude m_V of a star in the visual band of the electromagnetic spectrum is defined by the equation

$$m_V = -2.5 \log f_V - 14.1 \quad \text{(B9.1)}$$

where f_V is the flux of the star in the visual band, extending from 5000 to 6660 Å. Defining L_V as the luminosity in this part of the spectrum, we have

$$f_V = \frac{L_V}{4\pi D^2} \quad \text{(B9.2)}$$

Equations B9.1 and B9.2 can be combined to give

$$m_V = -2.5 \log \left(\frac{L_V}{4\pi D^2} \right) - 14.1$$
$$= -2.5 \log L_V$$
$$+ 5 \log D \text{ (pc)} \quad \text{(B9.3)}$$
$$+ 81.1$$

where the constant 81.1 comes from eliminating the 4π and changing D to parsecs.

The absolute magnitude of a star is its apparent magnitude at a distance of 10 pc, or, using equation B9.3,

$$M_V = -2.5 \log L_V$$
$$+ 5 \log 10 + 81.1 \quad \text{(B9.4)}$$

Equation B9.4 can be subtracted from equation B9.3 to give

$$m_V - M_V = -5 + 5 \log D \quad \text{(B9.5)}$$

Equation B9.5 holds for any wavelength region where the apparent magnitude m is defined. The constant in equation B9.1, however, differs with the wavelength region. A magnitude that refers to the flux added together over all wavelength regions is a *bolometric* magnitude and is directly related to the star's total luminosity.

faint stars. First-magnitude stars are the brightest, those that appear after sunset; third- or fourth-magnitude stars are the faintest you can see with the naked eye from most suburbs; and twenty-second-magnitude stars are the faint ones that can be photographed with contemporary telescopes.

Equation 9.1, the fundamental relationship between flux, distance, and luminosity, can be recast into magnitude language. The *apparent magnitude* of a star is a measure of its flux. The *absolute magnitude* of a star is its apparent magnitude at a distance of 10 pc (1 pc = 3.085×10^{18} cm; see Box 9.3). The absolute magnitude of a star is thus a measure of its luminosity (see Box 9.2 for additional mathematical details). The difference between the apparent and absolute magnitudes of a star is thus related to its distance, and this difference is called the *distance modulus.* You can measure flux and use equation 9.1 to relate flux to luminosity and distance, or you can measure apparent magnitude and use the equations in Box 9.2 to relate apparent magnitude to absolute magnitude and distance.

☆ **DISTANCE MEASUREMENTS** ☆

Distance is thus the key to relating an observable property, a star's brightness, to a fundamental stellar property, its power or luminosity. The most fundamental method used in measuring stellar distances is also used by surveyors to measure distances on the earth: triangulation. Observe the position of a star from two ends of the earth's orbit (Figure 9.4). These two position measurements define a long, slender triangle whose sides are the distance to the star and the diameter of the earth's orbit. This triangle can be solved by geometric methods (Box 9.3) to determine the distance to the star, if the apparent angular displacement and the radius of the earth's orbit are known. By convention, half the apparent angular displacement is referred to as the star's *parallax;* the more distant the star, the smaller the parallax. It turns out (Box 9.3) that if the parallax is measured in seconds of arc, the distance of the star can be neatly expressed as

$$\text{Distance (in parsecs)} = \frac{1}{\text{parallax}} \tag{9.2}$$

The unit of stellar distance, the *parsec* (pc), is 3.085×10^{18} cm. You can experimentally verify that as the angular shift becomes smaller, the distance becomes greater. Hold your finger quite close to your face and look at it first with one eye and then with the other, noticing the shift in relation to a distant wall. Now hold it at arm's length and observe the smaller shift.

The trigonometric parallax method for determining stellar distances is far easier to describe than it is to put into practice. The parsec turns out to be a rough measure of the average distance between stars; the Alpha Centauri system, the nearest known stellar system, is 1.33 pc away and has a parallax of 0.75 sec of arc. It is difficult to measure 1 sec of arc, the diameter of a dime $2\frac{1}{2}$ km away, and the parallaxes of all stars are smaller than that. Practitioners of the parallax-measuring art, astrometrists, have been able to hone their observing procedures so that they can measure parallaxes as small as 0.01 sec of arc with reasonable accuracy; even so, this degree of precision only allows distance measurements for stars nearer than 100 pc.

Methods of distance measurement become more approximate as stellar distances increase. If a particular star is located in a cluster of stars, it is possible to determine the distance to the whole cluster by looking for a star whose luminosity can be determined by its spectral similarity to a nearby star, with known distance and luminosity. We measure the flux from the cluster star and use the relationship between flux, distance, and luminosity

FIGURE 9.4   Measuring a star's position from two points in the earth's orbit allows us to measure the stellar distance by triangulation.

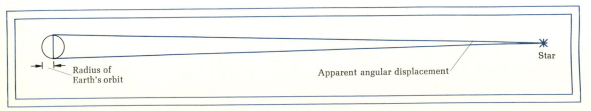

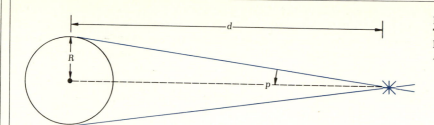

Box 9.3
Trigonometric
Parallaxes
and the Parsec[1]

Since the distance d in the figure is large compared to the radius R of the earth's orbit, one can analyze the geometric situation as though R were the arc of a circle. Thus R is some fraction of the circumference of a circle centered on the star, and p is a fraction of $360°$:

$$\frac{R}{2\pi d} = \frac{p(\text{degrees})}{360°}$$

$$= \frac{p(\text{sec of arc})}{360° \times 60 \times 60}$$

Solving the above equation for d, one obtains

$$d = \frac{R}{p(\text{sec of arc})}$$

$$\times \frac{360 \times 60 \times 60}{2\pi}$$

Since $R = 1$ AU and $(360 \times 60 \times 60)/2\pi = 206{,}264.8062$, one can conveniently define 1 *parsec* to be $206{,}264.8062$ AU, or 3.085×10^{18} cm, and obtain the simple relationship

$$d = \frac{1}{p} \text{ parsec}$$

[1] I thank Dr. John Warner for this simple and neat presentation.

(or between apparent and absolute magnitude if you wish) to measure its distance. Because the stars in the cluster are all close together in space, the distance to the whole cluster and thus to any star in the cluster is then known.

★ 9.2 TEMPERATURES AND RADII ★

The luminosity of a star is, then, one of the stellar properties that can be experimentally measured. But what features of a star determine its luminosity? A star is luminous because each square centimeter of its surface emits radiation. A star's luminosity thus depends on its surface area; the bigger the star, the more radiation it will emit. The luminosity also depends on how intensely radiation is emitted from each square centimeter of stellar surface. The intensity of the radiation emitted from each unit of surface area depends on the star's temperature; a hot star produces much more radiation than a cool one. Thus the luminosity of a star increases with increasing stellar temperature and size.

The relationship between luminosity, temperature, and radius described in the last paragraph can also be cast into mathematical language. A hot surface of temperature T emits radiation at a rate given by σT^4, where σ is just a number, 5.67×10^{-5} if the energy rate is expressed in ergs per second per square centimeter. Since the number of square centimeters on a star is its surface area $4\pi R^2$, where R is its radius, luminosity of a star is

$$\text{Luminosity} = 4\pi(\text{radius})^2 \, \sigma(\text{temperature})^4 \qquad (9.3)$$

The radius of a star is a very difficult quantity to measure, because stars appear as pinpoints of light, not disks, even in large telescopes. While sophisticated techniques developed in the last few years can determine the radii of some near and bright stars, the relationship between temperature, radius, and luminosity must be used to measure the sizes of most stars. If the temperature and luminosity are known, equation 9.3 can be used to measure the radius. We can determine the temperature of a star by examining its color or scrutinizing its spectrum.

☆   STELLAR COLORS   ☆

Hot stars are blue, and cool stars are red. Blue photons are high-energy photons that come from high-temperature gases where individual atoms zip around at high speeds; low-energy red photons come from cooler gases, where the atoms are more sluggish. It is thus possible to measure the temperature of a star by measuring its fluxes in the blue and red regions of the electromagnetic spectrum and computing the ratio of these fluxes. A cool red star will emit more red light than blue light, while a hot blue star will emit less red light and more blue light (Figure 9.5).

We use the language of magnitudes to express the relative amounts of red and blue light coming from a star. The *color index* of a star is the difference between its red and blue magnitudes, and convention dictates that

$$\text{Color index} = \text{blue magnitude} - \text{red magnitude} \qquad (9.4)$$

We measure blue magnitudes and red magnitudes by isolating a particular part of the spectrum with a filter placed in front of the radiation detector in the telescope; this filter allows only photons of the desired color to reach the detector. Different researchers use different types of filters and determine different color indices; Figure 9.5 shows one set of filters in common use. The color index most often measured is the *B-V* color index (for Blue-Visual). A blue star has a higher blue flux than a visual flux, its blue magnitude is a smaller number than its visual magnitude, and its color index is a negative number. A cooler star like the sun is brighter in the visual range than in the blue range, and its color index is a positive number. (Remember that magnitudes run backward.)

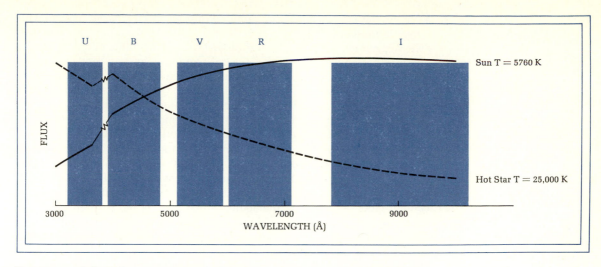

FIGURE 9.5 Colors of stars are most often measured on the UBV (Ultraviolet, Blue, Visual) system. Filters are used to isolate the parts of the spectrum shown by the shading when the flux from a star is measured. Also shown are the Red and Infrared filters used to measure very cool stars. By adding more filters, infrared astronomers have extended this system to almost 10μ.

☆ SPECTRAL CLASSIFICATION ☆

The spectrum of a star provides another indication of its temperature. Color measurements isolate broad bands in the electromagnetic spectrum of a star, whereas a photograph of a star's spectrum provides a detailed picture of the distribution of light intensity as it depends on wavelength.

Chapter 8 describes how observations of the sun's spectrum were used to determine the abundances of chemical elements in the solar atmosphere. Hydrogen is the most common constituent of the sun, but because only a few hydrogen atoms are in an energy state that can absorb photons of visible light, the hydrogen lines in the sun's spectrum are quite weak. The atoms in hotter stars zip around with higher energies, and energetic collisions between atoms can kick hydrogen atoms into higher energy states capable of producing absorption lines. In the relatively cool sun, only 1 of 10^8 hydrogen atoms is in the energy state that can absorb photons of visible light; in a hotter star, with a temperature of 10,000 K, 1 in 10^6 atoms (100 times as many) are in this high energy state. As a result, the absorption lines produced by hydrogen are considerably stronger in hotter stars, and observation bears this out.

The appearance of a stellar spectrum, then, provides an indication of a star's temperature. Because most stars have more or less the same abundances of chemical elements in their surface layers, the presence or absence of the lines of certain elements is an indication of the surface temperature of a star. Stellar spectra have been grouped into *spectrum classes*, labeled with letters (Figure 9.6). The order is *O, B, A, F, G, K, M*, from hottest to coolest. (The sexist mnemonic "Oh Be A Fine Girl Kiss Me" is often cited as a means of remembering the order of the spectral sequence; you can probably invent a memory crutch less reminiscent of the Victorian era.) These spectrum classes have been subdivided by numbers, in the sense that a G2 star like the sun is hotter, and more like an F star, than Capella, a G5 star.

The spectral sequence is a temperature sequence because temperature determines whether a particular element can absorb visible photons. Take hydrogen, for example. In the very cool class-M stars, with temperatures of about 3000 K, so few hydrogen atoms are in the excited state that produces visible absorption lines that these lines are unobservable. The spectra of M stars are dominated by molecular absorption bands. M-star atmospheres contain large quantities of molecules, since there is not enough energy to break the molecules apart. At higher temperatures, in the K stars, molecules disappear because atoms collide with high enough energies to disrupt them; and hydrogen lines begin to enter the picture because more hydrogen atoms are in the first excited state. Hydrogen absorption lines become progressively stronger through the G, F, and A stars. In the A stars, hydrogen lines dominate the spectrum (Figure 9.6). At still higher temperatures, atoms collide with such energy that the electron is stripped off the hydrogen atom, and many hydrogen atoms become ions, weakening the line in the B and O stars. A variety of theoretical calculations can reproduce the behavior of hydrogen lines in stars of differing temperatures, and thus derive a relationship between temperature and spectral class.

These two sections have shown how the message of starlight can be decoded. Measure the brightness of a star, determine its distance, and its lu-

FIGURE 9.6 Photographs of the spectra of stars in the principal spectrum classes. (Yerkes Observatory photograph)

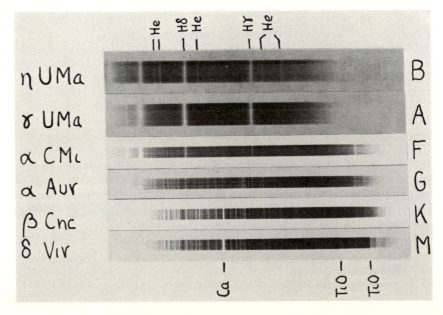

Box 9.4 An
Analysis of Sirius A

Temperature: The temperature of the optically brightest component of the brightest star in the sky could be determined from its *B-V* color index of −0.01 or from its spectrum classification as an A1 star. Sirius is so bright that a more detailed method can be used. Its flux has been measured over a wide range of wavelengths, from the infrared through the visible to the ultraviolet (measured by space satellites), and computer programs then matched this flux to that from a model star to fix its temperature at 9970 K, the value in Table 9.1.

Luminosity: Sirius is close enough to the earth so that its dis-

tance can be measured as 2.67 pc by parallax methods. Since the flux over all wavelengths has been measured, equation 9.1 gives its luminosity as 9.57×10^{34} ergs/sec, or 25.0 times the luminosity of the sun.

Radius: The relation between temperature, luminosity, and radius described in the text (numerically, equation 9.3) gives the radius of Sirius A as 1.16×10^{11} cm, or 1.67 times the radius of the sun. Sirius is near enough so that its radius can be directly measured by sophisticated techniques, and it is encouraging that the direct measurement agrees with this value.

minosity can be calculated. A color measurement or a determination of a star's spectral class fixes its temperature. Because a star's luminosity is determined by its temperature and radius, the size of this distant ball of gas can be measured even though it is just a pinpoint of light in the telescope. Box 9.4 illustrates the application of this analytical procedure to Sirius, the brightest star in the sky.

★ 9.3 CLASSIFYING STARS ★

Take another look at Figure 9.1. The first two sections of this chapter have shown how astronomers can explain the wide diversity of stellar brightnesses and colors in terms of stellar sizes, temperatures, and distances. But the attempt to pull order out of chaos is not complete; a long list of stellar temperatures and sizes is still a little confusing. Table 9.1 summarizes the properties of a number of representative stars. (These stars are all found in the winter sky, many of them in Orion; Figure 2.8 illustrates their locations in various constellations.) The temperatures of these stars run all the way from Sirius B's 32,000 K to Betelgeuse's 3250 K. (The bright star Sirius is actually a binary system; the star emitting the most visible radiation is called Sirius A, and the fainter star is Sirius B.) Luminosities run from 4.7×10^5 solar luminosities (Epsilon Orionis) to 0.06 solar luminosities (Sirius B); Sirius B is also the extreme in size at 0.0078 solar radius, versus 1100 solar radii for Betelgeuse. There are stars with even more extreme values of temperature,

TABLE 9.1 PROPERTIES OF STARS IN THE WINTER SKY

STAR	APPARENT VISUAL MAGNITUDE	ABSOLUTE VISUAL MAGNITUDE	DISTANCE (Parsecs)
Epsilon Orionis	+1.50	−6.99(?)	500(?)
Rigel	+0.13	−7.03	270(?)
Regulus	+1.35	−0.58	24.3
Sirius A	−1.46	+1.41	2.67
Procyon	+0.37	+2.65	3.50
Sol	−26.8	+4.77	0.000005
Capella[a]	+0.09	−0.42	12.6
Epsilon Eridani	+3.74	+6.15	3.30
Aldebaran	+0.80	−0.79	20.8
Betelgeuse	+0.4 var	−6.11(?)	200(?)
Sirius B	+8.4	+11.27	2.67

[a] Capella is a double star. The magnitudes and colors are for the combined system; the temperature, radius, and luminosity are those of the brighter and cooler component. The distances of Epsilon Orionis, Rigel, and Betelgeuse are estimated, so the absolute magnitudes, radii, and luminosities of these three stars are approximate.

SOURCES: D. Hoffleit, *Yale Catalogue of Bright Stars,* 2nd ed. (New Haven: Yale University Press, 1964); A. D. Code, J. Davis, R. C. Bless, and R. Hanbury Brown, "Empirical Effective Temperatures and Bolometric Corrections for Early Type Stars," *Astrophysical Journal* 203 (1976): 417–434; C. W. Allen, *Astrophysical Quantities,* 2nd ed. (London: Athlone, 1964). Additional sources on specific stars: *Capella:* W. D. Heintz, "Parallax and Motions of the Capella System," *Astrophysical Journal* 195

radius, and luminosity; the extremely hot, cool, large, small, luminous, or dim stars are so strange that their properties are very poorly measured.

The wide variety of stars in Table 9.1 illustrates the dilemma of someone attempting to interpret stellar structure. With big stars and little stars, hot stars and cool stars, luminous stars and dim stars, how many of them are like the sun inside and how many of them have some completely different internal make-up? If most of the stars in the sky are indeed like the sun inside, as the stellar-interior model of Chapter 8 would seem to indicate, the stars that are totally different need to be culled out of Table 9.1. What stars should be left out?

☆ THE HERTZSPRUNG-RUSSELL ☆
DIAGRAM

A table is a useful way to summarize a body of knowledge, but it is a poor way to organize that knowledge. In the early part of the twentieth century, Einar Hertzsprung and Henry Norris Russell (Figure 9.7) independently

TABLE 9.1 Continued

B-V COLOR INDEX	SPECTRAL CLASS	TEMPERATURE (K)	RADIUS (Solar radii)	LUMINOSITY (Solar units)
−0.18	B0	24,800	37	4.7×10^5
−0.05	B8	11,550	74	8.9×10^4
−0.11	B7	12,210	3.63	2.7×10^2
0.00	A0	9,970	1.67	2.5×10^1
+0.42	F5	6,510	2.07	7.0
+0.62	G2	5,760	1.00	1.00
+0.81	G5, G0	5,200	14.1	1.3×10^2
+0.87	K2	5,000 (est.)	0.7	2.8×10^{-1}
+1.55	K5	3,780	61	6.9×10^2
+1.85	M2	3,250	1100	1.2×10^5
−0.03	White dwarf	32,000	0.008	5.8×10^{-2}

(1975): 411–412; A. K. Dupree, "Ultraviolet Observations of Alpha Aurigae From Copernicus," *Astrophysical Journal* 200 (1976): L27–L32. *Aldebaran* and *Betelgeuse*: H. M. Dyck, G. W. Lockwood, and R. W. Capps, "Infrared Fluxes, Spectral Types, and Temperatures for Very Cool Stars," *Astrophysical Journal* 189 (1974): 89–100; A. J. Wesselink, "Surface Brightnesses in the U, B, V System with Applications on M_v and Dimensions of Stars," *Monthly Notices of the Royal Astronomical Society* 144 (1969): 297–311. *Sirius B*: H. L. Shipman, "Sirius B: A Thermal X-Ray Source?" *Astrophysical Journal* (*Letters*) 206 (1976): L67–L70; K. D. Rakos and R. J. Havlen, "Photoelectric Observations of Sirius B," *Astronomy and Astrophysics* (in press).

FIGURE 9.7 The American astronomer Henry Norris Russell, one of the two astronomers who discovered the usefulness of the Hertzsprung-Russell diagram as a way of classifying stars. (Yerkes Observatory photograph)

thought of a scheme that made a good deal of sense of the bewildering diversity of stellar properties. They plotted the temperatures and luminosities of various stars on a graph, with a temperature indicator on one axis and a luminosity indicator on the other. Each star was assigned a location in this *Hertzsprung-Russell diagram,* according to these two properties. The stars listed in Table 9.1 are plotted in such a diagram in Figure 9.8.

A number of features of the Hertzsprung-Russell (or H-R) diagram are worth consideration. If one wishes to produce an H-R diagram that any astronomer will recognize, temperature must increase to the *left* along the horizontal axis, and luminosity must increase upward. Any temperature indicator (color index or spectral class) can be plotted along the horizontal axis, and any luminosity indicator (absolute magnitude in any wavelength band) can be plotted along the vertical axis.

The chief value of the H-R diagram is its clear classification of stars into different groups. The largest stars are those that are luminous but at the same time red, and these *red giants* fall into the upper right corner of the diagram. The smallest stars are quite dim in spite of their high temperatures; they are found at the lower left and are called *white dwarfs.* In Figure 9.9, an H-R diagram where the axes are logarithmic plots of luminosity and temperature, lines of constant radius are drawn, showing the increase in stellar size from lower left to upper right. The approximate locations of the white-dwarf

FIGURE 9.8   A Hertzsprung-Russell diagram of the stars listed in Table 9.1, plotting their luminosities and temperatures.

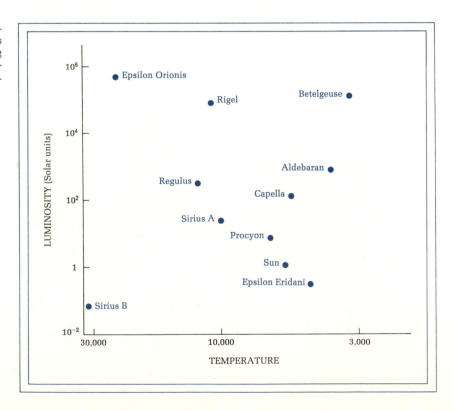

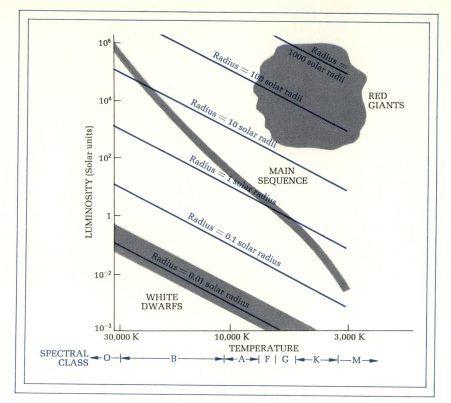

FIGURE 9.9 A Hertzsprung-Russell diagram, showing where stars of different sizes are found. The largest stars are the *red giants*, in the upper right corner, and the smallest are the *white dwarfs* in the lower left. The *main sequence*, consisting of stars comparable in size to the sun, is in the middle.

and red-giant stars are shown. This drawing demonstrates the utility of the H-R diagram as a star sorter.

The last few pages have discussed a number of different stellar properties and demonstrated how the H-R diagram presents them graphically. All the stellar properties listed in Table 9.1 are reflected in the H-R diagram of Figure 9.8, save the distance of the star, which is not a fundamental stellar property. You might want to stop and look at the graph for a while, being sure that you understand why each star is where it is. Several of the review questions at the end of this chapter are designed to help you understand the relations among the radius, temperature, and luminosity of a star.

☆   **THE MAIN SEQUENCE**   ☆

Between the domain of the red giants in the upper right corner and the home of the white dwarfs at the lower left lies a well-defined middle ground where most stars in the sky are plotted. The stars in Figure 9.8 were selected as illustrating the extremes of stellar properties; the concentration of stars along the diagonal of the H-R diagram running from lower right to upper left becomes apparent when a more typical sample of stars is plotted. Figure 9.10

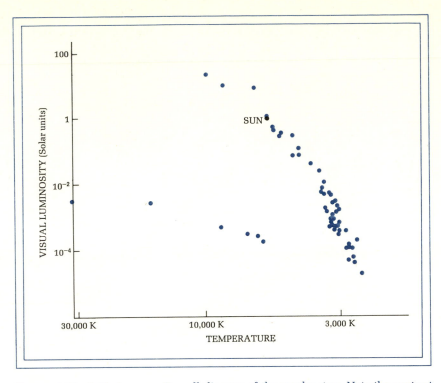

FIGURE 9.10 A Hertzsprung-Russell diagram of the nearby stars. Note the contrast with Figure 9.8; most stars are on the main sequence, and many are less luminous than the sun (color). Here the visual luminosity, rather than the total luminosity, is plotted, underestimating the total luminosities of the cool M-type stars that radiate mostly in the infrared. [Luminosities are from P. van de Kamp, "The Nearby Stars," *Annual Review of Astronomy and Astrophysics, 9* (1971), p. 105. Temperatures for hot stars are from the scale of Table 9.1; temperatures for the cool stars are from the scale of G. J. Veeder, *Astronomical Journal, 79* (1975), p. 1067; and temperatures for the white dwarfs are from H. L. Shipman, *Astrophysical Journal, 177* (1972), pp. 723–743, and from J. Liebert, *Astrophysical Journal* (in press).]

shows an H-R diagram of stars within 5 pc of the sun. This band in the H-R diagram, separating the white dwarfs from the red giants, is called the *main sequence.*

Why is there a main sequence of stars? Stars on the main sequence are those that have interior structures quite similar to the sun, and whose masses increase from lower right to upper left. The massive main-sequence stars must have great central pressures to hold up the enormous weight of their envelopes, and this central pressure comes from a high central temperature. Thanks to the high central temperature, nuclear reactions proceed rapidly in the cores of these massive stars in the upper left corner of the diagram, and they are hot and luminous. Tiny stars have low central pressures, low central temperatures, and low reaction rates; as a result they are less luminous and cooler; they are found in the lower right of the diagram. Figure 9.10 shows that the dim, small stars are quite common in our neighborhood. Intuitive expectations about the main sequence are borne out by detailed calculations that show that stars with the same structure as the sun but different masses would indeed fall along the main-sequence band in Figures 9.9 and 9.10.

The H-R diagram is the tool that allows us to bring order into the chaos that is the tremendous stellar zoo. It allows us to sort stars easily into those that are fundamentally like the sun, the main-sequence stars, and those that are radically different, the white-dwarf and red-giant stars. The main sequence is a group of stars similar in structure but differing in mass. A direct measurement of the masses of main-sequence stars would provide more insight into their properties. But it is difficult to measure the mass of a single, isolated star with high precision. Two stars orbiting each other, locked together by gravity, form a binary star, and the masses of binary stars can be measured.

★ 9.4 BINARY STARS AND ★
STELLAR MASSES

Closer examination of the apparently single stars in the sky reveals that many of these pinpoints of light are double or multiple systems, containing two or more stars orbiting around each other. It is these binary stars that allow us to determine a small number of stellar masses. In a binary star system, just as in the solar system, the gravitational attraction of the two stars is balanced by their motion, so that the two stars neither fall together and collide nor fly apart into space.

☆ KEPLER'S THIRD LAW REVISITED ☆

Because gravity is produced by mass, analysis of the motion of binary stars results in a determination of the mass needed to balance that motion. The heart of double-star analysis is Kepler's third law, the mathematical formulation of the balance between motion and gravity. This law was discussed in connection with orbits in the solar system in Chapter 3, and was derived for the simple case of circular motion of a small planet in Box 3.2:

$$\text{(Mass of } A + \text{mass of } B) \times \text{(period)}^2 = \text{(orbital radius)}^3 \qquad (9.5)$$

Here the masses of the two orbiting objects A and B are measured in solar masses, the period is measured in years, and the orbital radius is measured in astronomical units.

The mathematics in equation 9.5 should not obscure the essential physical phenomenon: the balance between orbital motion and gravity. This balance may be clarified by a few examples. First apply the equation to the earth-sun system. The mass of the earth is 0.000003 solar mass, so the combined mass of the two orbiting objects is essentially 1 solar mass. The period of the earth's orbit is 1 year, and the orbital radius is 1 AU, so the equation is satisfied. It should be satisfied in a binary system where the combined mass of the stars is 1 solar mass, the orbital radius 1 AU, and the system's period 1 year.

Now suppose a binary system exists with a shorter period—half a year, for example—but with the same separation between stars. The stars have less time to complete one circuit, and so they move faster. The gravitational force between them, and hence their mass, must be greater to balance this faster motion; equation 9.5 indicates that the combined mass of the system must be 4 solar masses to achieve a balance.

Consider another system, with a period equal to the earth-sun system's 1 year but with a separation of 2 AU. Again the stars must move faster, and the greater separation further weakens the gravitational force binding them together, so once again a greater mass is needed to achieve a balance. Equation 9.5 indicates just how much greater the mass must be: 8 solar masses.

We can apply Kepler's third law quite easily to *visual binaries*, pairs of stars that can be split by the telescope. If a visual binary is near enough so that its parallax can be measured, observers can carefully plot the changing positions of the two stars and measure the period and size of the binary system's orbit. We can split more distant binaries, or closer binaries, only with the spectrograph; these are called *spectroscopic binaries*, since the two disks cannot be separated with the telescope. In a few spectroscopic binaries, the orbits of the two stars are arranged so that one star passes behind the other when the pair dances around in its orbit. At such a time, the light from the whole system is dimmed because one star is eclipsed; these systems are called *eclipsing binaries*.

How can the spectrograph split a double star? The spectrum of the pair of stars shows a split because the motions of the stars around each other cause light from the approaching star to be shifted to slightly shorter wavelengths, and light from the receding star to be shifted to slightly longer wavelengths. Where two stars are visible, both spectra can be seen; where only one star is visible, the shift in the spectrum shows the motion. It is the *Doppler effect* that causes the shifting spectrum.

☆ THE DOPPLER EFFECT ☆

We can best understand the Doppler effect by considering the musical pitch of the siren of a speeding police car (Figure 9.11). Suppose that the car has a siren turned to middle C. The police officer in the car will hear middle C coming from the siren. Someone on the right, ahead of the police car, will hear a higher musical pitch because the sound waves are compressed in wavelength in the direction of the car's motion. Someone on the left, behind the car, will hear a lower pitch because the waves are stretched out. The shift in musical pitch is caused by a shift in the wavelength of the sound.

In the same way, light waves from a moving star (Figure 9.12) are shifted. If the star is approaching the observer, the wavelength is shortened and the spectrum is *blueshifted*. A receding star will have a *redshifted* spectrum. The magnitude of the shift depends on the velocity of the star; the faster it moves, the more the waves are either piled up or stretched out, and the

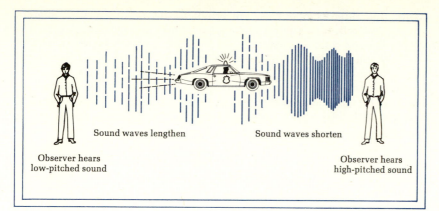

Sound waves lengthen

Sound waves shorten

Observer hears
low-pitched sound

Observer hears
high-pitched sound

FIGURE 9.11 Stationary observers hear Doppler shifts from a moving police car; the siren has a higher pitch when the car approaches the listener. (Adapted from Harry L. Shipman, *Black Holes, Quasars, and the Universe.* Copyright © 1976 by Houghton Mifflin Company, p. 89.)

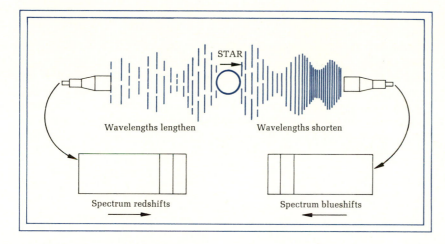

STAR

Wavelengths lengthen

Wavelengths shorten

Spectrum redshifts

Spectrum blueshifts

FIGURE 9.12 Starlight shifts in wavelength because of the Doppler effect. A star moving toward the observer (right) has wavelengths in its spectrum shortened by its motion, while a star moving away from the observer (left) has wavelengths in its spectrum lengthened. (Adapted from Harry L. Shipman, *Black Holes, Quasars, and the Universe.* Copyright © 1976 by Houghton Mifflin Company, p. 89.)

greater the shift. Mathematically, if the speed of the star is only a small fraction of the speed of light, this relationship is

$$\frac{\text{Shift in wavelength}}{\text{wavelength}} = \frac{\text{object velocity}}{\text{wave velocity}} \qquad (9.6)$$

Symbolically, if λ is the wavelength, $\Delta\lambda$ the wavelength shift, v the object velocity, and c the wave velocity, the Doppler shift is given by the simple expression $\Delta\lambda/\lambda = v/c$, for small shifts.

★ 9.5 THE SEXTUPLE STAR CASTOR ★

Castor, the second brightest star in Gemini, looks like only one star to the naked eye, but it is actually a sextuple system. A telescopic view splits

Castor into three components, the bright Castor A, the nearby, slightly fainter Castor B, and a faint, distant companion Castor C. However, the system is more complex than that. All three types of binary star are represented in this system: Castor A and B form a visual binary (as does the A-B pair with the distant Castor C); A and B are each spectroscopic binaries; and C is an eclipsing binary (Figure 9.13). The analysis of the Castor system discussed in this section demonstrates how the precepts of the previous section are applied to the analysis of binary stars and the determination of stellar masses.

☆   THE VISUAL BINARY CASTOR A-B   ☆

The motion of Castor A and Castor B in their orbit around each other was first detected by William Herschel in 1804. Herschel was the first person to realize that binary stars were physically associated with each other. Measurements of the changing positions of Castor A and Castor B over centuries reveal that the orbital period is 380 years and that the orbital radius is 93 AU. The application of Kepler's third law, equation 9.5, to this system reveals the combined mass of the system as 5.5 solar masses. If the two stars divided the masses equally, both stars would gyrate around their common center of mass in similar orbits. Measurements show that Castor A is a little more sluggish, since it is the more massive, less mobile star. Detailed observations set the mass of Castor A at 3.2 solar masses, while the dimmer Castor

FIGURE 9.13 The triple star system of Castor. Through a telescope three stars are visible: Castor A, Castor B, and the distant Castor C. (Through most telescopes the field will be a mirror image of that shown here.) Each of these three stars is a binary, with Castor A and Castor B spectroscopic binaries, and Castor C an eclipsing binary. (Details from H. A. Rey, *The Stars*, published by Houghton Mifflin Company. Copyright © 1952, 1962, 1967, 1970 by H. A. Rey.)

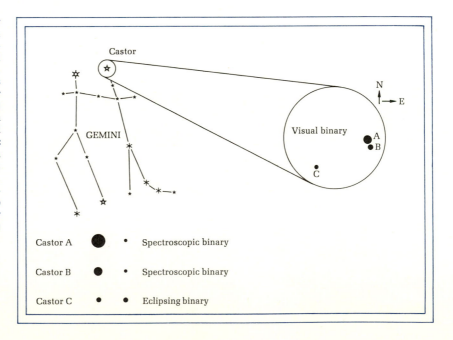

B has a mass of 2.3 solar masses. Were Castor a less complex and less interesting system, the analysis would be finished, because the orbital motion of the distant Castor C around the A-B pair is imperceptible. But the spectrograph reveals more complexity.

☆ **THE SPECTROSCOPIC BINARIES** ☆
CASTOR A AND CASTOR B

Observations of the spectra of Castor A and Castor B reveal that the dark lines in their spectra shift back and forth in wavelength over periods of several days (Figure 9.14). We cannot see the spectra of the companions to these stars, but the periodically changing Doppler shift in the spectra reveals the motion of the brighter components of the system in their orbits. Ideally, we would like to apply Kepler's law to these systems as well, but the Doppler effect only tells us how fast the stars are moving along our line of sight, toward us or away from us. There may be some additional motion across our line of sight that the spectrograph would not detect. A reasonable guess is that the orbits of the spectroscopic binaries are oriented in the same plane as the orbit of the Castor A-B system, just as satellite orbits in the solar system are in the plane of the ecliptic. Such a presumption produces masses of 3.0 and 0.2 solar masses for the two components of Castor A, and 2.0 and 0.3 for the two components of Castor B. These numbers are not absolutely reliable, because we had to guess the orbital orientation of these systems in order to derive them.

☆ **THE ECLIPSING BINARY CASTOR C** ☆
(YY GEMINORUM)

The distant, unimposing Castor C is the component of the system that bares most of its secrets. It is a spectroscopic binary like Castor A and Castor B, but the spectra of both of its components are visible; they are both cool main-sequence stars of class M1. Castor C is also an eclipsing binary, which

FIGURE 9.14 The spectrum of the brightest component of Castor—Castor A—showing that it is a spectroscopic binary. The two spectra were obtained several days apart, and the changing motion of the visible star is shown by the changing Doppler shift in the star's spectrum. (Lick Observatory photograph)

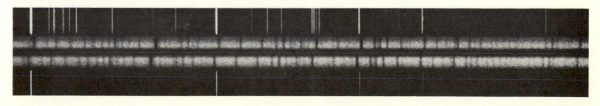

accounts for its listing in the variable star catalogue as YY Geminorum. Figure 9.15 illustrates an eclipse of this star, showing the diminution of light and the shift in the spectral lines of each companion. The orbit is a tight one, for the stars are only 0.018 AU, or 3.8 solar radii, apart.

Thanks to the eclipses, we know that the plane of Castor C's orbit is along our line of sight and that Doppler-shift measurements pick up the entire motion of each star. The Doppler shift gives the orbital velocity as 125 km/sec. Since we know it takes 0.81 days for the system to make a complete circuit, we can use the familiar relation Distance = rate × time to find the distance around the orbit of these two stars. Now Kepler's third law can be applied, because we know the orbital size and its period. The equal Doppler shifts of the two stars indicate that their masses are equal, and Kepler's law gives each as 0.58 solar masses.[1]

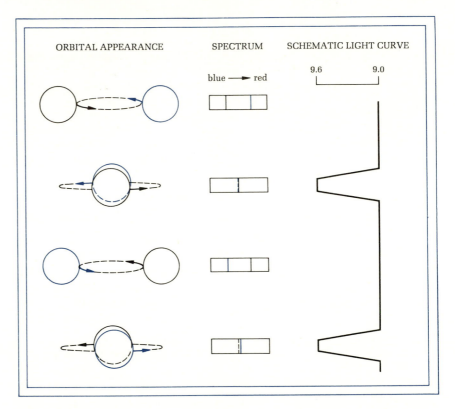

FIGURE 9.15 Orbital appearance, spectral-line shifts, and light variations of the eclipsing binary YY Geminorum. The Doppler shifts in the spectrum show the changing motions of the two stars toward and away from us; the eclipses occur when one star passes behind the other. One star is shown in color for clarity; as far as we can tell, the two stars are identical.

[1] In applying Kepler's third law to binary stars, the proper orbital size to use is the distance between stars, in this case 0.018 AU.

Eclipsing binaries are useful in other ways as well. A measurement of the duration of an eclipse, combined with a knowledge of the speed of the orbiting star, produces a measurement of the size of each star. Each component of Castor C has a radius of 0.6 solar radius. This radius measurement can only be made for eclipsing binary stars.

Castor C shares one of the other properties of eclipsing binaries: It is peculiar. Many cool M main-sequence stars are flare stars (recall Chapter 8), and Castor C is no exception. Flares can disrupt the nice eclipsing-binary light curve shown in Figure 9.15. Castor C also shows other variations in its light curve, and a recent analysis indicates that these secondary variations are due to giant starspots on its surface, similar to the spots that other flare stars have. Eclipsing binaries have proved to be less comparable to the sun than one might like, because so many of them show peculiarities of one sort or another.

★ 9.6 TESTING STELLAR ★
INTERIOR THEORY

This chapter has shown that careful measurements of a wide variety of stellar phenomena can result in a fairly complete knowledge of stellar properties. With the aid of the H-R diagram, we can sort out the red giants and white dwarfs, leaving a collection of main-sequence stars that should be much like the sun. We can then extend the test of our picture of the solar interior by seeing whether it can explain the properties of these stars as well. Much of the confidence that we have in standard stellar-interior theory is due to the success of this model in explaining the characteristics of main-sequence stars.

Main-sequence stars have the same basic interior structure as the sun. Their principal feature is that they are roughly chemically homogeneous from center to surface; there is some hydrogen gas at all points in their interior. Their energy source is central hydrogen fusion. The differences among them stem from their differing masses. The massive, hot, luminous stars are at the upper end of the sequence, in the upper left corner of the H-R diagram, and the small, cool, dim ones are in the lower main sequence, at the lower right. The differing masses cause differences in the central temperatures of the stars located at various points on the sequence.

While main-sequence stars are generally similar in their interior structure, there are minor differences between the small stars of the lower main sequence and the massive stars of the upper main sequence. Stars larger than the sun rely primarily on the CNO Bi-Cycle for energy production. Further, massive stars have such hot interiors that radiation cannot carry energy away from their core fast enough; convection currents of hot gas, swirling in their cores, transport this energy toward the surface.

But these are differences of detail, not differences of substance. The principal difference among main-sequence stars of different temperatures and luminosities is the difference in their masses, according to the theoretical picture of what the main sequence is. The correctness of the theory can be assessed by answering the question observationally: Is the main sequence indeed a sequence of differing masses, and does the run of stellar mass along the main sequence proceed in agreement with the theoretical calculations? While there are many ways to answer this question, the simplest is to graph the masses and luminosities of main-sequence stars and see if observationally derived masses of binary stars correspond to the theory, Figure 9.16

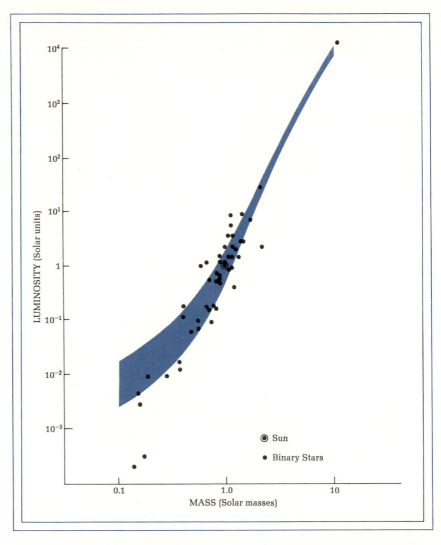

FIGURE 9.16   The mass-luminosity relation. The black dots show masses and luminosities derived from observations of nearby binary stars; the color band shows the location of the theoretical mass-luminosity relation calculated from our theory of stellar structure. The coincidence of observations and theory supports the theory.

shows a plot of some fairly recent analyses of stellar masses; the agreement of the theoretically predicted relation between luminosity and mass (the solid line) with the observations is quite good. Each dot on that graph represents a star, and the theory seems to be able to explain the relationship between the mass and luminosity of that star.

Figure 9.16 summarizes why astronomers spend long, sometimes cold, nights at the telescope, cursing the balky equipment and hoping for that one last data point on the light curve of an eclipsing binary. Satisfying as it is to find one's name listed as a source for an entry in a catalogue of stellar masses, radii, and luminosities, understanding comes from the insight that these measurements provide into the internal structure and evolution of stars. When the H-R diagram was first discovered, long hours were spent deciding whether stars moved up or down the main sequence in the course of their evolution. Observation of stellar properties combined with theoretical models indicates that such a question was pointless; stars remain on the main sequence in one place as long as they have a structure similar to the structure of the sun. The position of a star on the sequence is fixed by its mass. The agreement of theory with observation shows that we understand the internal structure of main-sequence stars.

In the first chapter, three big questions that astronomers ask were posed: Where is it? What is it? How does it evolve? The first two questions have been answered by stellar astronomers: Stars are balls of gas at tremendous distances. We have even been able to understand the interior structure of one group of stars, the main-sequence stars. But how do they evolve? Where did main-sequence stars come from? What happens to them? These are the questions that the next three chapters of Part 3 will focus on.

SUMMARY

The stars are balls of hot gas that look like pinpoints when seen through a telescope. Careful measurement and analysis of the light coming from these stars can determine the fundamental properties of these stars: their radii, temperatures, and luminosities, provided their distances are known. The H-R diagram can be used to sort stars into groups, showing which stars are on the main sequence, which are red giants, and which are white dwarfs. We can measure the masses of some stars, those in binary systems, and it is the measurement of these masses that confirms our basic picture of the internal structure of main-sequence stars.

This chapter has covered:

The measurement of the flux from a star and the relationship between flux, luminosity, and distance;

The methods used to measure stellar distances;

The relationship between stellar radius, luminosity, and temperature;

The use of spectral classification and color measurements as temperature indicators;

The H-R diagram, its use, and its organization;

The determination of masses of binary stars;

The application of binary-star techniques to the star Castor; and

The confirmation of our theoretical picture of main-sequence stars by analysis of stellar properties.

KEY CONCEPTS

Absolute magnitude
Apparent magnitude
Binary star
Color index
Detector
Distance modulus
Doppler effect

Eclipsing binary
Flux
H-R diagram
Luminosity
Magnitude
Main sequence
OBAFGKM
Parallax

Parsec
Photometer
Red giants
Spectrum classes
Spectroscopic binary
Visual binary
White dwarfs

REVIEW QUESTIONS

1. In the following pairs of objects, which is the brighter? Explain your answers.

a. a sixth-magnitude star and an eighteenth-magnitude star;

b. a star with a flux of 2×10^{-5} ergs/cm²/sec and a star with a flux of 2×10^{-7} ergs/cm²/sec;

c. a star with a luminosity of 1 sun at a distance of 1 pc and a star with a luminosity of 1 sun at a distance of 10 pc;

d. a star 10 pc away with an absolute magnitude of 5 and a star of 10 pc away with an absolute magnitude of 1.

2. Review the parallels between the descriptions of the relationship among brightness, luminosity, and distance in the language of flux and in the language of magnitudes.

3. Which of the following pairs of stars is hotter? Briefly explain your answers.

a. a star with a color index of 0 and a star with a color index of 1;

b. a G star and an O star;

c. a blue star and a red star.

Questions 4 to 6 refer to Table 9.1 and the Hertzsprung-Russell diagram of Figure 9.8. Be sure that you could answer the questions if you were given either the table or the diagram.

4. Sirius B and Epsilon Orionis have roughly the same temperature. Which is more luminous, and what physical characteristic makes them different?

5. Aldebaran and Rigel have roughly the same radius. Which is more luminous, and why?

6. Betelgeuse and Aldebaran have roughly the same temperature and roughly the same brightness. Betelgeuse is 40 times larger than Aldebaran. Which is further away?

7. Make up your own questions similar to questions 4 to 6, and answer them.

8. How could a blind person measure the speed of a passing police car?

9. What is the distinction among the three types of binary stars? Which components of the Castor system form each type of binary?

10. If binary stars did not exist, how well confirmed would our model of a main-sequence star's internal structure be?

FURTHER READING

Batten, A. H. *Binary and Multiple Systems of Stars.* Elmsford, N.Y.: Pergamon, 1973.

Meadows, A. J. *Stellar Evolution.* Elmsford N.Y.: Pergamon, 1967.

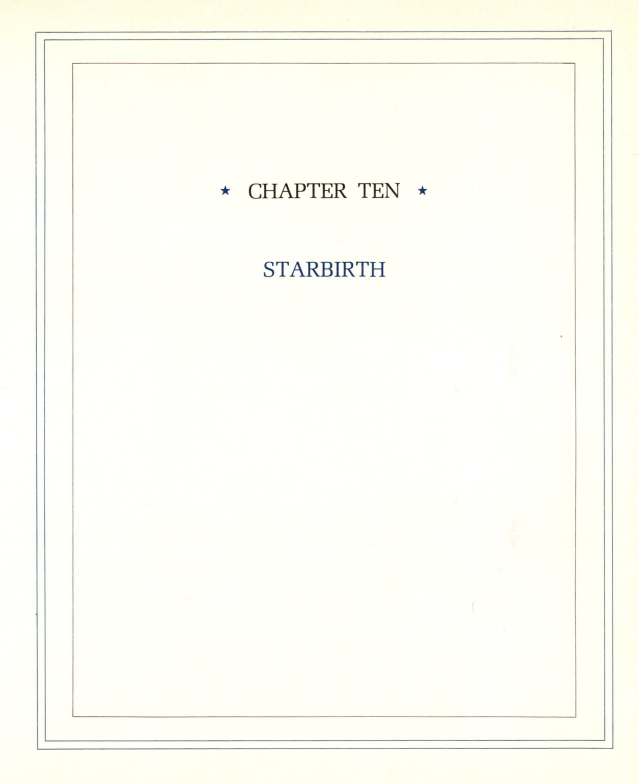

★ CHAPTER TEN ★

STARBIRTH

Chapters 8 and 9 present a complete theoretical model of a star in the prime of its life—a main-sequence star. Where did this star come from? Observations show that the space between the stars is filled with gas, clumped together in clouds of various shapes and sizes. Stars form from these interstellar gas clouds, when gravity within them becomes strong enough to draw a large number of gas molecules together into a small ball, a star. But how does this event occur?

Investigations of various components of the interstellar medium are slowly converging toward a tentative picture of the formation of a star. An important accompaniment to the formation of one star, our sun, was the formation of a planetary system.

★ MAIN IDEAS OF THIS CHAPTER ★

Interstellar space contains a great deal of gas and dust, generally aggregated together into various clouds.

Condensations of gas and dust in interstellar space are the sites of the formation of new stars.

Stars are formed by the gravitational contraction of interstellar gas clouds.

The formation of a planetary system was a natural event; although the details are not understood, it seems that many stars, when they are born, have planets around them.

★ 10.1 INTERSTELLAR GAS ★

Stars and their associated planetary systems are very small compared to the vastness of interstellar space. Two stars in interstellar space are like two people who have the entire planet earth all to themselves; the distances between them are 10^7 times their sizes. Compared to interstellar space, the solar system is quite crowded. This space is not just a void; it is filled with gas. This gas is important in star formation. Small clumps of interstellar gas are continually condensing and forming new stars, and an understanding of exactly how stars form requires a reasonably accurate picture of the gas from which they were born (Figure 10.1). We do not yet have a comprehensive picture of the interstellar medium; it contains a bewildering variety of gas clouds of different types. This chapter begins with a description of the different kinds of objects found in the interstellar medium, and the last two sections present the present state of our knowledge of how stars and planets are born.

FIGURE 10.1 The Orion nebula, a large cloud of glowing gas. At the center of this object are a number of types of gas clouds that represent various stages in the formation of a star. Sections 10.1 and 10.2 discuss the events taking place in this object; Figure 10.8 maps the clouds seen by radio and infrared astronomers. See also Color Photo 11 for a shorter-exposure photograph that shows more detail in the central portions. (Lick Observatory photograph)

☆ HYDROGEN GAS ☆

The most abundant element in interstellar space is the most abundant element in the universe: hydrogen. Radio astronomers detect hydrogen in interstellar space, and they have discovered that about 5 percent of the mass of our galaxy exists in the form of large clouds of neutral hydrogen gas, called *H I regions* (pronounced "aitch-one"; H I is the spectroscopist's name for neutral hydrogen). The density of these clouds is quite low: less than 10 hydrogen atoms per cubic centimeter, only 1/100,000 of the density of the earth's atmosphere 1000 km above the surface. Such clouds, vacuums by terrestrial standards, are difficult to observe since they neither reflect nor absorb visible light.

Hydrogen atoms radiate away some of the clouds' energy at a wavelength of 21 cm, in the radio part of the electromagnetic spectrum; as a result, the discovery and analysis of H I regions is the business of radio astronomers. A hydrogen atom consists of two particles—a proton and an electron, each one a spinning bit of electrical charge (Figure 10.2). A spinning charge acts like a small magnet, so one can represent a hydrogen atom as a pair of magnets, each with a north pole and a south pole. Opposite magnetic poles attract, so that the hydrogen atom is in a lower energy state if its magnets are pointed in opposite directions.

In a cold interstellar cloud, most hydrogen atoms will be found in the lower energy state. A few atoms will collide with others and be kicked up into the higher energy state, flipping their spins. Most of the time, an atom will get rid of the excess energy in another collision, flipping its spin again, but sometimes the energy will be radiated away as a photon with a frequency of 1421 megahertz (MHz) or a wavelength of 21 cm. A cloud of hydrogen gas will thus tend to radiate at this frequency, and a watching radio astronomer can observe the 21-cm radiation coming from an H I cloud.

Such 21-cm observations of cool H I clouds in the galaxy have been made for 20 years now, and a reasonable picture of these regions of interstellar space has been extracted from the observations. H I regions vary widely in size, density, and temperature. A typical cloud contains about 300 solar masses of gas, is 10 pc across, and has a temperature of 70 K. These clouds are often quite irregular in shape. The analysis of this component of the interstellar gas is somewhat difficult, because a radio astronomer has no

FIGURE 10.2 A hydrogen atom contains a spinning electron and a spinning proton. Each can be regarded as a magnet with a north and south pole. The energy of the atom depends on the alignment of their spins. The 21-cm photon, emitted when an atom drops from the high-energy state to the low one, is a key tool in probing the interstellar medium.

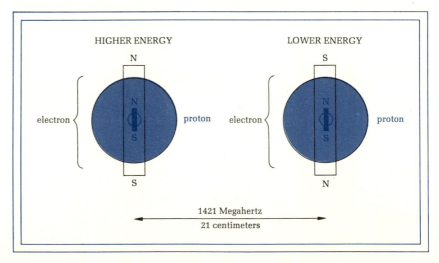

way of directly measuring the distance to an H I region. Is it a small cloud close to us or a large one further away? Radio observations provide only an incomplete picture of interstellar gas, and other means of observing interstellar gas must complete the picture.

☆ **INTERSTELLAR ABSORPTION LINES** ☆

Interstellar gas atoms jumping from one energy state to another can leave their footprints in another way. An astronomer observing the spectrum of a distant star sees absorption lines in the stellar spectrum, but sometimes some additional absorption lines are observed. Starlight passing through an interstellar cloud can be absorbed by a waiting gas atom that steals a photon from the beam of starlight and jumps to a higher energy state. The existence of the interstellar medium was first suspected because of these additional absorption lines.

The visible part of the electromagnetic spectrum contains *interstellar absorption lines* from trace constituents of the interstellar gas, atoms such as calcium and sodium. Measurements of the Doppler shifts of these lines and their intensities allow us to map the clouds lying between us and a wide variety of stars. The typical cloud speed is 10 km/sec, and there are generally about seven or so clouds between us and a star 1 kiloparsec (kpc) away.

Observations of interstellar absorption lines in the ultraviolet part of the electromagnetic spectrum reveal the true complexity of the interstellar gas. Hydrogen atoms in interstellar space are generally in their lowest energy state, and they can only absorb high-energy ultraviolet photons. Hydrogen molecules, too, are good absorbers of ultraviolet photons. Thus the mapping of the most abundant constituent of interstellar space requires observations of ultraviolet radiation from stars—observations that can only be made from a satellite orbiting above the blanket of the earth's atmosphere. Such observations, made with the Copernicus satellite, show that in some directions the average density of hydrogen is well over 1 atom/cm^3, while in others the density of neutral hydrogen is less than 0.01 atom/cm^3. This wide disparity shows that some parts of interstellar space are well filled with interstellar hydrogen, while others are quite empty.

What fills interstellar space between the hydrogen clouds? A possible constituent is a very-high-temperature interstellar corona first discovered when ultraviolet absorption lines from O VI were observed by the Copernicus satellite. (O VI is the spectroscopist's term for an oxygen atom with five electrons missing; such atoms only exist when the temperature is near several hundred thousand degrees.) What heats the interstellar medium to such high temperatures? Theorists are now puzzling over these new observations that further complicate the picture of the interstellar medium.

It is very difficult to untangle the different components of the interstellar medium with observations of 21-cm radiation and absorption lines, because the telescope observes all the components of the medium that are along the line of sight. Fortunately, other observational techniques can be used to study individual interstellar gas clouds.

★ 10.2 BRIGHT GAS CLOUDS ★

Some clouds of interstellar gas that are found near hot, bright stars emit visible radiation. These clouds are among the most photogenic objects in the sky. Perhaps the best known interstellar gas cloud is the Orion nebula (Figure 10.1, Color Photo 11), the middle "star" in Orion's sword. These objects are sometimes called *gaseous nebulae*; the word *nebula* is a relic of nineteenth-century astronomy and is a catchall term that was once used to describe anything that looks fuzzy when seen through a telescope. These gaseous nebulae are best seen from very dark skies, but Orion, the nearest and brightest, is quite rewarding even in a light-polluted urban sky. Figures 10.3 and 10.4 and Color Photo 12 show more of these beautiful objects. Pause and look at the pictures for a few minutes; in the remainder of this section we will analyze this particular type of gas cloud.

FIGURE 10.3   Messier 8, the Lagoon nebula, in Sagittarius. This is an H II region, a cloud of glowing ionized gas. (© by the Association of Universities of Research in Astronomy, Inc. Kitt Peak National Observatory.)

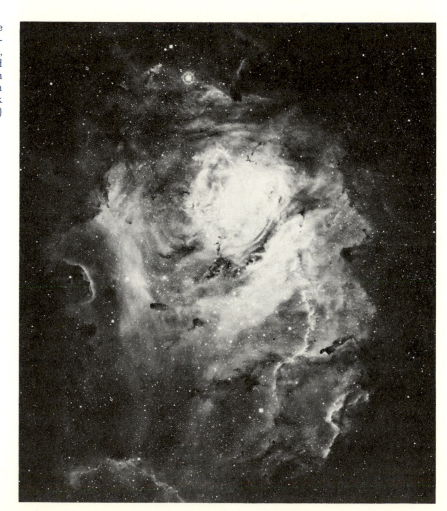

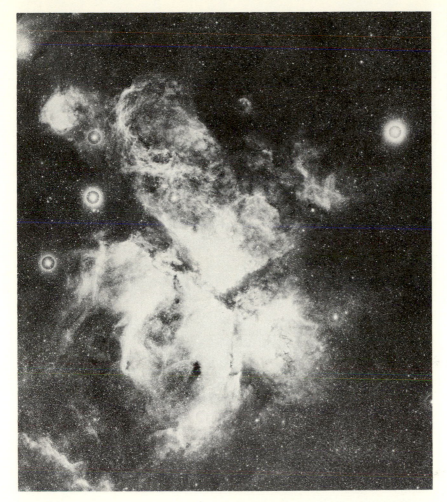

FIGURE 10.4   The Eta Carina nebula, one of the brightest H II regions of the southern sky. Note the dark areas, which are obscuring dust clouds. (Bart J. Bok, The Cerro Tololo Inter-American Observatory)

As usual, the first question to ask is how far away the gaseous nebulae are. The Orion nebula is associated with a cluster of stars that includes most of the stars in the Orion constellation. Comparison of the main-sequence stars in this cluster with others in our neighborhood produces a distance to the cluster and, hence, a distance to the nebula. The distance is 500 pc; from that, we can determine the diameter of the large glowing region in Figure 10.1 as 4 pc. The inner, extremely bright region shown in Color Photo 11 is somewhat smaller, only 1 pc in diameter. Most gaseous nebulae are similar in size to Orion.

The radical difference between a gaseous nebula like Orion and a star is illustrated in Figure 10.5, the spectrum of the Orion nebula. A stellar spectrum is a more or less continuous band of light, showing that the star is emitting radiation over a wide range of wavelengths. But in Orion, all the light from the nebula is concentrated at a few particular wavelengths, in *emission*

FIGURE 10.5 The spectrum
of the Orion nebula. Light
from the nebula is concentrated at a few specific
wavelengths, in emission
lines. These emission lines
are responsible for the colors
evident in Color Photo 11 of
the Orion nebula and Color
Photo 12 of the Trifid nebula. (Yerkes Observatory
photograph)

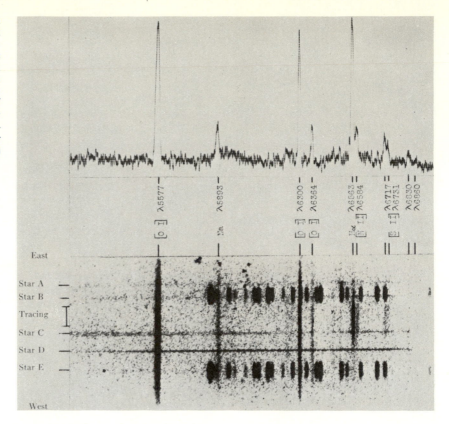

lines. The emission lines, concentrations of photons at particular wave-lengths or colors, are responsible for the vivid colors in Color Photos 11 and 12. The central region of the Orion nebula, overexposed to whiteness in Color Photo 11, looks green in a small telescope; the outer regions, visible in the color photograph, are red. What occurs in the Orion nebula to produce these vivid reds and greens?

The Orion nebula and other gaseous nebulae are huge clouds of ionized hydrogen gas. Some astronomers call these objects *H II regions,* after the spectroscopist's name for ionized hydrogen, H II. At the center of the Orion nebula are four bright, hot, blue stars, referred to as the "Trapezium" be-cause of their orientation. It is these stars that cause the nebula to glow.

How are the emission lines produced? The hydrogen line at 6562 Å, the line responsible for the red color of the upper left part of Orion, is a good example. Energetic photons from the hottest of the four Trapezium stars collide with hydrogen atoms in the nebula, knocking electrons away from atoms and producing hydrogen ions. An ion soon encounters another electron in the nebula, and if the encounter is close enough, the proton and electron *recombine,* forming a hydrogen atom once again. If the electron is in one of the atom's upper-energy states, it will drop down in energy, emitting a photon each time it descends the energy stairway (Figure 10.6). The energy, frequency, and wavelength of the emitted photon are determined by the energy intervals of the hydrogen atom. When the electron is in high-energy states, there is little difference between two successive energy levels, and it emits a radio photon. The vivid colors and the optical spectrum are produced further down the energy ladder; the 6562 Å red photon is produced when an electron drops from the third level to the second level. The green color of the Orion nebula is caused by emission lines from doubly ionized oxygen at 4958 and 5007 Å. Thus the optical radiation from the nebula is concentrated at particular wavelengths, determined by the energy-level structures of the atoms in the nebula. The same process of ionization and recombination occurs in a fluorescent light, which, like a gaseous nebula, has an emission-line spectrum.

Analysis of the emission lines and the continuous radiation from nebulae like Orion provides a fairly comprehensive picture of the structure of these objects. Most are about a few parsecs across, containing tens or hundreds of solar masses of ionized gas. Temperatures in these nebulae run around 10,000 K, and their density is a few hundred to a few thousand atoms per cubic centimeter. In all cases, a hot, luminous, blue, O-type star at the center of the nebula makes it shine.

The alert reader may have noticed that the shape of the Orion nebula is a bit irregular. Would it not be reasonable that all hydrogen atoms close enough to the Trapezium would produce emission lines by the sequential

FIGURE 10.6   The creation of emission lines in a gaseous nebula such as Orion. (a) A stellar high-energy photon strikes a hydrogen atom and ionizes it, removing the electron from the atom. (b) Electron and atom recombine. (c) and (d) As the electron cascades down the energy stairway, it emits photons. When it descends among the higher, closely packed energy levels, it loses little energy with each step and radiates in the radio range of the electromagnetic spectrum; when it descends from the third to the second level, it produces the red 6562-Å photons evident in Color Photo 11.

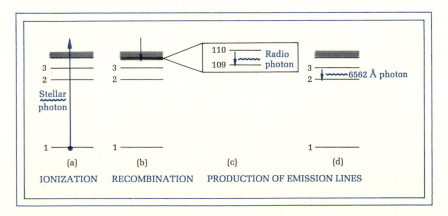

process of ionization and recombination illustrated in Figure 10.6? Such a supposition is quite reasonable. The dark bay at the lower right in Color Photo 11 and Figure 10.1 cannot be explained by the simple recombination picture above. Something is absorbing photons from that part of the nebula. Such dark absorption lines are seen in many other parts of the sky, and they are due to a ubiquitous component of the interstellar medium: *dust clouds*.

★ 10.3 DUST ★

Interstellar dust is the nemesis of any optical astronomer wishing to investigate star formation or the interstellar medium. Look ahead at Figure 13.1; notice how the Milky Way is slashed by black dust lanes. Dust is also evident in Figures 10.3 and 10.4, where the bright gaseous nebulae are obscured by dust in many places. Dust clouds also exist far from H II regions; Figure 10.7 shows a dust cloud with a rather unusual shape. At one time these clouds were thought to be openings or holes in the Milky Way, but now we know that they are simply gigantic aggregations of dust that obscure the visible light from the stars behind them. This dust hinders optical astronomers by absorbing visible light that would otherwise reach us. Since most distance-finding methods are based on optical astronomy, dust makes it difficult to map the Milky Way.

FIGURE 10.7   The Horsehead nebula in Orion. The bottom half of the photograph and the horsehead shape are regions of space where an aggregation of interstellar dust obscures the visible light from stars in distant parts of the Milky Way galaxy. At the top of the photograph, many more distant stars are visible. (Hale Observatories)

The dust clouds shown in Figures 10.3, 10.4, and 10.7 and in Color Photos 11 and 12 are known to be dust because these clouds scatter different colors of light unevenly. Dust particles tend to scatter blue light more easily than they scatter red light, and so red light can penetrate the dust but blue light cannot. Thus, light from distant stars appears redder when seen through a dust cloud. This phenomenon of *interstellar reddening,* perhaps more accurately described as interstellar "debluing," can be used to determine the amount of dust between us and a distant star. The same phenomenon occurs at sunset, when sunlight is scattered out of the sunbeam by the earth's atmosphere, making the sunset appear red (recall the discussion of sky color in Chapter 5).

These dust clouds can also be studied from the side in some cases. Suppose someone were observing a cloud like the Horsehead nebula from the side rather than from the front as we see it. If such a cloud were very close to a star, it would reflect a great deal of blue starlight and be visible as a reflection nebula. Just as the sky, produced by reflected sunlight, appears blue, a reflection nebula should also be blue. The bright stars in the Pleiades star cluster (Color Photo 13) are surrounded by blue reflection nebulae, a direct confirmation that dust scatters blue light more effectively than it scatters red light.

Astronomers have studied the composition and size of the interstellar dust particles by examining the way they scatter or absorb light at different wavelengths, particularly in the ultraviolet part of the spectrum, where the dust absorbs best. We compare two stars with similar spectra; if one of them is behind a dust cloud and the other is much closer to us, it is relatively easy to determine just what the dust does to starlight. Dust absorption peaks at a wavelength of 2200 Å, and such a peak is characteristic of graphite particles a few tenths of a micron in diameter. The rise in absorption toward 1000 Å indicates that silicate particles, particles with a composition similar to terrestrial rocks, are an important constituent of dust. Dust also emits in the infrared part of the spectrum, and silicates are indicated by strong emission at a wavelength of 10 μ.

★ 10.4 THE INTERSTELLAR MEDIUM: ★ A LARGE-SCALE VIEW

Observations of gas and dust thus provide a large-scale, bird's-eye view of the chaotic and diverse interstellar medium. Let's review what's there.

H I regions, vast clouds of neutral hydrogen, 1 to 10 pc across, observed by radio astronomers detecting the 21-cm line of neutral hydrogen;

The 10^5- to 10^6-*degree* coronal interstellar gas, observed by satellites detecting interstellar absorption lines;

H II regions like the Orion nebula, clouds of glowing ionized gas shining by transforming light from young hot stars at their centers into a dazzling array of colors and emission lines; and

Dust clouds, forbidding and mysterious, that hide the stars beyond and their interiors by absorbing visible light from stars.

How does this picture all fit together? A few years ago, most astronomers viewed the interstellar medium as a placid place, mostly filled with H I clouds emitting 21-cm radiation. Theorists predicted that an intercloud medium, some rarefied 10,000° gas, would fill the gaps between the H I clouds, but this intercloud medium has never been observed. The prevailing view was that the overall structure of the interstellar medium would not change as time passed.

Recent observations have provided a much more dynamic view of interstellar space, where the critical events are starbirth and stardeath. The birth or death of a massive star is accompanied by the emission of hordes of energetic ultraviolet or x-ray photons. Each one of these photons can ionize an interstellar gas atom. The heated, ionized gas expands, and the massive star has perhaps blown a bubble in the interstellar medium. The hot 10^6-degree gas may be what is left inside a bubble. The space between the bubbles is filled with irregularly shaped H I regions.

Don't take this model too seriously; it is a very tentative one and represents one way of interpreting recent discoveries. What is clear is that the interstellar gas is a chaotic, evolving place. Any attempt to characterize the interstellar medium as having a "typical" density, an average temperature, or any other unchanging property everywhere in space is doomed to failure. We are just beginning to uncover the events that cause the interstellar gas to expand, contract, and clump together.

★ 10.5 STELLAR NURSERIES ★

It is the clumping of the interstellar gas into relatively dense clouds that is the prelude to starbirth. Stars are continually forming in our galaxy; the young, massive stars at the center of the Orion nebula are less than 10 million years old. These stars formed when gravity caused a few tens of solar masses of interstellar gas to collapse from a cloud a few parsecs in diameter to a star a few million kilometers in diameter. The star is only about 10^{-8} times as large as the gas cloud it was born from.

How does starbirth occur? It is easy enough to say that gravity pulls matter together, and that stars form by gravitational collapse of interstellar clouds, but such a view lacks some very important details. What is it that determines when a cloud will collapse? Why do some clouds fragment into a few large, massive stars while others fragment into smaller stars? Do some clouds fragment into still smaller objects—planets—that can circle a newly formed star?

Astronomers are just beginning to attack some of the detailed questions concerning star formation. Optical observations shed little light on starbirth, because the interesting events take place in dense clouds from which light cannot escape. However, the longer wavelengths of the electromagnetic spectrum have proven to be a useful probe. Dust does not absorb infrared and radio radiation as effectively as it absorbs optical radiation, just as it absorbs short-wavelength blue light more than it absorbs long-wavelength red light. Our understanding of star formation has increased dramatically with the opening of these new regions of the electromagnetic spectrum. While infrared astronomers can observe thermal emissions from hot dust and gas, radio astronomers can observe radiation from molecules in these dark clouds. This section describes the different ways in which radio and infrared astronomers observe the dense clouds in which stars are born; the following section presents current ideas of how the diverse observations fit together into a scenario for star formation.

☆ RADIO OBSERVATIONS ☆
OF INTERSTELLAR MOLECULES

Radio astronomers have discovered a wide variety of molecules, compounds in which two or more atoms are chemically bound together, in interstellar space. Optical astronomers had discovered a few diatomic (two-atom) molecules that show interstellar absorption lines, but most molecules emit and absorb radiation best outside the optical part of the electromagnetic spectrum. A molecule can exist in different energy states when the atoms vibrate or when the molecule as a whole rotates in different ways. The energy difference between two rotational states of a molecule is generally on the order of 10^{-16} erg, and so the transition between two rotational states results in the emission or absorption of a photon in the short-wavelength part of the radio spectrum. A molecule can interact with a high-energy optical photon only by changing its electronic energy state, a rare occurrence in the cold, low-density regions of interstellar space. Consequently, the study of interstellar molecules is primarily the business of the radio astronomer.

In the 1960s, when molecular radio astronomy began, radio astronomers looked for diatomic molecules only. OH was discovered in 1963, but other hydrogen-containing molecules like CH and SH were unsuccessfully sought in the next few years. Since one atom will collide with another only once in 100 years in a typical H I region, no one suspected that three or more atoms could unite to form a more complex molecule under these conditions.

Yet a few groups of radio astronomers persisted, notably Charles Townes's team at Berkeley, and they extended the search to molecules with three and four atoms. Townes and his collaborators were rewarded when they discovered the 1.25-cm transition of ammonia in 1968. Ever since, radio astronomers have fiercely competed in the race to discover more and more molecules. At this writing some 40 molecules have been discovered; a partial list is given in Table 10.1.

Many of the compounds listed in Table 10.1 are rather common compounds on the earth. Of the diatomic and triatomic molecules, carbon monoxide (CO) is emitted by automobile engines, SO_2 is a common atmospheric pollutant, and H_2O is water. Of the more complex molecules, formaldehyde is familiar as embalming fluid to anyone who has been in a biology lab during a dissection experiment. Cyanides, compounds containing a CN bond, are often poisons, and these, too, are found in interstellar space. Perhaps the most intriguing compound, from a public relations viewpoint, is ethyl alcohol. The discoverers, a team from the University of Maryland and the University of Chicago, announced in their scientific paper that about 10^{31} fifths of it were spread over the galactic center. Since these 10^{31} fifths are

TABLE 10.1 A PARTIAL LIST OF INTERSTELLAR MOLECULES

2 atoms	H_2, OH, SiO, SiS, NS, SO, CH, CH^+, CN, CO, CS
3 atoms	H_2O (water), SO_2, HCN, H_2S
4 atoms	NH_3 (ammonia), H_2CO (formaldehyde)
5 atoms	HCOOH (formic acid), HC_3N (cyanoacetylene)
6 atoms	CH_3OH (methanol), $HCONH_2$ (formamide)
7 atoms	H_2CCHCN (vinyl cyanide), $HCOCH_3$ (acetaldehyde)
8 atoms	$HCOOCH_3$ (methyl formate)
9 atoms	$(CH_3)_2O$ (dimethyl ether), C_2H_5OH (ethyl alcohol, the active ingredient in alcoholic beverages)

distributed over 2×10^5 pc³, a galactic-center distillery would have to sweep a cube of space nearly 100,000 km on a side to collect enough ethanol to make a bottle of wine. Many of these interstellar molecules are part of the building blocks of living systems; while it is unlikely that any molecules would survive the rigors of star formation and be incorporated in planets, their abundances show that they can be formed under the relatively hostile conditions in interstellar space.

The frantic pace of molecular discoveries, accompanied by the race for priority among different teams, may well settle down in the next few years, for radio astronomers are now realizing the power of molecular radio radiation as a probe of the dark interiors of dense interstellar clouds. A molecule emits radiation when it collides with the most common interstellar molecule, molecular hydrogen (H_2), and is excited to a higher energy state. It can drop down to a lower energy state by emitting a photon that the radio astronomer detects, just as a hydrogen atom drops down to a lower energy state and emits a 21-cm photon. Some molecules, like carbon monoxide (CO), are relatively easy to excite and can be observed in relatively low-density regions, while others are more difficult to excite and are only observed in dense interstellar clouds where collisions are more frequent. Thus the types of molecules observed in a particular region of space can determine the density of that region. Further, the densest clouds, where molecules are most common, cannot be probed by the optical astronomer, for the dust that shields the molecules from destruction by stellar ultraviolet photons makes it impossible to see into these clouds. But it is here, in the densest clouds, that stars form. Molecular clouds are stellar nurseries.

☆ MOLECULAR CLOUDS ☆

Interstellar molecules are primarily found in dark, dense dust clouds, often located near the H II regions found around newly formed, hot, bright stars. Every molecular discoverer has a favorite hunting ground, and the Orion nebula and the H II regions near the center of the Milky Way galaxy have been the most productive sources of new molecular discoveries. The presence of molecules in dust clouds is no surprise; ultraviolet radiation from stars will destroy molecules unless they are hidden in dark dust clouds.

The Orion complex provides a fairly good example of the different types of molecular clouds that can exist (Figure 10.8). This complex is associated with the H II region, the Orion nebula, but is located behind it from our viewpoint. The largest cloud is an enormous, extended cloud, observed primarily in radiation from carbon monoxide molecules (CO). The large cloud is at least 8 pc across, with a density of about 10^3 H_2 molecules/cm³, inferred from analyses of the CO emission. This cloud contains a huge amount of material—at least 10^4 solar masses—in contrast to the 500 or so solar masses in the H II region.

The more exotic, more complex molecules in Orion are found in a far smaller, denser molecular cloud, about 0.3 pc across, located near the center of the visible nebula but behind it. This cloud is at least 100 times denser than the large CO cloud. It is here that collisions between the dominant H_2 molecules and the more complex, polyatomic molecules occur sufficiently often so that the large molecules are found in excited states that can emit radio photons. The separation between the large, relatively low-density clouds and the smaller, higher-density ones seen in Orion is typical. All these clouds are far denser than an average H I region observed at 21 cm.

At the inner core of the Orion complex is one of the most puzzling sources of molecular radiation, an interstellar *maser*. The word *maser* is an acronym

FIGURE 10.8 The molecular clouds associated with the Orion nebula. The gray region is the central bright H II region visible to optical astronomers and shown in Color Photo 11; the location of the four bright stars of the Trapezium is indicated by "T." The three molecular clouds, which lie behind the visible nebula, are indicated in color: the OH maser, the compact molecular cloud containing exotic molecules, and the CO cloud are of unknown extent, probably larger than that in the diagram.

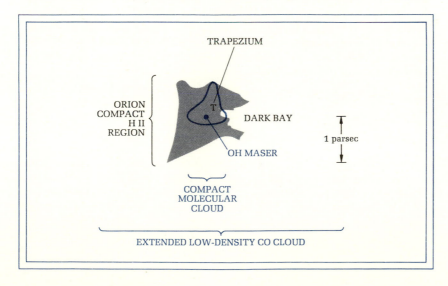

for microwave amplification by stimulated emission of radiation; the more familiar word "laser" is the same acronym, but the first word is "light." Molecular emission from a maser is so intense that it cannot be explained as the normal result of a collision between an OH molecule and a hydrogen molecule followed by the descent of the OH molecule to a lower energy state. Something must be dumping the OH molecule into an excited energy state, but what? Theoretical models, uncertain as they are, indicate that this maser must be very small, tens of astronomical units across, and very dense, with more than 10^8 atoms/cm^3, for the maser action to work.

☆   INFRARED OBSERVATIONS   ☆

Our analysis of the Orion complex has revealed three types of clouds emitting radiation from molecules: the huge, low-density cloud; the small, higher-density cloud in the middle; and the cloud responsible for the maser emission. Unfortunately, none of these clouds is visible optically, because dust is in the way. Again, the Orion complex can serve as a guide, since it is quite similar to other infrared sources associated with H II regions and molecular clouds.

The Orion region contains three small, intense infrared sources. One of them is identified with the compact H II region surrounding the Trapezium, the bright inner part of the nebula visible in Color Photo 11. Here, dust in the H II region is heated to a temperature of about 100 K. The infrared radiation that is seen is thermal emission from this hot dust. The dust in the Orion source emits more photons at a wavelength of 10 μ than a simple hot object would, and this emission is attributed to silicate compounds in the dust.

The other infrared sources in the Orion region are located at the center of the molecular cloud, near the OH maser. One of these sources is quite cool, with a temperature of about 70 K. The other, the Becklin-Neugebauer object (named for its discoverers) has a temperature of 500 K. Observations of this source seem to indicate that it is one of the best candidates for a protostar, a star that is forming right now. However, it is also possible that this object already exists as a star, with all its visible light emission being absorbed by a huge dust cloud that envelops it.

★   10.6 STAR FORMATION   ★

The preceding pages have described a number of different types of objects seen in stellar nurseries. The visible signs of star formation are H II regions, clouds of ionized gas surrounding hot, young stars. Radio astronomers observe molecular clouds, clouds of cool gas that range in size and density from the huge sources of CO emission to the small, dense OH masers. Presumably, these different objects represent different stages of the evolutionary sequence leading to star formation. How are they linked? Two astronomers active in this field have summarized the dilemma in a recent review article:

> How do stars form? In the days when observational astronomy was entirely optical, it was hoped that large new telescopes such as the 200" would lead us to the answer. Astronomers such as Bok and Herbig struggled valiantly amid the muck. But, alas, the interstellar dust could not be pushed aside, and try as they might, optical astronomers could barely penetrate the surface of vast cloud complexes in which new stars, in various stages of development, are deeply embedded. The spectacular rise of molecular radio astronomy and infrared astronomy has allowed us to pull back the dusty veil and view the grand spectacle as it unfolds. Whether we shall be clever enough to understand the message remains to be seen.[1]

We are not yet clever enough to understand the whole story, but a general consensus has developed among astronomers, and it agrees with the broad outlines presented in this section. This scenario for star formation, described below and sketched in Figure 10.9, is another working hypothesis, to be supported or shot down as more facts come in.

The starting point is an H I region of the interstellar medium, perhaps located in a giant cloud complex. Something, no one yet knows what, causes these clouds to become compact enough so that the prime agent in star formation, gravity, can compress them still further. Every atom in an H I region

FIGURE 10.9 Different types of objects representing different stages in star formation. Evolution proceeds from left to right, when gravity causes clouds to contract and to increase their density. Many of the details of this progression from interstellar cloud to star are not understood in detail.

OBJECTS	H I regions	Dark clouds	CO envelopes	Dense molecular clouds	Masers	Stars	H II regions
MASS (Solar masses)	300	100	10^4	100	10^{-6}?	1	300
DENSITY (atoms/cm^3)	10	10^3	10^4	10^5	10^8	10^{23} (interior) 10^{14} (surface)	10^5
TEMPERATURE (K)	70	10	20	30–100	10–1000	10^6 (interior) 4000 (surface)	10,000
 APPEARANCE	RADIO	OPTICAL	RADIO	RADIO	RADIO	OPTICAL	ORION

[1] B. Zuckerman and P. Palmer, "Radio Radiation from Interstellar Molecules," *Annual Review of Astronomy and Astrophysics* 12 (1974): 295.

feels a relentless pull toward the center, and it falls toward the middle of the cloud. The central region increases in density under the action of gravity. A sufficient density increase brings the dust grains in these clouds together, and a dark optical cloud like the Horsehead (Figure 10.7) may evolve.

When the density of one of these dark clouds reaches 10^4 particles/cm^3, collisions between hydrogen molecules and CO molecules occur often enough so that the CO molecules produce radio radiation. The CO clouds that have been observed are extremely large, containing 10^4 to 10^5 solar masses of material. Presumably, they evolved from large H I complexes, but this evolutionary sequence is one of the many details that are still obscure.

Gravity has not yet stopped, since at the center of these large clouds the densities are even higher. The dense molecular clouds, like the one observed in the Orion nebula, have densities of 10^5 or even 10^6 molecules/cm^3. These small clouds condense within the larger, extended CO envelopes, since gravity tends to pull matter together toward the center of any cloud. A number of infrared sources are often associated with these molecular clouds.

For the next stage, we must turn to the theorists. A number of investigators have made computer calculations of these early stages of stellar evolution. All calculations indicate that once the densities characteristic of these molecular clouds are attained, gas molecules will fall freely toward the center of the cloud, building up a dense ball of gas in the short time of 10^3 to 10^4 years. This dense ball of gas is generally referred to as a *protostar*. When it first forms, it is still embedded deep within the surrounding cloud, so it is extremely difficult to observe. The theoretical calculations differ when each theorist tries to determine details like the temperature of the protostar, but most indicate that the temperature rises from the 100-K temperature of the parent cloud to about 1000 K.

Have we observed protostars? This question is perhaps one of the most important unanswered questions about star formation. The OH masers are found in clouds that certainly look as if they should be associated with protostars (Figure 10.9); their densities and dimensions are between those of dense molecular clouds and those of stars. Yet the OH masers are not the stars themselves; they are found in clouds of dense gas, tens of astronomical units in size, hundreds of times larger than a main-sequence star. The best protostar candidates are objects such as the Becklin-Neugebauer infrared source associated with the Orion complex and a similar object in the constellation Ophiuchus. Most astronomers hope that these are protostars; it is also possible that they are stars that have *already* formed but that look much cooler than stars because they are hidden by the protective blanket of dust.

A star first becomes visible when the dust begins to blow away; how it appears depends on the mass of the newly born star. Massive stars are generally surrounded by massive clouds, and the star is well on the main sequence before the surrounding cloud dissipates. The massive main-sequence star is a strong producer of ultraviolet photons that can ionize the surrounding gas, producing an H II region like the Orion nebula.

A lower-mass star, surrounded by a smaller cloud, may well dissipate the cloud before it reaches the main sequence. Here we can once again make contact with the real world of observations after our venture into the uncertain domain of theoretical calculations that disagree with each other on details. Figure 10.10 shows a Hertzsprung-Russell diagram with some evolutionary tracks of stars in their pre-main-sequence phases. At some point on each track, the surrounding placental cloud is blown away and the star becomes visible. Two distinct classes of pre-main-sequence stars, the *T Tauri stars* and the *Herbig emission stars,* are shown on the diagram.

Both classes of stars share one characteristic of H II regions: Their spectra contain emission lines, indicating that these stars are surrounded by small clouds of ionized gas. In the T Tauri stars, measurements of Doppler shifts indicate that the ionized gas is flowing away from the star, being pushed

FIGURE 10.10   A Hertzsprung-Russell diagram showing, in black, the evolutionary paths followed by stars in their pre-main-sequence stages. In the dashed portions of the tracks, stars have not yet emerged from the placental clouds that formed them and, as a result, are invisible. The shaded regions, in color, show two types of stars observed in their pre-main-sequence stages. Very massive stars emerge from the surrounding dust clouds only after they reach the main sequence and, if sufficiently massive, are surrounded by visible clouds of ionized gas like the Orion nebula. Numbers along the main sequence show the mass of each star in solar masses.

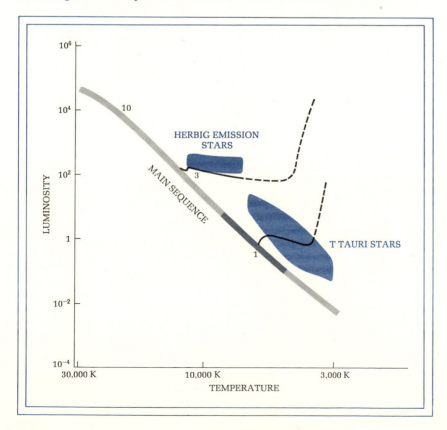

outward by radiation from the star. Herbig emission stars have a somewhat similar appearance, but they are found in a part of the H-R diagram that indicates they are the precursors of more massive main-sequence stars.

It is possible that in at least one case the emergence of a star from its placental cloud has been observed by astronomers. FU Orionis, a star in Orion, brightened by five magnitudes in 1936 and has since remained bright. The current interpretation of this event is that FU Orionis blew the surrounding dust cloud away and has emerged as a new star. Another star, V1057 Cygni, has brightened in a similar manner.

<h2>★ 10.7 FORMATION OF THE PLANETS ★</h2>

When the sun emerged from its placental cloud, it was probably a T Tauri star. Did the mass outflow blow all the surrounding material away? In the sun's case, it definitely did not; the planet we live on is a remnant of this placental cloud. Our current ideas regarding the origin of the solar system indicate that the planets were formed at the same general time as the sun. Many uncertainties still remain in our theoretical picture, but its overall form seems clear.

The problem of planetary formation is perhaps one of the oldest in the general field of stellar evolution. Pierre Simon de Laplace, an eighteenth-century French mathematician, first proposed the currently accepted theory in its broadest outlines. He visualized the beginnings of the planets in a rotating gas cloud, now called the *solar nebula*. The influence of gravity formed a small condensation in the center, which grew to stellar dimensions and eventually became the sun, following the scheme described in the last section. The remnants of the solar nebula then formed a disk, rotating in the same direction as the original cloud. The planets formed from this disk. This skeletal scheme is depicted in Figure 10.11.

Laplace's original scheme sounded reasonable, but like many theories there was no way to prove it true or false for almost 200 years. In a sense, it wasn't even a scientific theory, for it lacked the hallmark of a scientific model—confrontation of the theoretical model with the reality of observations. It was a nice idea, for it meant that the formation of planets would be a relatively frequent occurrence. But other theories could be possible. In the early twentieth century, two British astronomers, Chamberlin and Moulton, proposed an alternative theory, now discredited, that the planets were torn out of the sun by an encounter with a passing star. If the Chamberlin-Moulton theory were correct, planets would be rare because stellar encounters are very infrequent.

In the last decade or so, the skeletal picture originally proposed by Laplace has been fleshed out considerably. Theoretical pictures of the origin of planets have begun to make some contact with observations. While there are still many uncertainties, many unanswered questions, the general picture has made considerable progress in the long road from philosophical speculation to scientific model. Meteorites, in particular, have proven to be excellent probes of the solar nebula.

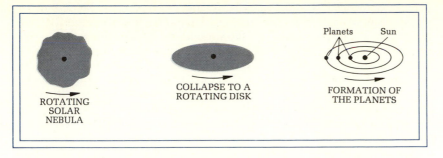

FIGURE 10.11   Origin of the planets. A rotating cloud collapses under the influence of gravity. The center condenses and eventually becomes the sun. The rest of the cloud collapses to form a disk, which eventually condenses into planets. Section 10.7 of the text adds some flesh to the skeletal picture shown here.

☆ **CHEMICAL CONDENSATION IN** ☆
THE SOLAR NEBULA

Perhaps the best understood process in the formation of planets is the growth of small, centimeter-sized particles out of the originally gaseous solar nebula. John S. Lewis and his collaborators of MIT have made extensive calculations of the kinds of compounds that form from a cooling solar nebula. The general picture is that the solar nebula near the sun was quite hot, heated by the radiation of the newly formed star. Further away from the solar nebula, the gas was cooler. Detailed calculations of solar-nebula models by Harvard's A. G. W. Cameron and his collaborators indicate that this general decrease of temperature with increasing distance from the sun is characteristic of a wide variety of solar-nebula models. The temperatures in Cameron's model are in general agreement with our observations of the bulk compositions of different planets.

In the inner solar system, temperatures were quite hot, ranging from 1500° near the orbit of Mercury to about 500° near the orbit of Mars. At Mercury's distance, the only elements that could condense were iron and high-melting-point compounds such as calcium and aluminum oxide. Mercury's high density agrees with this picture; Mercury is the densest of all the planets, and its density of 5.4 (Chapter 4) agrees with the density of material that would condense out of a high-temperature solar nebula. Venus, Earth, and Mars formed in somewhat cooler regions, where iron would condense not as a pure metal but as iron oxide. Earth and Venus are both relatively dense planets, but proper accounting for the compression of matter by the gravitational forces of these planets produces densities in agreement with the presumed condensation temperatures of 1000 K and 600 K. The theory can even account for the presence of other chemical compounds such as water on the earth's surface. One minor puzzle is the moon, for its composition indicates it did not condense at the same time and place as the earth.

271

In the outer solar system, the cooler temperatures of 150 to 200 K (not too much greater than the temperatures of molecular clouds; see Figure 10.9) indicate that at some distance from the sun ice should start to condense from the cooling gas. The model is more difficult to check here because far less is known about the outer solar system; but the icy composition of many of the satellites in the outer solar system (Chapter 6) does indicate that the outer parts of the solar system were quite cool. If the methane on the surface of Pluto, recently detected by infrared observations (Chapter 6), indicates that the interior of Pluto contains methane, it must have condensed at a temperature below 60 K.

The only direct samples we have of material from a rather different part of the solar system are the meteorites, and analysis of the composition of these objects provides still more insight into the temperatures in the solar nebula. Once again the question is, "What was the temperature at the point in the solar nebula where this object condensed?" Knowledge of the chemical reactions taking place in the *condensation* process allows us to conclude, for example, that the carbonaceous chondrites (Chapter 7), those samples of primitive solar-system material, condensed where the temperature was between 350 and 380 K. Water is chemically trapped in these minerals. If the temperature were too high, little water would be trapped; if it were too low, too much water would be trapped, and some ice would even have formed. This temperature agrees with the temperature that one expects for the asteroid belt.

Analysis of the chemical reactions that take place in the solar nebula provides us with some relatively firm facts about the conditions that prevailed when solid bodies were formed from the gas of the solar nebula. The distinction between rock objects in the inner solar system and icy objects in the outer solar system indicates that the inner solar nebula was hot and the outer solar nebula was cold. This difference parallels the difference between Jupiter's rocky inner satellites, Io and Europa, and the icy outer ones, Ganymede and Callisto (Chapter 6). It would be very difficult to account for this difference between icy and rocky objects with any other theory of the origin of the solar system, given the agreement of the model with detailed analyses of the composition of the primitive meteorites, the carbonaceous chondrites.

☆ GROWTH TO PLANETARY DIMENSIONS ☆

Thanks to the chemical condensation calculations just discussed, we can follow the evolution of the solar nebula as far as the growth of small, centimeter-sized solid bodies with some confidence that the model is correct. Planets, however, are far larger than these small bodies that can be produced by chemical condensation. How did they form?

Solid bodies cannot be held above the plane of the ecliptic, the equatorial plane of the rotating gas cloud, by the pressure of the gas in the solar nebula. Calculations show that they settled down to form a very thin disk, about 0.01 AU thick, in the plane in which the planets now orbit. This settling is a direct result of the rotation of the original cloud; centrifugal forces hold the orbiting particles outward in their plane of rotation. One of the dynamic regularities of the solar system, the fact that the planets all revolve around the sun in the same direction and in almost the same plane, is readily explained.

Calculations by several investigators, notably Cal Tech's Peter Goldreich and Harvard's William Ward, bridge the next gap—that between small solid bodies and asteroidal bodies. Again gravity is the contracting agent; they were able to show that a thin disk of small bodies would break up under the influence of its own gravity and quickly form kilometer-sized bodies.

The next gap, growing from dimensions of 1 km to about 10^4 to 10^5 km, is the most difficult one to cross at this time. The general picture is that the small planetesimals collide with each other and stick, gradually building up a planet made of 10^{12} of these small objects. Various investigators have tried to calculate just how and when this event would happen, and how long it would take. Problems arise: The calculations indicate that Uranus and Neptune could never grow to their present sizes in the 4.5 billion years that the solar system has existed. Yet the calculations are uncertain, and probably some forgotten factor can make Uranus and Neptune grow fast enough. It is here that some puzzling regularities are left unexplained. Why do all the planets that have not had their rotation slowed by tides rotate with periods between 10 and 30 hours? Why do they rotate in the direction in which they orbit the sun (except for Uranus)? What explains Uranus's peculiar motion? Why are the planets spaced according to Bode's law?

Another intriguing puzzle is the origin of the comets. Their icy composition clearly indicates that they were not formed in the inner solar system, where the earth is. But were they formed somewhere near Jupiter and then tossed out to their present location in the comet cloud tens of thousands of astronomical units away from the sun? Or were they formed out in the comet cloud where they now spend most of their time? Most researchers believe that they were formed in the comet cloud, almost a quarter of a light year from the sun, and are thus samples of protostellar material.

The uncertainties in our model for the evolution of the solar system should not obscure the achievements of the model as originally proposed by Laplace and fleshed out in the last two decades. The progression from rocky inner planets to icy outer ones is a logical consequence of the temperature distribution in the solar nebula, and detailed examinations of meteorites substantiate this picture. The fact that the planets all orbit in the same direction in the same plane is also explained without much difficulty. Gas clouds similar to the solar nebula should exist around other stars, and so our reconstruction of the origin of planets indicates that planetary systems should be rather common in the universe. Life exists on planets, so the prevalence of planets indicates that life sites do exist elsewhere. Do such planets actually support living beings? That question is reserved for Chapter 16.

SUMMARY

An investigation of the path of events that stars follow when they form by the gravitational collapse of interstellar gas clouds has revealed a wide variety of objects. Gas clouds of all sizes exist in interstellar space, and star formation follows a path from the comparatively rarefied H I regions, through the dense molecular clouds, onto the main sequence. While the early stages of star formation are beginning to emerge from the thick veil of dust that hides them from optical astronomers, the protostar stage, immediately preceding the main sequence, is not yet in sight. The ability of infrared and radio radiation to penetrate the interstellar dust has allowed astronomers working in these parts of the electromagnetic spectrum to uncover parts of the star-formation story. The formation of planets in our solar system was a natural event that has probably occurred elsewhere.

This chapter has covered:

The large-scale state of the interstellar medium;

The physical characteristics of H I regions, observed by radio astronomers, and glowing gaseous nebulae or H II regions;

The different types of molecular clouds that can exist;

Infrared sources that are associated with molecular clouds;

A hypothetical scenario for star formation; and

Another scenario describing the formation of the planets.

KEY CONCEPTS

Chemical condensation	H II region	Protostar
Dust clouds	Herbig emission star	Recombination
Emission lines	Interstellar absorption lines	Solar nebula
Gaseous nebulae	Interstellar reddening	T Tauri star
H I region	Maser	21 cm
	Nebula	

REVIEW QUESTIONS

1. Why is the wavelength 21 cm of special interest to radio astronomers?

2. What is the difference between an interstellar absorption line and a stellar absorption line? Are they formed by the same process or different processes?

3. Why does the observation of interstellar absorption lines from O VI, five-times-ionized oxygen, indicate the presence of a hot gas in interstellar space?

4. What is an emission line?

5. Suppose that none of the stars in the Trapezium produced any high-energy ultraviolet photons. Would the Orion nebula still be visible? Explain.

6. Why is the nebulosity around the Pleiades blue in Color Photo 13?

7. What is the origin of the hot interstellar gas?

8. Why are molecules not found in low-density H I regions?

9. Describe the three different regions in which molecular emission is produced by clouds associated with the Orion complex.

10. What is the best candidate for a protostar?

11. What does gravity do to cause star formation?

12. How do T Tauri and Herbig emission stars fit into star formation?

13. Why is it reasonable to find rocky planets in the inner solar system and icy planets in the outer solar system?

FURTHER READING

Cameron, A. G. W. "The Origin and Evolution of the Solar System." *Scientific American* 233 (September 1975): 32–41.

Chaisson, E. J. "Gaseous Nebulae and their Interstellar Environment." In *Frontiers of Astrophysics,* edited by E. H. Avrett. Cambridge, Mass.: Harvard University Press, 1976, pp. 259–351.

Strom, S. E. "Star Formation and the Early Phases of Stellar Evolution." In *Frontiers of Astrophysics,* edited by E. H. Avrett. Cambridge, Mass.: Harvard University Press, 1976, pp. 95–117.

★ CHAPTER ELEVEN ★

LATE STAGES OF STELLAR EVOLUTION

Most of the stars in the sky are main-sequence stars because this stage of stellar evolution lasts a long time. Yet from an evolutionary point of view, the main-sequence stage is relatively uninteresting, since the temperature and radius of the star scarcely change until the central hydrogen runs out. The drama of stellar evolution, evident in star formation, does not show up again until the central hydrogen is exhausted. Then the star's core contracts, its surface expands, and it becomes a red giant. The most luminous stars in an old star cluster such as Messier 3 (Figure 11.1) are red giants.

The evolution of low-mass stars like the sun can be followed a little further. These stars lose the huge envelopes characteristic of the red-giant stage and end their lives as *white-dwarf stars*, the tiny cores that formerly were the centers of red giants. One important group of giant stars, the *Cepheid variables*, have proven to be very valuable distance indicators because their light output varies regularly, in a way that is related to their luminosity.

★ MAIN IDEAS OF THIS CHAPTER ★

A star leaves the main sequence when its core contains no more hydrogen fuel.

Following the main-sequence stage, a star becomes a red giant when its luminosity and size increase.

The different fuels available in the core and gravity determine the evolution of red-giant stars.

Pulsating red-giant stars can be very useful as distance indicators.

FIGURE 11.1   The globular cluster Messier 3. The most luminous stars in this old star cluster are red giants, stars tens to hundreds of times larger than our sun. (Hale Observatories)

The late evolution of binary systems can differ greatly from the late evolution of single stars.

★ **11.1 LEAVING THE** ★
MAIN SEQUENCE

Look back at Table 9.1 and its accompanying H-R diagram, Figure 9.8. Many of the brightest stars in the winter sky—Capella, Rigel, Betelgeuse, and Aldebaran—are tens to thousands of times larger than the sun or other main-sequence stars. These stars are known as *red giants* because of their large size, and some extremely large and luminous stars are known as *supergiants* (Figure 11.2). In the opposite corner of the H-R diagram are the *white dwarfs*, tiny stars no larger than the earth. These two corners of the H-R diagram are filled by stars in very late evolutionary stages. What causes

FIGURE 11.2 *Left:* A Hertzsprung-Russell diagram, showing the locations of supergiants, red giants, the main sequence, and white dwarfs. *Right:* A comparison of the size of two giant stars, Capella and Aldebaran, and the sun.

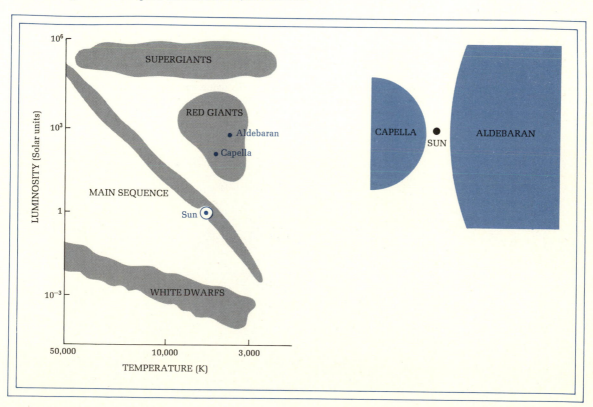

a star to leave the main sequence? In this section we examine the general trends of a star's evolution through the red-giant stage.

☆ CENTRAL-HYDROGEN EXHAUSTION ☆

Any main-sequence star—our sun, the red glowworm Epsilon Eridani, or the luminous Sirius A—is a ball of mostly hydrogen gas that sustains itself by fusing hydrogen in its center. Energy generated in the stellar core balances energy lost in the form of starlight (Figure 11.3). Now that the question is not just what is a star (Chapter 9), but how stars evolve, we ask whether this happy state of affairs can last forever. Eventually the hydrogen fuel that supports the central energy source will run out. When central hydrogen is exhausted, the star's condition changes drastically as it seeks some other energy source to replenish the energy lost as starlight.

The immediate cause of a star's migration off the main sequence is the abrupt contraction of the central core. The central fire flames out, and hydrogen fusion begins in a shell surrounding the center. This shell does not heat the center of the star, and the only way the center can sustain the necessary pressure is by contracting. Contraction of a mass of gas heats the gas, so the continued contraction of the central core can provide the heat needed to maintain the pressure that balances the weight of the envelope. The structure of the center of a star at this stage of evolution is shown in Figure 11.4.

FIGURE 11.3 Structure of a main-sequence star. Energy produced by hydrogen fusion in the core balances the outflow of starlight. Note that the core is called a *hydrogen-burning* core even though it is not chemical combustion, normally called "burning" in terrestrial contexts, that provides the energy. [Drawing based on a model by I. Iben, Jr., *Astrophysical Journal*, 147 (1967): 624.]

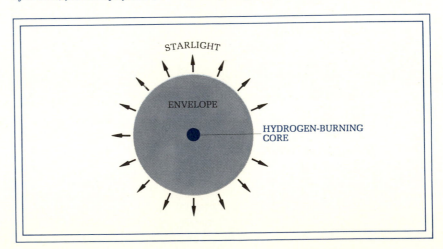

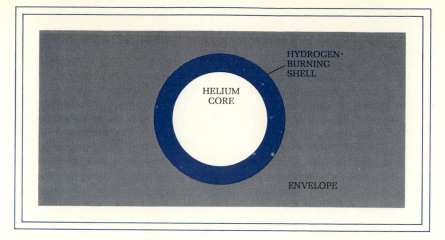

FIGURE 11.4   The structure of the central core of a star shortly after it leaves the main sequence. All energy is being generated in a shell surrounding the helium core; the core itself must contract to maintain the necessary central pressure. The core of a red giant is very small; on the scale of this drawing, the surface of the star could be several yards across. [Drawing based on a model by I. Iben, Jr., *Astrophysical Journal*, 147 (1967): 624.]

The contraction of the core of a star evolving off the main sequence is paradoxically accompanied by the expansion of the outer envelope. Box 11.1 describes the causes of this rather strange phenomenon. The star swells up tremendously; Figure 11.2 illustrates the proportions of some relatively modest giants in comparison to the sun. The increased core temperatures lead to an increase in the star's luminosity. The star, increasing both its size and luminosity as it leaves the main sequence, heads to the upper right corner of the H-R diagram, becoming a red giant like Aldebaran. (This phase of stellar evolution is called the "red-giant stage," even though some giant stars, like Capella, are not strictly red.)

☆   MAIN-SEQUENCE LIFETIMES   ☆

Stars remain in the main-sequence stage of stellar evolution for the largest part of their life cycles. Any particular stage of stellar evolution is characterized by the energy source that balances the outflow of energy as starlight; main-sequence stars are those that fuse hydrogen in their cores (Figure 11.3). In later stages of stellar evolution, hydrogen is no longer fused in the core to provide energy, and other energy sources come into the picture. The duration of any particular stage of stellar evolution is determined by the amount of available energy that a particular mode of energy production can supply and the rate at which that energy is used up.

Box 11.1 THE
EXPANSION OF A
RED GIANT

When the core of a red giant contracts, the envelope expands.
This curious behavior is caused
by the chemically inhomogene-
ous structure of a red-giant star,
which in all stages consists of a
core of helium or heavier elements surrounded by an envelope
of a lighter element, hydrogen.
Atoms in the core, containing
more mass, are heavier; gravity
acts on them more strongly. As a
result, the core must have a very
high pressure to hold these heavy
atoms up. This same high pressure pushes the lighter hydrogen
atoms (in the envelope) outward,
because the pressure is more than
enough to support the lightweight hydrogen atoms.

The expansion of a red giant is
accompanied by an increase in
luminosity. This increase in luminosity results from the increased core temperature and the
lowered envelope density. A
hotter core produces more radiation, which can flow through the
inner regions of the star. In a
main-sequence star, the thick envelope is a good insulator that
blocks the flow of energy through
the star, but a tenuous red-giant
envelope is a poor insulator, and
energy can flow more freely to the
surface. As long as the chemical
difference between the core of
heavy atoms and the envelope of
light hydrogen atoms persists, a
star remains large and luminous,
a red giant in the upper right
corner of the H-R diagram.

The length of time during which a star remains on the main sequence depends on the star's mass. A massive main-sequence star, 10 times as massive
as the sun, for example, is 6000 times as luminous as the sun. It is therefore
consuming its hydrogen fuel 6000 times more rapidly than the sun, because the energy supplied by hydrogen fusion in the center must exactly
balance the loss of energy observed as starlight and measured by the star's
luminosity. Even though a 10-solar-mass star has 10 times as much available
fuel as the sun has, it is consuming it so fast that it cannot stay on the main
sequence as long as the sun. Specifically, if we note that the main-sequence
stage lasts until about 10 percent of the mass of the star is transformed by fusion into helium, then the sun has 0.1 solar mass of fuel, and the 10-
solar-mass star has 1.0 solar mass of fuel. The larger star is consuming its
fuel at 6000 times the sun's rate, so it lasts on the main sequence for 1/600 of
the time that the sun does.

The differing main-sequence lifetimes of stars have a marked effect on the
appearance of stars in star clusters of differing ages. Consider the old globular cluster M 3 (Figure 11.1) and the young cluster, the Pleiades (Color Photo
13). The H-R diagrams of these two clusters are shown schematically in Figure 11.5. The Pleiades, a very young cluster, contains no red-giant stars, and
its main sequence extends to relatively high temperatures and luminosities.
In the 10^8 years that have passed since the Pleiades formed, few stars have
evolved off the main sequence. (Other young clusters contain a few red
giants.) The old cluster M 3, called a *globular cluster* because of its shape,

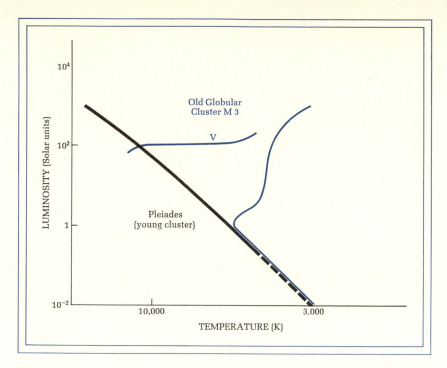

FIGURE 11.5   A Hertzsprung-Russell diagram of a young star cluster, the Pleiades
(Color Photo 13), and the old globular cluster M 3 (Figure 11.1), plotted schematically.
The dashed black line shows where the Pleiades main sequence would be if these
cool main-sequence stars had formed by this time. The Pleiades stars are about 10^8
years old, while M 3 is over 10^{10} years old. The V notes where the RR Lyrae variable
stars (section 11.4) are. This diagram illustrates that young clusters like the Pleiades
have few red giants and still contain hot main-sequence stars, while old clusters like
M 3 have no hot main-sequence stars and many red giants. [Source: Pleiades data from
R. I. Mitchell and H. L. Johnson, *Astrophysical Journal*, 125 (1957): 418, transformed
using the temperature scales of Figure 9.9; M 3 data based on A. Sandage, *Astrophysical Journal*, 125 (1957): 435, transformed by the temperature scales of Figure 9.9
and Table 9.1. The transformations change the original data, given as colors and
magnitudes, into temperatures and luminosities.]

has an extremely different H-R diagram; the main sequence terminates about
where the sun is. M 3 has a large number of stars in the red-giant regions of
the H-R diagram; the most luminous stars in this cluster are the red giants,
not the blue main-sequence stars.

The end of the main-sequence stage of stellar evolution is marked by the
exhaustion of the central hydrogen supply. We don't completely understand
the life cycle of a star after it leaves the main sequence, for many complex
factors enter any theoretical calculation. The next section describes, in
broad outlines, the evolution of stars through the red-giant stages. Section
11.3 can be skipped by readers in a hurry; it describes our current understanding of the details of red-giant evolution as they affect the future evolution of the sun.

★ 11.2 RED-GIANT EVOLUTION ★

☆ THE HELIUM-BURNING STAGE ☆

The first stage of red-giant evolution, immediately following the main-sequence phase, is a contraction of the helium core accompanied by a rapid movement of the star toward the upper right corner of the H-R diagram, caused by the star's increasing radius and luminosity. This rapid evolution comes to an end when the central core becomes hot enough to fuse helium. When the temperature reaches 100 million (10^8) K, about seven times the present interior temperature of the sun, three helium nuclei can collide and stick together, forming a carbon nucleus. This fusion reaction, like the fusion of four hydrogen nuclei or protons to make one helium nucleus, produces energy that can heat the central core and provide the central pressure necessary to balance the weight of the envelope. Figure 11.6 depicts the interior of a star at this stage of stellar evolution.

Red-giant evolution is the result of the interplay of two strong and opposing tendencies: nuclear fusion and gravitational contraction. Gravity is the relentless commander of stellar evolution, continually sending out its orders to contract. Every atom in the star is pulled toward the middle, tending to increase the central density. If there were no other player in the game, this contraction would last indefinitely and the star would end its life as a black hole, an object so dense that nothing, not even light, can escape. But there is another player in the red-giant evolution game: nuclear fusion reactions that provide heat and pressure to the central core. The pressure provided by the energy from fusion opposes the tendency of gravity to

FIGURE 11.6 A rough sketch of the core of a 1.5-solar-mass star in the core-helium-burning stage. On this scale, the surface of the star would be 2 meters across. While no models for the sun in this evolutionary stage are available, its interior should be similar. [Freely sketched from data of D. J. Faulkner and R. D. Cannon, *Astrophysical Journal,* 180 (1973): 435.]

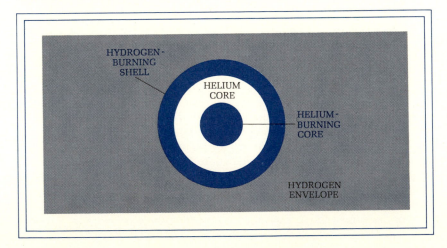

compress the core. As different fusion reactions are brought into play, the star's core postpones the orders of gravity in different ways. The helium-burning stage of stellar evolution represents one such postponement of the tendency of stars to contract under the effects of gravity.

☆ LATER STAGES OF ☆
RED-GIANT EVOLUTION

Eventually, the helium in a red-giant core is exhausted, just as the hydrogen became exhausted in the main-sequence stage. The core is now made of carbon, and it contracts just as the helium core contracted earlier. In some stars, the carbon core can become an energy source, when the temperature rises to the point at which carbon can fuse and form magnesium. Again, this energy source cannot support the star indefinitely. In some stars, the products of carbon fusion, magnesium and neon, can serve as the raw materials for further fusion reactions. How far fusion reactions progress depends on the temperature that the star's center can attain. Heavier nuclei contain more protons, which tend to keep the nuclei apart and inhibit them from fusing unless the temperature is extremely high.

Red-giant evolution thus can terminate when a star's core cannot become sufficiently hot to fuse the material that the core contains. Even the largest stars, with the highest central temperatures, cannot sustain their fusion reactions indefinitely, for there is no more energy to be extracted from atomic nuclei once the fusion reactions have converted the core material into iron. Iron is the end of the line as far as fusion is concerned, for the fusion of two iron nuclei does not release energy; it absorbs energy.

☆ GRAVITATION AND THE FINAL STATE ☆

So far in stellar evolution, from the main-sequence stage through the red-giant stage, the star has been kept from collapsing under its own weight by the pressure produced by the heat in its central core. Nuclear-energy generation has been able to replenish that heat. But once the nuclear fusion reactions end, the core must shrink under the weight of the overlying layers. The relentless command of gravity must be coped with in a more permanent way, because the temporary respite provided by nuclear energy is over. The final state of a star is determined by the way in which it copes with gravity.

In many stars, a different kind of pressure can successfully resist the tendency of gravity to compress the core. A long-lasting solution to the gravity problem requires a form of pressure that does not depend on heat, because there is no more fuel to provide heat to the stellar core. A kind of pressure that fulfills this requirement is *degeneracy pressure*, produced by an intrinsic resistance to compaction of the particles composing matter. Degenerate matter is very dense; under stellar conditions, a density of 10^6 g/cm^3 or several tons per cubic inch is required if degeneracy pressure is to provide significant resistance to compaction. A gas of degenerate electrons is like a box full of billiard balls; to compress the box, you would have to squash the billiard balls, not just push them closer together.

One way in which a star can end its life is to bring degeneracy pressure from electrons into play. The cores of low-mass red giants contain degenerate electrons that can resist further compression no matter how hot or cold the core is. The details are not understood, but it is clear that some red giants lose their outer envelopes in the later stages of red-giant evolution and leave the cores as their corpses. Such cores are called *white-dwarf stars*. These stars do not collapse any further, because degeneracy pressure resists gravity indefinitely. In spite of their high temperature, these stars have very low luminosities because of their small size. White-dwarf matter is rather extraordinary; a cupful of it would outweigh two dozen elephants (Figure 11.7).

The final stages of red-giant evolution are not always as quiet as the story of white-dwarf formation might indicate. No one knows quite how, but in at least some cases the cores of massive red-giant stars are compressed to the point at which the electron degeneracy pressure is overcome. Violence accompanies the end of these stars; they explode, blowing off their outer layers in a spectacular stellar funeral and they leave rather strange remnants, neutron stars or possibly black holes. In a neutron star, matter has been crushed so that individual electrons and protons no longer exist. Neutron stars are only about 10 km across and are 10^9 times as dense as white-dwarf stars. A black hole is an object that has been compressed so far that the strength of the gravity at its surface does not permit light to escape. Chapter 12 contains a more thorough description of some of these objects.

This section has described in rather general terms the important milestones of a star's life following the exhaustion of the hydrogen fuel that sustained it as a main-sequence star. These key events are summarized in Table 11.1. Not all stars go through all stages, and Table 11.1 should only

FIGURE 11.7 Matter in a white-dwarf star is unlike ordinary matter. A cupful of white-dwarf stuff would outweigh two dozen elephants. (Adapted from Harry L. Shipman, *Black Holes, Quasars, and the Universe.* Copyright © 1976 by Houghton Mifflin Company, p. 37.)

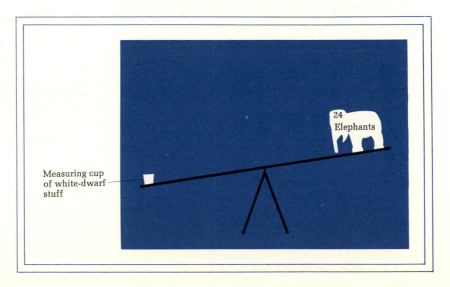

be considered a very general outline of the late stages of stellar evolution. It is probably only the massive stars that go through the more advanced stages of red-giant evolution. The next section describes the future evolution of our sun in as much detail as present theoretical sophistication allows, showing how the general descriptions in Table 11.1 fit into the life cycle of the sun.

★ 11.3 THE SOLAR LIFE CYCLE ★

The first two sections of this chapter describe the physical changes that take place in a red giant's interior while it travels along the inexorable path to death as a white-dwarf star, neutron star, or black hole. Theoretical astrophysicists have gone to great lengths to describe these physical changes in numerical terms and to calculate precisely what effects they have on the observable properties of red giants: the temperature and luminosity of the surface. We seek to touch base with observations of the real world so we can ask, "Do the nuclear-burning stages described in section 11.2 really occur? How long does each take? How massive must a star be in order to go through all of them?" Our understanding of red-giant evolution is in a tantalizing stage, since most of the major evolutionary stages have been sorted out in theory, but the details are only just beginning to be matched to the real world of observations.

TABLE 11.1   MILESTONES OF RED-GIANT EVOLUTION

Central-Hydrogen Exhaustion   A star no longer fuses hydrogen at its center. The core contracts (Figure 11.4), and the star increases in size and luminosity, heading toward the red-giant region of the H-R diagram (Figure 11.2).

Helium Burning   Three helium nuclei fuse to form a carbon nucleus, producing energy in the process. Stars in this stage have two central energy sources—helium fuses in the stellar center and hydrogen fuses in a shell surrounding the center (Figure 11.6).

Later Stages of Nuclear Burning   After the helium supply in the stellar center is exhausted, some stars can use carbon as a nuclear fuel, fusing carbon nuclei to form neon and magnesium. It is possible that some stars can fuse matter still further, carrying fusion reactions all the way to iron, after which no more energy can be extracted from a fusion reaction.

The Final State   In low-mass stars, electron degeneracy pressure stops further compaction of the core when densities reach 10^6 g/cm^3 (Figure 11.7). The star loses its envelope, and the core remains as a white-dwarf star (section 11.4). White-dwarf stars collapse no further and simply cool.

In high-mass stars, the core collapses catastrophically (in some stars at least), the star explodes, and the final state is a neutron star (about 10 km across) or a black hole.

We best understand the evolution of lightweight stars like the sun, and so this section describes the future evolution of the sun in some detail. While we cannot look 5 billion years into the future to see this evolution unfold, there are other stars in the sky, and points of contact between the calculations and the observation of these other stars will be delineated along the way. Evolution through the red-giant stage is somewhat different for stars of differing masses, and these differences, too, will be noted.

☆ BEFORE THE RED-GIANT STAGE ☆

The sun started as an interstellar gas cloud. This cloud fragmented and collapsed under the influence of gravity when its density reached the point at which gas atoms were pulled toward the center. As described in Chapter 10,

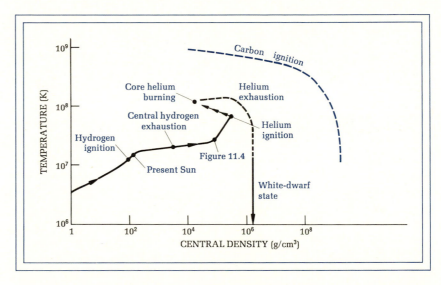

FIGURE 11.8 Evolution of the temperature and density of the sun's center. Gravity tends to compress and heat the center, pushing it to the upper right in the diagram. Section 11.3 describes the various evolutionary phases in detail, and Figure 11.9 illustrates the sun's path through the H-R diagram that corresponds to the interior evolution illustrated here. [Sources: I. Iben, Jr., *Astrophysical Journal*, 147 (1967): 624, for the points up through helium ignition: J. Faulkner and R. Cannon, *Astrophysical Journal*, 180 (1973): 435, for the core helium-burning point; my own estimates for the tracks between helium ignition and core helium burning and between core helium burning and the white-dwarf stage, since such tracks have not been calculated and the curves are roughly sketched; the density in the white-dwarf stage is from the models of T. Hamada and E. Salpeter, *Astrophysical Journal*, 134 (1961): 683, assuming a remnant mass of 0.5 solar masses.

once the cloud started to collapse, a massive ball quickly formed in the middle. Temperatures and densities at the center of this ball increased, pushing the embryo sun upward and to the right on the temperature-density graph of Figure 11.8. Some time before the interior became hot enough to fuse hydrogen, the remnants of the gas cloud were blown away, and the sun first appeared on the H-R diagram, probably somewhere near the beginning of the evolutionary track shown in Figure 11.9. Small bits of the gas cloud remained in orbit around the sun as planets.

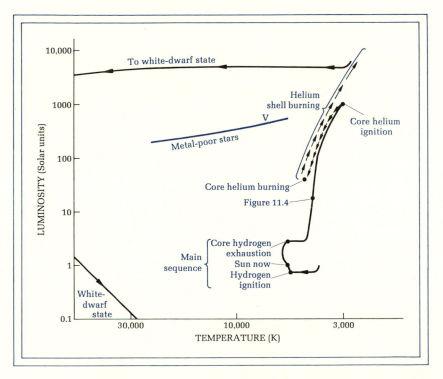

FIGURE 11.9 Evolutionary track of the sun through the H-R diagram. The marked points on the track are also noted in the interior-evolution diagram, Figure 11.8, and in section 11.3 of the text. The location of the helium-shell-burning portion of the track is very uncertain; calculations have not been carried to this point, and the track is based on the locations of stars that we believe are similar to the sun. The general behavior should be correct, but the details may not be. [Sources: Pre-main-sequence from I. Iben, Jr., *Astrophysical Journal,* 141 (1965): 993; main sequence to "Figure 11.4" from I. Iben, Jr., *Astrophysical Journal,* 147 (1967): 624; core helium burning from D. J. Faulkner and R. D. Cannon, *Astrophysical Journal,* 180 (1973): 435; helium shell burning from B. M. Tinsley, *Astronomy and Astrophysics,* 20 (1972): 386, and B. M. Tinsley, *Astrophysical Letters,* 9 (1971): 105; the position of the sun at core helium ignition and stages following helium shell burning from B. Paczynski, *Acta Astronomica,* 20 (1970), no. 2.]

A few tens of millions of years after the sun first emerged from its prestellar cocoon, temperatures at the center reached 14 million degrees and hydrogen fusion began. The establishment of a central energy source enabled the sun to remain in a relatively stable state for a long time. Now, some 4.7 billion years after the beginning of hydrogen fusion, the central temperature and density are scarcely different from the central conditions when hydrogen fusion began, as Figure 11.8 shows. The sun is now just a little hotter and a little brighter than it was at the beginning, as shown by its changing position on the H-R diagram of Figure 11.9. However, this change is extremely small in relation to the changes that will occur when the sun enters its giant phase.

The sun will remain on the main sequence for another 5 billion years or so, creeping upward from its present position as its luminosity increases slightly. This increasing luminosity will be caused by a slight increase in the temperature and density of the central core, shown in Figure 11.8. Temperatures on the earth will rise slightly but will remain well within the range of tolerable temperatures for billions of years. Some time before the solar hydrogen runs out, temperatures on the earth will rise to the point at which the oceans will boil. Just when this will happen depends on the extent to which the greenhouse effect (Chapter 5) can increase the earth's surface temperature; of course, some regions of the earth will be steamier than others. Detailed calculations including the greenhouse effect in an approximate way indicate that we have 4.5 billion years to go before the earth ceases to be habitable.

A little over 5 billion years from now, 10 billion years after the sun arrived on the main sequence, the central hydrogen source will run out, and the sun will readjust its structure rather drastically. The slow, stately, billion-year time scale of main-sequence evolution will speed up considerably. A few hundred million years after its hydrogen is exhausted, the sun will have climbed up in the H-R diagram and will be 10 times as luminous as it is now. This point is marked as "Figure 11.4" on Figures 11.8 and 11.9; at that time, the sun's interior will (according to the calculations) resemble Figure 11.4. Five hundred million years will pass while the sun climbs up to the tip of its evolutionary track, into the red-giant region, and reaches the helium-ignition phase.

☆ HELIUM BURNING ☆

When the sun's core becomes sufficiently dense and hot, the fusion of helium nuclei in its interior will temporarily halt the compression. Since the core is degenerate when helium starts to fuse, helium ignition abruptly expands the core, sending it on a track shown by the arrows in the interior evolution diagram of Figure 11.8. When the sun reaches the point labeled "core helium burning," its interior will resemble Figure 11.6. If the sun had far fewer heavy elements, it would move far to the left in the H-R diagram and be located somewhere along the color line labeled "metal-poor stars" in the H-R diagram of Figure 11.9. The H-R diagrams of globular clusters contain this horizontal branch of stars in the core-helium-burning stage and studies of these horizontal branch stars have provided an important observational checkpoint for our understanding of red-giant evolution.

More massive stars follow similar evolutionary tracks in the H-R diagram to this point, rising to high luminosities and high locations in the diagram before igniting helium. They then move back down, and somewhat to the left, when helium burning begins in the core. These stars often move far higher in the H-R diagram; the very luminous red giants and supergiants must be very massive stars, although we have no direct measurement of their masses. Note that the sun, at this point, is not even 100 times as luminous as it is now.

☆ **LATER STAGES OF NUCLEAR BURNING** ☆

For the sun, helium fusion will be the end of the line. When the helium runs out in its core, a helium-burning shell will be established, and soon much of the helium core will be fused to carbon. The core-helium-burning stage will last about 100 million years; it is not yet certain how much more time will pass before the sun dies peacefully as a white dwarf.

More massive stars, stars larger than 3 or 4 solar masses, can become hot enough in their interiors to fuse carbon, reaching the dotted line labeled "carbon ignition" in the interior evolution diagram of Figure 11.8. Middleweight stars, with masses between 4 and 10 solar masses, are degenerate when carbon burning starts, and it may be that explosive carbon burning leads to an explosion of the star's outer layers. The heavyweight stars, heavier than 10 solar masses, can remain nondegenerate all the way through the final stages of nuclear burning, when the fuel is processed to become iron, the end of the line for fusion as an energy source.

The calculation of later stages of nuclear burning is extremely complicated because red giants in this stage lose mass from their surfaces. At this time in its life, a red giant consists of a tiny core, not much larger than the earth, surrounded by a vast envelope about 0.2 AU in radius. The force of gravity that binds the envelope to the core is very weak, and there are a number of ways in which this mass can be pushed away from the core and off into space. The tenuous connection between the envelope (the star that we see) and the core makes it extremely difficult to calculate just what the sun will look like in these late stages. The best guess is that once helium is exhausted, the sun will move up in the H-R diagram, becoming larger and more luminous, along a track something like the dashed line in Figure 11.9.

Some writers refer to the stage of helium burning in a shell as the "second red-giant stage." It is not certain just how large and luminous the sun will be at this point; some stars in globular clusters and in other old stellar groups are known to have radii larger than 1 AU. If the sun becomes that large, the earth will be orbiting inside the solar surface. Such an event sounds ominous, but at this stage the sun will be less than 1/1000 as dense as sea-level air. Effects of this second red-giant stage on the earth remain to be calculated, but life on earth will have long since been cooked to death by the sun's tremendous luminosity. The mass lost by the sun in this stage will provide a strong stellar wind, 10^2 to 10^7 times as intense as the present solar wind.

Some stars eject a shell of gas in the final stages of their red-giant evolution. Some type of instability at the base of the red-giant envelope causes the entire envelope to be puffed off and to expand into space. These envelopes are visible as *planetary nebulae,* luminous patches of gas that shine by recombination radiation. The hot star at the center, visible in some of the planetary nebulae shown in Color Photo 14, ionizes the gas atoms, and the recombining atoms concentrate the light from the nebula in emission lines that produce the vivid colors shown in the photo. The name "planetary nebulae" comes from the green color and roughly circular shape of these objects, by which they resemble the outer planets Uranus and Neptune.

The cores of planetary nebulae are the predecessors of white-dwarf stars, the final state of low-mass stars like the sun. They are found far to the left, at the high-temperature end of the H-R diagram. These hot, blue, tiny balls rapidly contract to the white-dwarf state, in which they are no larger than the earth. At the end of its life, the sun will be one of these tiny balls; the earth, fried by the high luminosity of the sun in its red-giant stage, will be frozen as it circles this tiny white sphere, about 7 or 8 billion years from now.

★ 11.4 VARIABLE STARS ★

The red-giant stage of stellar evolution is of interest because it is in this part of the H-R diagram that many peculiar and interesting stars are found. Stars that are abnormal in some way always fascinate astronomers. The question here is, "Just why is this star strange?" In some cases, the curiosity results only in the discovery of an abnormal star, but in other cases these peculiarities can be the keys to some astronomical puzzle. The particularly interesting stars in the red-giant region of the H-R diagram are the variable stars, stars that change the amount of light they put out.

The British amateur astronomer John Goodricke discovered the first variable star in 1782, when he noticed that the eclipsing binary Algol, the second brightest star in Perseus, did not remain constant in brightness. Goodricke, William Herschel, and the nineteenth-century astronomer Friedrich Argelander discovered more of these stars, and Argelander performed the characteristic first step in astronomical analysis when he published a catalogue of these variables in 1844. But what caused the variation in brightness? Argelander's catalogue remained only a list of curiosities for a long time.

☆ CEPHEID VARIABLES: A NEW ☆
COSMIC YARDSTICK

The advent of spectroscopy in the late nineteenth century proved to be a key to finding the cause of the variations of these strange objects. Doppler shifts in the stellar spectra were measured, and it became clear that many of the variables were eclipsing binaries (Chapter 9). But some variables, notably the *Cepheid variables,* were still puzzling. The Cepheids, named for the bright Cepheid variable Delta Cephei, varied regularly with periods ranging from a few days to a few months; Doppler shift measurements showed that the Cepheids moved in a way that could not be reconciled with an eclipsing binary model. Analyses of the Doppler shifts indicated that these stars pulsated, growing larger and smaller, hotter and cooler, over their cycle.

The most important piece of the Cepheid puzzle was discovered by Henrietta Leavitt at Harvard in 1912. Neither she nor anyone else suspected that her patient, painstaking inspection of a series of photographs of the Magellanic clouds would result in the discovery of a very important method for determining the distance to a group of stars. She was simply compiling a list of the variables in the Magellanic clouds, the two nearest galaxies to our own that are only visible from the Southern Hemisphere (Color Photo 16). Her discovery was a combination of luck and skill. She was lucky because all the variables she found were the same distance away from us, in one stellar group. The relationship among brightness, luminosity, and distance (discussed in Chapter 9) ensured that the bright stars, the ones that produced the largest images in the photograph, were also the luminous ones. She did not have to sort out the stars that are bright only because they are close to us. Leavitt's skill was in picking a regularity out of the chaos of her collection of variables. The Cepheid variables with the longest periods were also the brightest, most luminous ones. Plotting this regularity on a graph, she discovered a well-defined relationship between the period and luminosity of a Cepheid variable (Figure 11.10).

The *period-luminosity relationship* illustrated in Figure 11.10 has proven to be a unique and useful cosmic yardstick. The period of a Cepheid variable

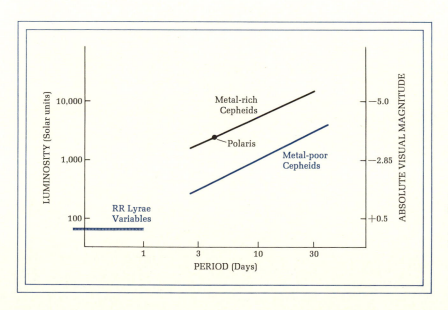

FIGURE 11.10 Cepheid variables show a characteristic relation between period and luminosity. The longer the period, the more luminous the Cepheid is. However, an attempt to determine the luminosity of the given Cepheid must take into account the difference between metal-rich Cepheids (black) and metal-poor Cepheids (color). [Sources: Metal-rich Cepheids from A. Sandage and G. A. Tammann, *Astrophysical Journal,* 167 (1971): 293; others from C. W. Allen, *Astrophysical Quantities* (London: Athlone, 1973), p. 216.]

is something that can be easily measured, and a measurement of the period can be entered on Figure 11.10 to obtain the luminosity of the Cepheid. (In practice, the color of the Cepheid provides additional information, since there is a slight difference between the luminosities of Cepheids of similar periods and different colors.) The flux from the Cepheid can also be determined, by measuring the apparent brightness of the star. The flux from a star, however, depends on the star's luminosity and its distance (Figure 9.3; equation 9.1); a luminous star will be bright if it is nearby, and dim if it is far away. Since equation 9.1 provides us with a numerical relationship among three quantities and we know two of them, the flux and the luminosity, the third, the distance to the Cepheid, can be readily determined. Box 11.2 illustrates how this is done.

The color curves in Figure 11.10 illustrate the care that must be taken if the distance to a Cepheid variable is to be determined with precision. Three types of variables are illustrated in that graph. The black curve shows the period-luminosity relationship for the metal-rich classical Cepheid variables, the ones that Henrietta Leavitt first mapped in the Magellanic clouds. One of the color curves shows the period-luminosity relationship for metal-poor Cepheid variables. The curve labeled "RR Lyrae variables" gives the luminosity for the RR Lyrae class of variable stars, a group of stars related to the Cepheids and also found in metal-poor star clusters. The RR Lyrae variables are located on the horizontal branch in globular clusters, and their position is indicated in the H-R diagrams of Figures 11.5 and 11.9.

Cepheid variables have been used to measure the distances to very distant stellar groups, including galaxies like our own, the Milky Way. When Edwin Hubble first made such a measurement, the distinction between metal-rich and metal-poor Cepheids was not known, and as a result Hubble believed that the distant galaxies were 2.5 times closer than we now believe them to be. Since Cepheid variables are very luminous stars, they can be photographed in extremely distant galaxies, and they can thus be used to determine the distance of any stellar group closer than 5 million pc.

☆　**MIRA VARIABLES**　☆

Another class of variable stars found in the red-giant region of the H-R diagram is the class of long-period variable stars, also named for the brightest and most remarkable member of this class, the star Mira in the constellation Cetus. Mira, the "wonderful" star, at its brightest is roughly a third-magnitude star, one of the brightest stars in this rather dim region of the sky. Later on in its 332-day cycle, it fades to ninth magnitude, becoming so dim that it is invisible. Its location is marked with a circle on the map of the autumn sky, Figure 2.7. In the late 1970s its maxima are well timed, occurring late in the year, so this star is visible, at maximum, in the evening autumn sky. The 332-day cycle will produce maxima in early November 1978, early October 1979, and early September 1980, with the maximum occurring about one month earlier in each succeeding year. Amateur astronomers have contributed greatly to variable-star work by keeping track of the variations of Mira variables; the variations are so large that the precision of photoelectric

Polaris, the north star, is a Cepheid variable with a period of 3.97 days. It does not vary too much; its variations, unlike those of most Cepheids, are imperceptible to the eye. Its period and brightness can be used to determine its distance. The distance calculation can be done in two ways, depending on whether the language of magnitudes or fluxes is used.

Flux language: The black curve of Figure 11.10 indicates that Polaris has a luminosity 2500 times the solar luminosity. Since the solar luminosity is 3.9×10^{33} ergs/sec, Polaris's luminosity is 9.8×10^{36} ergs/sec. A measurement of the flux from Polaris indicates that, in all wavelengths, 5.9×10^{-6} ergs of radiant energy cross each square centimeter of telescope mirror surface every second. With equation 9.1, the relationship between flux, luminosity, and distance,

$$\text{Flux} = \frac{\text{luminosity}}{4\pi(\text{distance})^2}$$

one can write

$$5.9 \times 10^{-6} = \frac{9.8 \times 10^{36}}{4\pi(\text{distance})^2}$$

Rearrange and determine the distance as 3.6×10^{20} cm, or 120 pc. At such a distance, the trigonometric parallax of Polaris would be 0.008 sec of arc, too small to be measured.

Magnitude language: Figure 11.10 provides the absolute visual magnitude M_V of Polaris when the period of this variable is entered on the graph. The apparent visual magnitude m_V of Polaris is 2.5, measured at the telescope. Equation B9.5 is equivalent to equation 9.1 and reads

$$m_V - M_V = -5 + 5 \log D$$

where D is the distance to Polaris. This equation thus means that

$$2.5 - (-2.85) = -5 + 5 \log D,$$
$$\text{or} \quad 5 \log D = 10.35 \text{ or}$$
$$\log D = 2.07$$

giving a distance of 120 pc.

BOX 11.2 MEASURING THE DISTANCE TO A CEPHEID VARIABLE

photometry is not needed in determining a period for a star. (If you are interested in this work, the American Association of Variable Star Observers coordinates amateur efforts in this field. Their address, along with the addresses of other amateur astronomy organizations, is listed in Appendix E).

The *Mira variables* are the most regular of a diverse class of red-giant variable stars. A number of similar stars vary semiregularly or irregularly, again with time scales of months. Betelgeuse, the bright star in Orion's right armpit, is an irregular variable, changing its brightness by a magnitude or so from one year to the next. R Corona Borealis, a star just visible to the naked eye in that circlet of stars, occasionally fades to magnitude fifteen, one ten-thousandth of its maximum brightness. It is not yet clear whether this diverse collection of red-giant variables will remain a bestiary of peculiar stars or whether something like a period-luminosity relation will be discovered, making these stars a useful distance indicator like the Cepheids.

There are three physical causes of variations in the luminosity of a star. The previous section describes pulsating stars, most of which are found in the red-giant region of the H-R diagram. Some stars are eruptive variables, suddenly increasing in brightness; these stars in the evolutionary stages that follow the red-giant stage will be discussed in connection with violent star-death in Chapter 12. The third type of variable star, the eclipsing binary, was discussed in Chapter 9, but most red-giant eclipsing binaries have a rather peculiar evolutionary past. A massive red-giant star can swell up to become as large as the binary system itself. In many binary systems, the stars are only a few astronomical units apart, and the radius of a massive red giant can extend to 5 or even 10 AU in some cases. The evolution of a red giant in a close binary system is affected by the presence of the secondary star.

To follow the evolution of a close binary system, we start with two stars, a massive one *A* and a smaller one *B*, orbiting each other. When both stars are on the main sequence, their separation far exceeds their size. Figure 11.11 is an attempt to illustrate this but, for normal main-sequence stars, millimeter-sized dots would have to be tens of centimeters (several inches) apart to convey a correct impression of the scale. When star *A*, the more massive component, evolves off the main sequence, it becomes much larger in its normal approach to the red-giant stage. However, it cannot increase its size indefinitely, for there is a point at which the gravitational attraction of star *B* will pull mass from star *A* (Figure 11.11*b*). Matter pulled from star *A* falls onto star *B*, forming a disk around star *B* and ultimately increasing *B*'s mass. Soon *B* consumes *A*'s envelope, and it then becomes the more massive component of the system.

The binary system now can follow one of three evolutionary paths, depending on its mass. Systems containing low-mass stars follow the track shown in the top panel of Figure 11.11*c*. Star *A* becomes a rather unusual type of star, a *subgiant*—a low-mass, very dim, small red giant. The binary system Algol is probably one of these. In a more massive binary, the core of star *A* will shed its envelope and appear as a *helium star*, a ball of pure helium that converts helium to carbon at its core. Still a third scenario is possible if the two stars are extremely close together. They can evolve to a still stranger state, a *contact binary system*, in which both stars have expanded as far as they can and each adds mass to the other. Contact binary stars are often surrounded by gas streams.

The advanced evolution of close binary systems would be just an interesting curiosity were it not for the role of these systems in verifying the late stages of stellar evolution. One evolved close binary system probably contains a black hole, the remnant of a massive star that was unable to overcome gravity at the end of its life cycle. We believe that this black hole evolved from a helium star in a system such as that depicted in the middle panel of Figure 11.11*c*. A thorough investigation of this system and its peculiarities has prompted a resurgence of interest in the properties of close binary systems.

This chapter has treated the late stages of stellar evolution, carrying stars to the point at which they run out of fuel. While we do not understand the

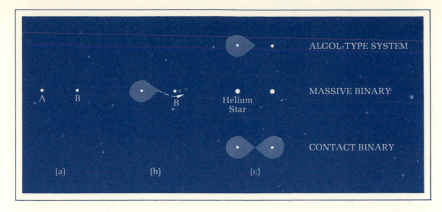

ALGOL-TYPE SYSTEM

MASSIVE BINARY

A   B

B   Helium
Star

CONTACT BINARY

(a)   (b)   (c)

FIGURE 11.11   Evolution of a close binary system. In (a), both stars are on the main sequence. In (b), the more massive star has evolved to the red-giant stage and has swelled up to the point at which the gravitational attraction of star B pulls part of the envelope from star A. The eventual state of the system is depicted in (c). Three results can occur, depending on the mass and separation of the two stars.

short evolutionary stage between the end of the red-giant sequence and star-death, two and probably three types of stellar corpses have been observed. Stellar corpses, white dwarfs, neutron stars, and black holes are the subject of the next chapter.

SUMMARY

When a star runs out of central hydrogen, its core contracts, its envelope expands, and it becomes a red giant. While the details of red-giant evolution depend on the star's mass, the rate at which it loses mass, and various core processes that are not fully understood, the general picture is the same for all stars: The core continues to contract, and various chemical elements are used as fuel for the fusion processes that supply the energy lost as sunlight by the enormous, luminous red giant. Some red giants vary in their luminosity; the Cepheid variables are sufficiently regular in their variation so that the variational period determines their luminosity and hence their distance. Other types of red-giant variables are not as well understood.

This chapter has covered:

A star's departure from the main sequence, prompted by the exhaustion of hydrogen in its central core;

The different fuels that are used to provide energy to a red giant;

The role of gravity in determining the final state of a star;

The future evolution of the sun;

The different types of variable stars found in the red-giant region of the H-R diagram; and

The evolution of binary star systems.

KEY CONCEPTS

Cepheid variables
Degeneracy pressure
Globular cluster

Mira variables
Period-luminosity
relationship

Planetary nebulae
Red giants
White dwarfs

REVIEW QUESTIONS

1. Why do massive stars last only a short time in their main-sequence stage?

2. What is the single most important distinguishing characteristic of a main-sequence star?

3. How would red-giant evolution differ if the fusion of helium nuclei did not liberate energy?

4. What is degeneracy pressure, and how does it differ from the pressure produced by heat?

5. What major evolutionary stages will the sun go through before it dies as a white-dwarf star?

6. Why are Cepheid variables useful as a cosmic yardstick?

7. The two components of Alpha Centauri are 23 AU apart, and neither is more massive than 1.06 solar masses. Will mass exchange be important in the future evolution of this system?

FURTHER READING

Strohmeier, W. *Variable Stars*. New York: Pergamon, 1972.

Several articles by Otto Struve, published many years ago in *Sky and Telescope*, describe individual peculiar stars. While the details may be out of date, these articles have no parallel as a stellar bestiary. These articles are listed below with their dates of issue.

Variable Stars
"Variable Stars and Stellar Evolution," April and May 1950
"Some Unusual Short-Period Variables," April 1955
"The Pulsating Star RR Lyrae," June 1962
"Mira Ceti," September 1954

Close Binary Stars
"The Story of Epsilon Aurigae," March 1962
"The Story of U Cephei," April 1963
"The Spectrum of Beta Lyrae," July 1957
"Contact Binaries," April 1961

★ CHAPTER TWELVE ★

STARDEATH

A living, evolving star changes its size and luminosity in response to the changing interplay between gravity and the pressure at its center. When all energy sources are exhausted, this interplay ceases and the star dies. Many stars leave remnants at the end of their evolutionary cycle, because another kind of pressure, degeneracy pressure, can keep the star from collapsing. These remnants—white-dwarf stars and neutron stars—are tiny, dense objects, far smaller than the stars that produced them. Some stars may just continue to collapse, becoming so dense that nothing can escape the pull of their gravity, not even light. Such an object is called a black hole. White-dwarf stars and neutron stars are known to exist in the real world, and black holes may have been discovered in recent years.

★   MAIN IDEAS OF THIS CHAPTER   ★

White-dwarf stars, packing one solar mass into a volume no larger than the earth, are the final state of low-mass stars and evolve only by cooling, not by changing their size.

Neutron stars are far smaller than white-dwarf stars and are observed as pulsating radio sources, or pulsars.

Dying stars can explosively increase in brightness, as novae or supernovae.

A black hole can form when a dying star becomes so compact that gravity prevents anything from escaping from its surface.

Black holes may have been discovered in binary x-ray sources.

★   12.1 THE END OF STELLAR EVOLUTION   ★

The evolution of a star through the red-giant stage comes to an end when the steady generation of nuclear energy in the star's core ceases to play a role in balancing the forces in its interior. To this point, the star's tendency to collapse under the influence of gravity has been forestalled by the pressure of the hot gas in the core. With the exhaustion of the central energy source, the hot gas must cool, and it may collapse if no other effects counter gravity. Stars that end their lives with small cores can maintain their core pressure when matter in the core becomes degenerate. Degenerate electrons or neutrons are packed so tightly that it takes additional force to crush them further. But a star that reaches the final state with a very massive core cannot resist gravity and must collapse indefinitely, becoming a black hole.

Only one of the roads to stardeath is well understood from a theoretical point of view. Section 11.3 shows how a low-mass star like the sun, at the end of its evolution as a red giant, could shed its outer envelope and leave a hot core as a remnant. It is uncertain whether all low-mass stars follow this quiescent evolutionary sequence, but a number of objects that are cooled-off red-giant cores have been found. These objects, the *white-dwarf stars* (Figure 12.1), constitute one of the better understood endpoints of stellar evolution.

More violent events occur when massive stars die. In the 1930s, Cal Tech astronomer Fritz Zwicky discovered that some stars temporarily became as luminous as an entire galaxy, in *supernova* explosions. The violent events at the core of a supernova are still not understood in detail, but recent work has discovered the remnant of a stellar core (after the supernova) as a *neutron star,* an object in which the mass is compressed into a 10-km sphere. Neutron stars, too, were recognized as possible stellar endpoints in the 1930s, in

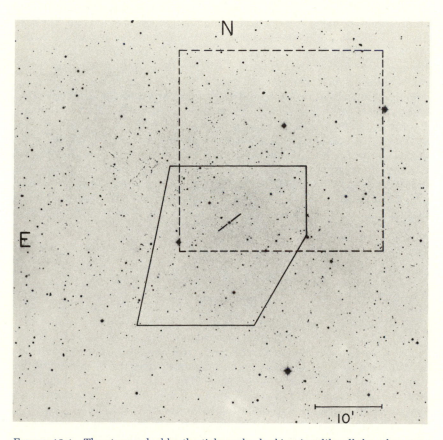

FIGURE 12.1   The star marked by the tick marks, looking just like all the other stars in this photograph, is 100 times smaller and 1 million times denser than a main-sequence star like the sun. This star, HZ (Humason-Zwicky) 43, is one of the hottest known stars in the sky, a white dwarf with a temperature between 55,000 and 70,000 K. (From M. Lampton, B. Margon, F. Paresce, R. Stern, and S. Bowyer, *Astrophysical Journal Letters,* 203, L71–L74, published by the University of Chicago Press. Copyright © 1976 by The American Astronomical Society. All rights reserved. The photograph is from the Palomar Observatory Sky Survey. Copyright © by the National Geographic Society-Palomar Observatory Sky Survey. Reproduced by permission from the Hale Observatories.)

an investigation that J. Robert Oppenheimer and his student George Volkoff published in 1939 just before the start of World War II. Oppenheimer himself did no further work on neutron stars, since he directed the Manhattan Project that built the first atomic bomb at Los Alamos. It was only in the 1960s that neutron stars were discovered as pulsating radio sources. In 1939, Oppenheimer and another student, Hartland Snyder, argued that some stars might become *black holes,* objects so dense that gravity would not allow anything, not even light, to escape from their surfaces. It is probable, but not yet certain, that these theoretical objects may also have been found in the real world by x-ray astronomers.

The investigation of the violent events that occur when middleweight and heavyweight stars die is an extremely active area of astronomical research. In such an active field, solid facts and concrete theories well supported by observations are few and far between, and controversy is extensive. The distinction between the model world of the theorist and the real world of the observer is one that must be kept in mind, for it is one thing to postulate a strange object like a black hole and another thing to find that one really exists in the cosmos. The material covered in this chapter is on the cutting edge of astronomical research, and you should be aware that these strange, bizarre objects are not at all well understood. Can it be that the white-dwarf star marked in Figure 12.1 is only as large as the earth, while the other stars in the photograph are 100 times larger? Is a pulsar really a star that is as dense as the nucleus of an atom? We believe so; the identification of white-dwarf stars is extremely secure, and the identification of pulsars no less so. The black hole, perhaps the strangest of all stellar final states, can only be observed indirectly, and its identification in the real world, though probable, is not yet certain.

★   **12.2 WHITE-DWARF STARS**   ★

☆   **DISCOVERY**   ☆

White-dwarf stars, the final state of stars like the sun, were first discovered as real objects in the first quarter of the twentieth century. In 1910, Henry Norris Russell, one of the originators of the H-R diagram, asked Harvard Observatory's director E. C. Pickering to look up the spectral classes of all the stars on Russell's parallax program. Russell's curiosity about the companion to 40 Eridani moved him to ask about that star, and he was startled to find that it was of spectral class A. Such a low-luminosity star must be incredibly tiny to be as hot as an A-type spectrum would suggest. Russell remembered 40 Eridani B, and when he first published an H-R diagram this white-dwarf star was on it, noted as a peculiar object and a possible spectral misclassification. Four years later, W. S. Adams, at the other end of the continent on Mount Wilson, obtained the spectrum of Sirius B and found that it, too, was a tiny hot star.

For 10 years no one believed that Sirius B and 40 Eridani B were really as small as the spectra seemed to indicate. Both these stars have known masses, since they are in binary systems, and their masses are 0.98 and 0.42 solar masses, respectively. But their small size gives them a volume one-millionth that of the sun, and so the observations indicated the existence of objects in which matter was compressed to a density 1 million times greater than the density of ordinary terrestrial rock. The English astrophysicist A. S. Eddington describes his feelings:

> The message of the Companion of Sirius, when it was decoded, ran: "I am composed of material 3,000 times denser than anything you have come across; a ton of my material would be a little nugget that you could put in a matchbox." What reply can one make to such a message? The reply that most of us made in 1914 was—"Shut up. Don't talk nonsense."[1]

The astronomical community only realized in 1926 that the message of the starlight from the white-dwarf stars was not nonsense. In that year, R. H. Fowler made the necessary link between observation and theory and successfully modeled white-dwarf stars as balls of degenerate matter. In the discovery of white-dwarf stars, observation preceded theory. In neutron-star and black-hole studies, the order was reversed, for we had a fairly concrete theoretical picture of these objects before observational confirmation of their existence appeared.

☆   PROPERTIES   ☆

Sirius B (Figure 12.2) is the nearest white-dwarf star, and its properties are representative of most white dwarfs. Analyses of the spectra of Sirius B, obtained with difficulty because of the bright visible light from Sirius A, indicate that Sirius B, with a temperature of 32,500 K, is considerably hotter than Sirius A, but its small size gives it a total luminosity of 0.061 times the luminosity of the sun. Its radius is 0.0078 solar radii, less than the radius of the earth (Figure 12.3). Its spectrum shows only hydrogen lines, resembling the A-type spectrum of Sirius A. At this high temperature, helium would produce absorption lines if it were present with the same abundance as is found in other stars, so we deduce that the surface of Sirius B is depleted of helium. Sirius B is surprisingly difficult to see in a telescope, since it is overshadowed by the glare of Sirius A, 10,000 times brighter in the visual part of the electromagnetic spectrum. Very good observing conditions are needed to split this pair. Sirius B is as far away from Sirius A as it will ever be in the late 1970s; since its orbital period is 50 years, it will be some time before it is as easy to see as it is now.

Main-sequence stars run the gamut from low-temperature, low-mass, low-luminosity stars in the lower right corner of the H-R diagram to the massive blue stars in the upper left. Red-giant stars do their job of populating the upper right corner of the diagram with all kinds of objects: ordinary red

---

[1] A. S. Eddington, *Stars and Atoms* (New Haven, Conn.: Yale University Press, 1927), p. 50.

FIGURE 12.2 This photograph shows two stars in a binary system. The bright one is Sirius A, a main-sequence star about twice as hot as the sun, and the brightest star in the sky. The fainter star, Sirius B, is 100 times smaller than Sirius A, and is much dimmer even though it is hotter. (Lick Observatory photograph)

FIGURE 12.3 White-dwarf stars are, in some cases, smaller than the earth, even though they are 1 million times as massive.

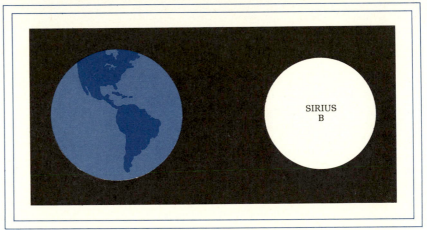

SIRIUS
B

giants, Cepheid variables, supergiants, and so forth. White-dwarf stars are spread over the lower left corner of the H-R diagram, and they show various spectral peculiarities and differences. While most have spectra showing that their atmospheres are pure hydrogen, a few have atmospheres of almost pure helium. Ten white-dwarf stars have very strong magnetic fields, more than 1000 times as strong as the magnetic field of the sun. Their temperatures vary from 4000 K, the temperature of the coolest known white dwarf, to greater than 55,000 K. HZ 43, the white-dwarf star shown in Figure 12.1, is quite inconspicuous in the optical part of the spectrum. Its high temperature, between 55,000 and 70,000 K, was only confirmed when a new type of telescope on the Apollo-Soyuz mission made the first extensive observation of the sky in the extreme ultraviolet part of the electromagnetic spectrum, at wavelengths between 100 and 1000 Å. This tiny star is the brightest object in the sky in this spectral range; its high temperature produces hordes of energetic extreme-ultraviolet photons. Other extremely hot stars are known as the nuclei of planetary nebulae.

☆ EVOLUTION ☆

The evolution of white-dwarf stars is far simpler than the evolution of stars in the main-sequence or red-giant part of the H-R diagram. Because white dwarfs no longer generate energy in their cores by nuclear reaction, they cool off when they emit energy in the form of starlight. They begin as sizzling little stars like the hottest ones (such as HZ 43) and cool slowly, taking a few million years to become as cool as Sirius B. When their temperatures drop, their luminosities decrease too, but their radii do not change. Thus their evolution in the H-R diagram consists of following a track of constant radius, sliding downward and to the right, growing cooler and dimmer. After a few billion years, they reach temperatures of a few thousand degrees and become very difficult to discover among the hordes of cool main-sequence stars that populate that part of the H-R diagram. As they cool, their internal structure remains the same. Degeneracy pressure can support a white-dwarf star no matter how cold it becomes, and so it remains the same size throughout its cooling period.

White-dwarf stars constitute perhaps the only type of final stellar state whose origin is reasonably well understood. Section 11.3 follows the evolution of the sun from protostar to white dwarf; most white-dwarf stars probably formed as described there, when a red-giant star lost its envelope as a planetary nebula and left a small core. Chapter 11 demonstrates that stars in binary systems evolve somewhat differently from single stars. White-dwarf stars in binary systems also show some peculiarities, and they serve as an introduction to the role of violent explosions in the late stages of stellar evolution.

☆ NOVAE AND CATACLYSMIC VARIABLES ☆

A sidelight to the white-dwarf story is the *nova* phenomenon, a sudden burst of light from a binary system containing a white-dwarf star. The word *nova* means "new" in Latin, and the original name for these objects was *nova stella* (new star) since they were first discovered as stars that appeared where no stars had been seen before. A nova is not, however, a genuinely new star, for in most cases a careful search of previous photographs of the sky reveals a faint, cool star that explosively increased in brightness when it became a nova.

Perhaps 40 novae occur every year in our own Milky Way galaxy, most of which are too far away to be readily discovered. About once or twice every decade, a nova becomes bright enough to be seen with the naked eye, as was a dramatic nova that flared up in Cygnus in late August 1975. When sunset occurred in Japan on August 29, 1975, Kentaro Osada, a Japanese amateur astronomer, noticed that the tail of Cygnus, the Swan, contained a rather bright extra star. The news was immediately flashed to the Bureau of Astronomical Telegrams in Cambridge, Massachusetts, where the sun had just risen and no one could confirm the existence of this new nova. But Osada's was only the first of hundreds of discovery reports that came in that day; observers around the globe hastened to report this brightening nova as the line of sunset swept westward across Asia. Hundreds of amateur and professional astronomers independently discovered this nova. Nova Cygni 1975 was a very fast nova, reaching maximum brightness only half a day after its

FIGURE 12.4   The variation with time of the brightness of Nova Cygni 1975. Only the visual-magnitude estimates are plotted, along with the photographic prediscovery images obtained from the meteor patrol photographs taken by Ben Mayer, an amateur astronomer. (From P. G. Young, H. C. Corwin, Jr., J. Bryan, and G. de Vaucouleurs, *Astrophysical Journal*, 209, pp. 882–894, published by the University of Chicago Press. Copyright © 1976 by The American Astronomical Society. All rights reserved.)

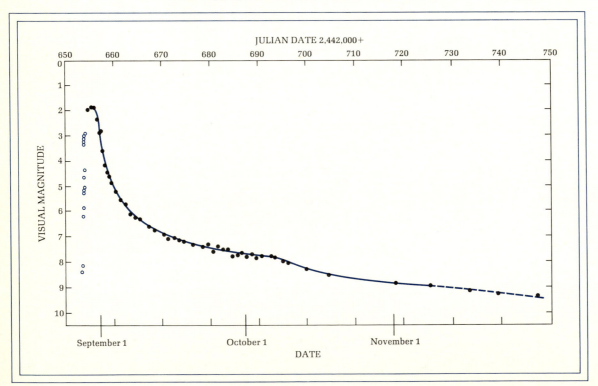

discovery (Figure 12.4). Its fast increase in brightness was paralleled by a rapid decline, since it faded below naked-eye visibility by September 10. Most novae go through their cycle more slowly, taking days to rise and weeks to decline. Nova Cygni is also unusual in that the original star is too faint to appear on photographs of the sky taken before its outburst. Most novae erupt from binary systems containing a white-dwarf star and a red main-sequence star, and it is surprising that neither member of the Nova Cygni system was visible before the outburst.

A number of variable stars resemble novae since their variations are marked by explosive increases in brightness. These objects, called "dwarf novae" or *cataclysmic variables,* flare more frequently than the novae do. The most active dwarf novae can flare up once every 2 weeks. If the novae and cataclysmic variables are considered as a group, they stretch from the most active dwarf novae, those that double or quadruple in brightness every 2 weeks or so, to the full-fledged novae, which increase their brightness 10^8 times about every 100 years. Ordinary novae do explode more than once.

What causes this dramatic stellar explosion? While they differ in important details, both the novae and the cataclysmic variables can be explained with the same general model, illustrated in Figure 12.5. The outbursts probably are caused by the accumulation of mass in the vicinity of a white-dwarf star in a close binary system. The two stars are so close to each other that the white dwarf pulls matter from the red main-sequence star. This matter is heated as it swirls toward the white-dwarf surface. When the temperature of the accreted matter reaches a few million degrees, hydrogen fusion begins and energy is released explosively. In some cases, a gas shell can be driven away from the system. Both the novae and cataclysmic variables occur in

FIGURE 12.5 Nova explosions occur in binary systems when mass transferred from a red-giant star to a white-dwarf star becomes hot enough to expand explosively. This illustration depicts the system when the white-dwarf star accumulates matter between nova outbursts.

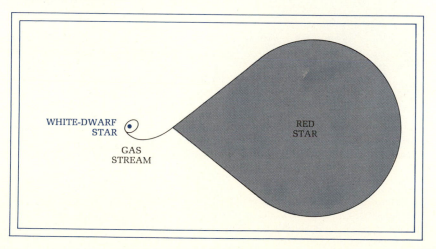

WHITE-DWARF
STAR

GAS
STREAM

RED
STAR

close binary systems; they are a special route that white-dwarf stars can take in their evolution.

★ 12.3 SUPERNOVAE ★

Dramatic as it is, a nova is not the most energetic stellar explosion known. Some stars can become as luminous as an entire galaxy when they explode and die; these titanic stellar funerals are called *supernovae* to distinguish them from the more frequent but less spectacular nova explosions. A nearby supernova explosion would be the most spectacular event likely to occur in a person's lifetime. This dying star would probably be visible in the daytime sky and might even be brighter than the full moon if the supernova were close enough to us. Several galactic supernovae have been observed in recorded history. Although we expect that a supernova explodes in our galaxy at least once every few decades, most are hidden from us by the thick veil of interstellar dust that blocks so many parts of the sky from our view (recall Chapter 10).

Our understanding of supernovae is based almost exclusively on their discovery in other galaxies, distant stellar systems like the Milky Way (Figure 12.6). Several observatories patiently photograph other galaxies week after

FIGURE 12.6   The spiral galaxy NGC 4303 is similar in many respects to the Milky Way galaxy. In this photograph, the arrow points to a supernova in NGC 4303, a stellar explosion that made the star as luminous as the entire galaxy it is in. If such an explosion were to occur in our neighborhood of our galaxy, it might be visible in the daytime sky. (Lick Observatory photograph)

week, and several supernovae in distant galaxies are discovered each year. The suddenness of a supernova explosion makes these objects rather difficult to study, but some of the brighter ones have been fairly well analyzed.

A supernova explosion is the most violent event that can occur in stellar evolution. The total amount of energy released in a supernova explosion is about 10^{50} ergs, equal to the entire energy output of the sun in its main-sequence evolution. This energy is all released in a very short time; the surge of visible light lasts a few days or weeks. However, the explosive event that caused the supernova to expand may last only a few hours or even minutes. What causes this sudden eruption? What makes a star explode, spewing its interior all over interstellar space? No one knows, yet.

In the stellar funeral marked by a supernova explosion, a sizable fraction of the star's mass is ejected outward, forming a gas cloud, a *supernova remnant*, that expands into interstellar space. Since much of this material comes from the star's interior, its chemical composition may be considerably different from the chemical composition of the gas cloud from which the star originally formed. The ashes of various nuclear reactions at the star's core can be transported outward, away from the star, and can change the chemical composition of the interstellar medium surrounding the supernova.

Perhaps the most conspicuous supernova remnant known is the *Crab nebula*, shown in Color Photo 15. The filaments of this glowing gas cloud looked like crab legs through the telescope of Lord Rosse, who named the object. It is the remnant of a supernova that the Chinese observed in 1054 A.D. The total mass of gas visible in Color Photo 15 is several solar masses, and so presumably nearly half the star was thrown out in the supernova explosion. Spectra of this gas cloud show that the filaments are glowing by recombination, the process responsible for the emission from the Orion nebula and planetary nebulae. All these objects show the characteristic vivid colors of emission lines, produced when high-energy radiation from some central object ionizes gas atoms that then cascade down their energy levels as electrons and ions recombine.

The recombination radiation in the Crab nebula shows that somewhere in the core of the nebula a source of high-energy photons provides the energy needed to ionize gas atoms in the filaments. The source of ionizing photons is a main-sequence star in Orion, a hot white-dwarf star in planetary nebulae. But what is it in the Crab nebula? A possible hint is provided by the blue S-shaped gas cloud at the center of the nebula, shown in the color photo and identified in Figure 12.7. This funny gas cloud is entirely different from the wispy filaments; the difference is suggested by its smooth shape and confirmed by its spectrum. While the filaments produce emission lines, the blue cloud has a continuous spectrum in which emissions are not concentrated at any particular wavelength. Further, the radiation from the blue cloud is strongly *polarized*. Ordinary thermal radiation from a dense ball of gas such as a star is not polarized, for the electric and magnetic fields in unpolarized thermal radiation vibrate in all directions. The electric and magnetic fields in the radiation from the blue cloud in the Crab tend to vibrate in particular directions, showing that this portion of the Crab is producing radiation in a rather unusual way (Figure 12.8). The blue cloud is not simply a cloud of hot gas. But what is it?

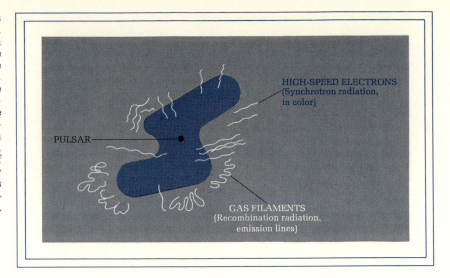

FIGURE 12.7 The various parts of the Crab nebula. Compare this illustration with Color Photo 15. Three types of emissions are produced in this object. Emission lines come from the filaments, synchrotron radiation comes from the blue cloud of high-speed electrons, and pulsed emission comes from the pulsar. While these three forms of radiation are related, they come from different objects and are produced by different mechanisms.

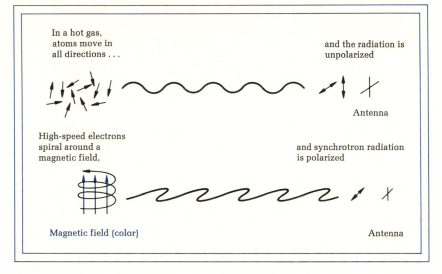

FIGURE 12.8 Synchrotron radiation is produced by high-speed electrons spiraling around a magnetic field; it is polarized.

The Russian astronomer Iosif Shklovsky discovered the source of the blue, high-energy radiation from the Crab in 1954. He remembered that particle accelerators, or synchrotrons, emit a ghostly blue, polarized light when they are turned on. A particle accelerator contains charged particles—electrons, for example—moving in a tight circle around a strong magnetic field. The magnetic field controls the direction in which the electrons move, and hence the direction in which they radiate. Electrons in a distant radio antenna will move in this same direction when they pick up radiation from spiraling charged particles. Shklovsky analyzed the radiation from the Crab, and the similarities between the radiation of the blue cloud and the blue light from synchrotrons were so great that the radiation from the blue cloud was confirmed as *synchrotron radiation.*

Thus our analysis of the Crab has led along a logical chain. The emission lines produced by the gas filaments need a source of high-energy photons to ionize the gas atoms. Synchrotron radiation is emitted at all wavelengths in the Crab nebula, from the radio to the x-ray part of the electromagnetic spectrum, and it is emitted in sufficient quantities to ionize the gas filaments. This radiation does not come from the filamentary emission lines, but from the blue S-shaped cloud.

What produces the high-energy electrons that power the blue S-shaped cloud? We may have solved the mystery of where and how the Crab produces its radiation, but this critical question remains unanswered in the analysis. To answer it, we must penetrate to the heart of the Crab. Analysis of the steady optical light from the two central stars in the Crab did not produce any insight into the nature of the source of the high-speed electrons; the answer to the question came from a totally unexpected discovery, *pulsars.*

★ 12.4 NEUTRON STARS AND PULSARS ★

☆ DISCOVERY ☆

Anthony Hewish was not setting out to discover pulsars when he conceived, planned, obtained funding for, and built a radio telescope at Cambridge University in the middle 1960s. This telescope was built to observe the rapid variations of radio sources caused by fluctuations in the interplanetary medium. A graduate student, Jocelyn Bell, was responsible for examining the miles of strip charts produced by this telescope and for finding the twinkling radio sources. She discovered that some sources observed by the telescope behaved very peculiarly, emitting their radio output in regularly spaced bursts rather than as a uniform hiss. Further investigation confirmed that these radio sources were indeed from some celestial object and not from a malfunctioning piece of laboratory equipment. These pulsating radio sources were named *pulsars* soon after their discovery was announced, and the discovery of a pulsar in the heart of the Crab nebula led to the solution of two problems at once. Investigation of the nature of pulsars showed that the pulsar was the source of the high-speed electrons.

☆ THE NATURE OF PULSARS ☆

When a pulsar was discovered in the Crab nebula, it became the target of an intense burst of research activity. Soon this pulsar was discovered to emit pulses in the optical region of the electromagnetic spectrum as well as the radio range; Figure 12.9 shows photographs of the pulsar in its on and off

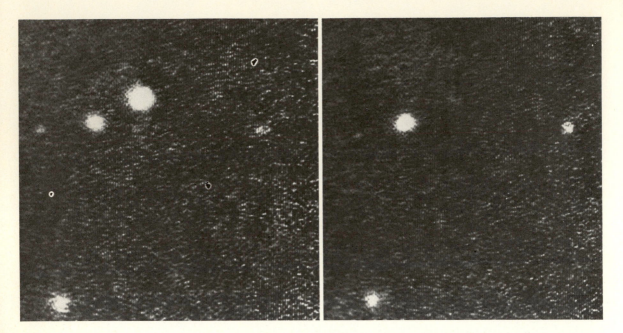

FIGURE 12.9   Television photographs of the pulsar in the center of the Crab nebula. In the left photograph, the pulsar is in its on state and visible, since a pulse is being beamed toward us. In the right photograph, the pulse beam is pointed somewhere else in space, and the pulsar is not visible. (Lick Observatory photograph)

states. Careful measurements of the pulses from the Crab and from other pulsars revealed that the pulse rate slowed down very, very gradually. A rotating object would show such a slowdown if it were losing energy. The importance of the discovery of the Crab pulsar was that we knew a great deal about the central object from studies of the whole nebula. Earlier measurements and calculations showed just how much energy had to be supplied by the central energy source in order to power the Crab.

Cornell University astronomer Thomas Gold put all the facts together in early 1969. The known rate at which the Crab pulsar slowed down provides an indication of the characteristics of the central object, since we know how much energy is lost from the rotating core. Gold's calculations indicated that his earlier pulsar model was correct, that a pulsar is an unbelievably small object, a *neutron star*. These hypothetical objects, proposed by Oppenheimer and others in the 1930s, were at last discovered in the real world. A larger object would lose too much energy when its rotation period increased, so the energy source of the Crab had to be a neutron star.

Neutron stars form when stellar matter is compressed to very high densities. When matter reaches a density of 10^6 g/cm^3, electrons become degenerate and start pressing against each other. Further compression of the matter forces the electrons to become smaller and smaller. When the density exceeds 10^8 g/cm^3, the electrons can shrink no longer, but some electrons find refuge in atomic nuclei by combining with protons to form neutrons. Further compression increases the number of neutrons in atomic nuclei by forcing more and more electrons to the nuclear refuge. At a density of 10^{12} g/cm³ the nuclei can take no more neutrons, so the neutrons escape from the atomic nuclei and are free. At typical neutron-star densities of 10^{15} g/cm³, a microscopic view of matter shows degenerate neutrons, neutrons packed so tightly that they can be squeezed no more. It is possible that still stranger states of matter exist at higher densities. Figure 12.10 schematically depicts matter at such densities.

A neutron star is an extraordinarily tiny object. Those who find white-dwarf densities (tons per cubic centimeter) hard to imagine will boggle still further at neutron-star densities of billions of tons per cubic centimeter. The cupful of white-dwarf stuff, which outweighs two dozen elephants in Figure 11.7, could be replaced by a billionth of a cupful, a small vial of a few thousandths of an inch across, which would still balance the scale. I find it very difficult to visualize matter at such densities. A neutron star is 1000 times smaller than a white-dwarf star, no larger than an average city. It is this kind of an object that powers the Crab.

☆   ANALYSIS OF NEUTRON STARS   ☆

Neutron stars were discovered in the real world when their identification with pulsars was pinned down as a result of analyses of the Crab nebula. But

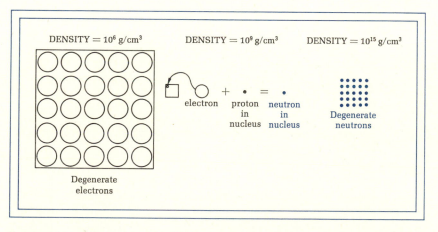

FIGURE 12.10   The appearance of matter at different densities. At white-dwarf densities of 10^6 g/cm³, degenerate electrons are closely packed (left panel). When this matter is compressed to a density of 10^9, or when the left panel is squashed so it is as small as the tiny square in the center panel, electrons combine with protons in atomic nuclei to form neutron-rich nuclei. At still higher densities of 10^{15} g/cm³ characteristic of neutron stars, matter is degenerate neutrons (right panel).

315

we do not stop when the neutron star displaces the white-dwarf star, in the stellar record book, as the smallest known stellar object. We can observe neutron stars only by looking at the pulses they emit in their pulsar stage, and so the analysis of neutron stars depends on precise measurement and timing of the pulses produced by the pulsars. The first question to ask is where the pulses come from.

Pulsar pulses are quite short but extremely regular in their timing. A pulsar actually pulses for an average of 3 percent of its pulse cycle. Many pulsars have two pulses per cycle, one weaker than the other. While the pulses always arrive on schedule, the strength of a pulsar can vary enormously from one pulse to another; some pulsars almost shut off for a few minutes at times. Although pulsars have been observed very intensively in the years since their discovery, no one has yet been able to classify their properties in a simple way.

A number of theorists have asked where pulsar pulses come from, and, while no one model is unambiguously correct, all the pulsar models share some general properties. The pulsar is seen as a rotating, magnetized neutron star whose magnetic axis is tilted with respect to the rotation axis (Figure 12.11). Such a tilt is not uncommon; the earth's magnetic poles are 900 miles away from the rotation poles. Electrons escape from the star's surface more easily near the magnetic poles, forming beams of charged particles. The rapid rotation of the pulsar sweeps these particle beams around and, in some way that is not understood, accelerates the particles and produces beams of electromagnetic radiation. When we observe the pulsar from a great distance, a beam sweeps by us once every time the pulsar turns on its own axis. When the beam points at us, the pulsar is on; when it is pointed somewhere else, the pulsar is off. The rapid rotation of the pulsar produces an on-off cycle that we see as pulses of radiation.

FIGURE 12.11 Pulsar pulses are produced by the rotation of a magnetized neutron star. The magnetic axis is not aligned with the rotation axis, and pulse beams aligned with the magnetic axis sweep around the sky as the pulsar rotates.

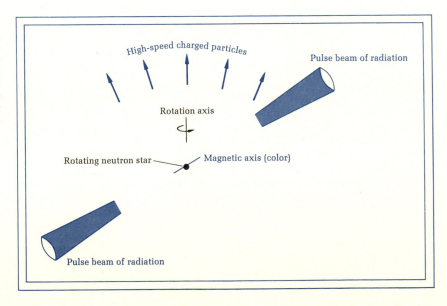

The trouble with this theoretical description of a pulsar is that it is a vague story, not a solid theory. Figure 12.11 is a generalized model, and it involves a certain amount of guesswork as to exactly where the beams of radiation are produced around the pulsar. Do the beams originate close to the pulsar surface, or far from the pulsar? Most of the pulsar's energy output is in the form of high-speed charged particles, not in the form of radiation. Where are these particles produced? Why are the intensities of pulses so variable from one pulse to the next? These questions remain unanswered.

Evolving pulsars generally slow down, because they are continually losing energy and producing high-speed charged particles. (Occasionally an adjustment of the crust of the star will produce a sudden speedup or "glitch.") All pulsars but one run down so that their periods double in a million years or so. No pulsars have been found with periods longer than 4 sec, so presumably once a pulsar becomes that slow the pulses fade altogether or become unobservably faint.

☆ **PULSAR EVOLUTION AND ORIGIN** ☆

Since the internal structure of a neutron star is fixed by the pressure of degenerate neutrons, pulsars do not expand or shrink once they have formed. Neutron stars just cool off, as white-dwarf stars do. They are visible as pulsars for about 10 million years, and then their rotation becomes sufficiently slow so that the pulses fade away. A nonpulsating neutron star would not be visible as a luminous ball because its small size would make it extremely faint.

The origin of pulsars is far less certain than their evolution. There is strong evidence for some kind of association between pulsars and supernovae, since the Crab is known to contain a pulsar, and a Southern Hemisphere pulsar, PSR 0833-45, is associated with a supernova remnant. Although two historical supernovae, Tycho's supernova of 1572 and Kepler's of 1604, did not produce observable pulsars, Figure 12.11 indicates that it is very possible for us to be outside the path of the pulsar's pulse beam and thus not see the pulses. More precise estimates indicate that we can observe something like one-tenth of all pulsars. Thus the evidence indicates that at least some pulsars are born in supernova explosions, the violent death throes of stars.

The working hypothesis currently fashionable in astronomy is that the supernova-pulsar sequence marks the end of the evolutionary path of middleweight stars, stars with masses between, say, 4 and 10 solar masses. Counts of the number of pulsars and analyses of their positions in the Milky Way support this hypothesis. We have fairly strong evidence that low-mass stars make white-dwarf stars, and there are some far weaker arguments that very massive stars are unlikely to make pulsars. The principal difficulty with this hypothesis is that so far no theoretical model of a supernova explosion has been able to explain both the high amount of energy released in the explosion and the production of a neutron star in it—either not enough energy is produced or the explosion blows up the star completely, leaving no remnant. This hypothesis for the genesis of pulsars is a working hypothesis, to be supported, destroyed, or modified by further research.

Neutron and white-dwarf stars are the final state of stars that have managed to find a way to overcome the tendency of gravity to contract the central core of a star. Degeneracy pressure from neutrons or electrons can maintain these stars indefinitely, since the central pressure will be the same whether the star is a sizzling, newborn stellar remnant or a cold object formed billions of years ago. But degeneracy pressure can only support a small stellar core. Electron degeneracy pressure cannot hold up a white-dwarf star more massive than 1.4 solar masses; an overly massive core will squash the electrons themselves and collapse the star. The maximum mass of a neutron star is open to debate; most experts place it at around 1 solar mass, and it is definitely less than 3 or 4 solar masses. While many stars can form such small cores by shedding mass before they progress to the final state, the most massive main-sequence stars known would have to shed almost all their material to stay within the limiting mass of stable degenerate stars. A star ending its life with a massive core will collapse indefinitely, leaving a black hole as its remnant.

★ 12.5 BLACK HOLES IN THEORY ★

The first person to contemplate the ultimate influence of gravity was Pierre Simon de Laplace, early architect of the theory of the origin of the solar system (Chapter 10). Laplace used an erroneous theory of both gravity and light, but he did come to the key conclusion. Compress an object sufficiently and gravity at the surface of that object will become so strong that nothing, not even light, can escape. Associated with the gravitational attraction at the surface of any object is an *escape velocity*, the velocity that a projectile must exceed if it is to escape from the gravitational field of the object. A *black hole* is the state in which a mass is compressed so much that the escape velocity at its surface exceeds the speed of light. Anything that falls into the hole is trapped, since no material object can travel faster than the speed of light and escape from the hole.

While Laplace did realize the extremes to which gravitational attraction could be pushed, he was unaware of the importance of the speed of light as a limiting velocity. A correct description of black holes did not come until Einstein formulated his theory of general relatively in 1916. Newton's theory of gravity is only a good approximation where gravitational fields are weak, and in a black hole gravity is a very strong force. Although numerous alternatives to Einstein's theory of gravity have been proposed since 1916, most predict black-hole properties in accordance with Einstein's theory. One should keep in mind that the theoretical description of a black hole does push Einstein's theory far beyond the domain in which it has been experimentally tested, the motions of objects in the solar system under the comparatively weak gravitational fields found there.

The key feature of a black hole is the *event horizon*, the boundary in space where the escape velocity exceeds the speed of light. A body compressed so that its surface lies within the event horizon forms a black hole. The event horizon is a spherical surface with a radius that depends on the mass of the object inside the hole; the radius of that sphere in kilometers is three times the mass of the object in solar masses. Evidently, a star must be compressed rather drastically if a black hole is to form; a black hole of 1 solar mass is about one-third as large as a neutron star of similar mass.

A black hole is a cosmic funnel, and the event horizon is a cosmic turnstile. Anything that falls within the event horizon will never come out again, for no rocket engine can accelerate it to a speed greater than the speed of light. It is in this sense that a black hole is a giant funnel in the cosmos (Figure 12.12). Once something falls in and is swallowed up, an outside investigator has no way of determining what kind of object fell into the hole. All we know is that the black hole gained a small amount of mass and increased slightly in size.

Near the event horizon, space and time are rather distorted by the hole's gravity, as the funnel diagram (Figure 12.12) indicates schematically. One of the most important consequences of Einstein's description of gravity was his realization that space and time are interdependent. He called the space-time arena in which objects move the "space-time continuum," and his description of gravity was one of an influence that distorts the space-time continuum around a mass. The details are rather complex, but one of the strangest of all black-hole properties is its effect on something falling into it

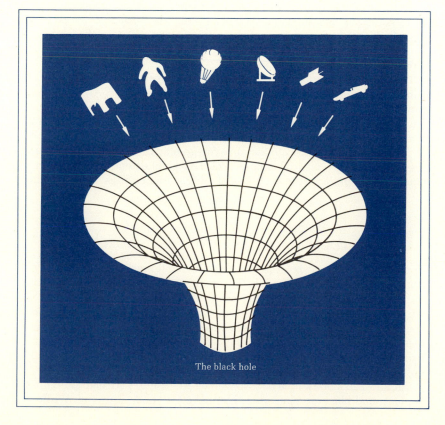

The black hole

FIGURE 12.12 Everything falling into a black hole loses its identity. You cannot know whether it was a space probe or a television set that fell in. (Adapted from Remo Ruffini and John A. Wheeler, "Introducing the Black Hole," *Physics Today* (January 1970): 31. © American Institute of Physics.)

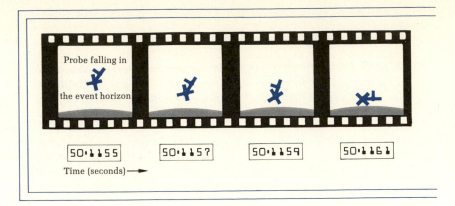

FIGURE 12.13 A "movie" showing how a distant observer would see a space probe approach a black hole. As the space probe falls in, its motion appears to freeze at the event horizon. The numbers are meant to apply to the fall of a rocket ship toward

(Figure 12.13). Someone watching a space probe falling toward a black hole will see the probe's motion appear to stop near the event horizon, for gravity distorts the passage of time near a black hole very severely. To us, on the outside, it will look as though time has come to a stop near the event horizon. But to someone on the space probe, time will march on as the probe falls inexorably into the black hole.

The future of a black hole is as simple as the future of the other stellar remnants, the degenerate neutron stars and white-dwarf stars. All a black hole can do is eat and grow. Black holes formed from the collapse of massive stars are very, very tiny on the cosmic scale, and interstellar space is quite empty. Even the most optimistic assumptions about the rate at which a black hole accumulates matter indicate that an average black hole will take five times the age of the universe to double in size. To a great extent, the evolution of a black hole is similar to the evolution of other stellar corpses: It just sits there.

★ 12.6 BLACK HOLES IN THE REAL WORLD ★

The black-hole story is a good story. Such strange objects are very intriguing, and they have struck the fancy of a number of science fiction writers. But do black holes really exist? This question has been one of the most exciting and controversial astronomical questions of the last few years, because there is now good evidence to believe that black holes really do exist in the real world. Most astronomers guardedly accept the existence of real black holes; at the present time the evidence indicates that real black holes are a probability, not a certainty.

But how can a black hole be discovered? Black holes are black and the sky is black, and black on black does not make for good observing. While it is

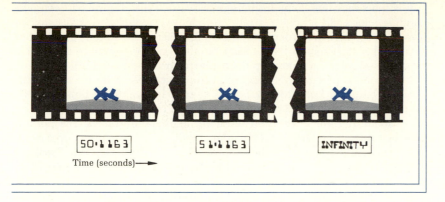

Time (seconds) ⟶

a 10-solar-mass black hole. (From Harry L. Shipman, *Black Holes, Quasars, and the Universe.* Copyright © 1976 by Houghton Mifflin Company, pp. 74—75.)

possible, in principle, to observe the effect of a black hole's gravity on the paths of photons traveling near it, the practical observation and discovery of black holes by this method is rather unlikely. But if a black hole were to exist in a binary system, the companion star would orbit the black hole, and we could detect the influence of the hole's gravity on the motion of the companion star. A critical question would still remain: Is the invisible object in a binary system a black hole or something else? In one binary system, *Cygnus X-1*, the evidence is quite strong that the system contains a black hole. There are several other binary systems where the evidence is not so clear.

☆ **CYGNUS X-1: THE BEST BLACK-HOLE CANDIDATE** ☆

As its name indicates, Cygnus X-1 is a source of celestial x rays. Precise location of the x-ray source in 1971 indicated that the x rays were coming from a system containing an otherwise unremarkable hot, blue supergiant star about 2500 pc away from the earth. A series of spectra of this star showed that the blue supergiant's spectrum produced Doppler shifts that alternated, the characteristic signature of a spectroscopic binary star (recall Chapter 9). These Doppler shifts showed that the star alternately moved toward and away from us when it orbited around an invisible companion. Since only one object in the system has a spectrum, the mass of each member of the system could not be fixed from the available information. However, a variety of analyses of the system indicate that the invisible companion must have a mass of at least 8 solar masses. Such an object is too big to be a normal neutron star or a white-dwarf star—but must it be a compact object at all? Is it not a normal star? There are several spectroscopic binaries with massive, invisible components. Why is Cygnus X-1 a good black-hole candidate?

It is the x rays that come from Cygnus X-1 that make it the best available candidate for a black hole. Figure 12.14 shows a model of this system. Gas streams from the blue supergiant toward the black hole. In the current view, this gas is simply matter that is lost from the blue supergiant by the normal processes of mass loss and is pulled toward the hole by the hole's gravity. The important part of this process is the formation of an *accretion disk* around the black hole. Matter falls downward toward the hole, swirling around and becoming hotter and hotter. In the inner region of the accretion disk, about 200 km away from the hole, temperatures reach the point at which x rays are emitted. While other models for the x-ray source are of course possible, it is very difficult to produce the high x-ray luminosity of Cygnus X-1 (about 10^{37} ergs/sec or 2000 times the optical luminosity of the sun) in any other reasonable way.

At the present time, the most controversial part of the Cygnus X-1 model depicted in Figure 12.14 is the argument that the object at the center of the accretion disk is a black hole. Accretion disks can form around any type of stellar remnant—white-dwarf star, neutron star, or black hole. The evidence indicates that the Cygnus X-1 system contains an accretion disk surrounding a stellar remnant and an 8-solar-mass companion. Devil's advocates seeking to avoid a black hole in the Cygnus X-1 system have argued that the 8-solar-mass companion may not be the object at the center of the accretion disk, that the blue supergiant has *two* companions—a neutron star at the center of the disk and an 8-solar-mass main-sequence star. It is difficult,

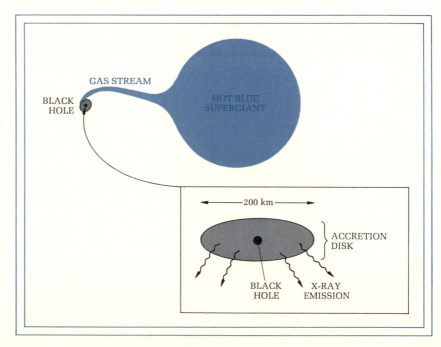

FIGURE 12.14   A model of the Cygnus X-1 system. The top panel shows a large-scale view of the system; mass lost from the blue supergiant is pulled toward the black hole and forms an accretion disk around the hole. The bottom panel shows the inner part of the accretion disk. X rays are emitted by matter in the accretion disk swirling around the black hole.

GAS STREAM

BLACK HOLE

HOT BLUE SUPERGIANT

200 km

ACCRETION DISK

BLACK HOLE

X-RAY EMISSION

almost impossible, to see how such a system could have evolved. Others argue that degenerate neutron matter could exhibit rather bizarre behavior and make an "obese" neutron star with a mass of 8 solar masses, but some of the fundamental principles of physics would have to be violated if such objects were to exist. The conclusion seems inescapable. The blue supergiant in the Cygnus X-1 system orbits a black hole. Black holes exist in the real world.

Yet must Cygnus X-1 be a black hole? Reflect on the last two paragraphs and note the number of times that the reasonableness of plausibility of an argument was appealed to. The model of Cygnus X-1 illustrated in Figure 12.14 is the most reasonable model of the system, but is it the only possible model for the system? Time will tell.

Perhaps the best way to settle the Cygnus X-1 controversy would be to discover another similar system. Opponents of the black-hole model for Cygnus X-1 can always argue quite correctly that this particular system is pathologically peculiar; nature may have conspired to cleverly disguise a neutron star (or something else) as a black hole. The existence of such a clever disguise in several different systems is quite difficult to accept. Thus the discovery of several other similar systems might help support the case for a black hole in Cygnus X-1.

☆ BINARY X-RAY SOURCES ☆

Perhaps one of the most productive research ventures was the launching of a small x-ray telescope on a satellite orbiting above the earth's atmosphere in 1969. Because x rays are blocked by the earth's atmosphere, x-ray telescopes need to be placed above it, and the expense of a satellite is offset by the fact that the satellite can continuously observe the sky. This satellite, named Uhuru, has discovered 161 x-ray sources, about 100 of which are located in the Milky Way galaxy. About 10 of these sources are supernova remnants, and many if not all of the others are probably binary systems containing a star in its final state—white-dwarf star, neutron star, or black hole. Figure 12.14, the Cygnus X-1 model, is a generalized model for such a system, although for many of the binary x-ray sources the stellar member of the system may not be a blue supergiant, and the object at the center of the accretion disk may not be a black hole.

Several of the binary x-ray sources are very similar to Cygnus X-1 in their x-ray characteristics. Three of these—named SMC (Small Magellanic Cloud) X-1, 3U 1700-37, and 3U 0900-40 (3U for the Third Uhuru catalogue)—are known to be identified with blue supergiants quite similar to the supergiant in Cygnus X-1. These other systems are not as good black-hole candidates as Cygnus X-1, though, because the mass of the x-ray source is in each case between 1 and 4 solar masses. While neutron stars are probably not that large, none of these systems has an x-ray source that is so massive that it must be a black hole.

Two binary x-ray sources, Centaurus X-3 and Hercules X-1, contain pulsating x-ray sources, or x-ray pulsars. The systems resemble Cygnus X-1 in their overall characteristics, but the companion stars are less massive and cooler, and the center of the accretion disk contains a neutron star (probably) or a white-dwarf star (possibly). A third category of low-mass objects has two members, Scorpius X-1, the brightest x-ray source in the sky, and Cygnus X-3. These objects may well be analogous to the dwarf novae. Many of the x-ray sources discovered by Uhuru have not been identified with any star yet, so we do not know exactly what they are.

A most puzzling class of objects are the *transient x-ray sources*, objects that suddenly emit a burst of x- or gamma-rays. These objects were first discovered by the Vela satellites, a group of space probes designed not to do astronomy but to discover gamma-ray blasts from clandestine atomic bomb explosions on the earth. Other satellites confirmed the Vela discoveries.

The role of serendipity in science appeared once again when the Vela satellites discovered that, several times a year, some celestial objects flare up in the gamma-ray part of the electromagnetic spectrum. These bursts are extremely variable, and objects have since been discovered to produce bursts of x rays too. One gamma-ray burst may have come from the Cygnus X-1 system. If these bursts come from systems like the binary x-ray sources, the peak luminosity of the objects is 10^{36} to 10^{40} ergs/sec, comparable to and perhaps greater than the steady luminosity of Cygnus X-1. About 30 distinct models for these sources have been proposed, but none has been accepted by the research community as being necessarily correct.

☆ THE FUTURE OF X- AND GAMMA-RAY ASTRONOMY ☆

A great deal of information about the late evolution of stars in close binary systems has come from x-ray astronomy. It was the x rays from Cygnus X-1 that led to its probable identification as a black hole. More objects like it need to be found. This field is currently one of the most exciting areas of astronomical research. The successful launching of the High Energy Astronomical Observatory (HEAO) should produce much more progress.

A number of fields, such as x-ray astronomy, are expensive fields in which to investigate, and this expensiveness has led to a change in the style of astronomical research. The HEAO, costing about $300 million, is very expensive and cannot possibly be funded by private donors. Only the federal government can support such a venture; on a federal scale, $300 million is an expensive but possible item in the national budget. As a comparison, $300 million equals the cost of two B-1 bomber planes, and the entire B-1 program would cost $23,000 million, almost 100 times more than HEAO. Astronomers can no longer be cloistered on mountaintop observatories like those which were endowed by wealthy individuals at the end of the nineteenth century. They must enter the world of science politics in Washington. While a great many fields, such as variable-star work, require only modest resources, an astronomer who has to go above the atmosphere to observe a part of the electromagnetic spectrum must depend on the government—ultimately on the taxpayer—for support.

Dying stars must finally cope with the tendency of gravity to collapse their cores. Neutron stars and white-dwarf stars are supported by degeneracy pressure, the pressure created by the close packing of particles of matter. Both of these types of stellar final states are very small, no larger than planets (white-dwarf stars) or cities (neutron stars). These objects are known to exist in the real world.

A star that cannot resist the pull of gravity becomes a black hole. For a long time, black holes were just theoretical objects, but recent work has indicated that they probably exist in the real world.

This chapter has covered:

The physical properties and evolution of white-dwarf stars;

The explosive variable stars, the novae and supernovae;

The discovery of pulsars;

The identification of pulsars as neutron stars, their properties, and evolution;

The properties of black holes; and

The discovery of black holes in binary systems, with Cygnus X-1 as the best black-hole candidate.

KEY CONCEPTS

Accretion disk	Escape velocity	Pulsar
Black hole	Event horizon	Supernova
Cataclysmic variable	Neutron star	Synchrotron radiation
Crab nebula	Nova	Transient x-ray source
Cygnus X-1		White-dwarf star

REVIEW QUESTIONS

1. Briefly describe the differences between a white-dwarf star and a neutron star.

2. Can a white-dwarf star ever become a black hole? If so, under what circumstances? If not, what eventually happens to it?

3. What is the difference between a nova and a supernova?

4. Three different types of objects exist in the Crab nebula. What are they, and what type of radiation do they emit?

5. What is the importance of the Crab nebula in the study of stardeath?

6. How does a pulsar evolve?

7. What is the event horizon?

8. Summarize the arguments leading to the discovery of a black hole in Cygnus X-1.

FURTHER READING

Gingerich, O., ed. *New Frontiers in Astronomy.* San Francisco: Freeman, Part VI, "High-Energy Astrophysics."

Golden, F. *Quasars, Pulsars, and Black Holes.* New York: Scribners, 1975.

Shipman, Harry L. *Black Holes, Quasars, and the Universe.* Boston: Houghton Mifflin, 1976.

SUMMARY OF PART THREE

The life cycle of a star is a continuing interplay between gravity and other forces. A star is born when gravity causes a diffuse interstellar cloud to fragment and condense into small globs of gas, which become hot enough in their cores to sustain nuclear reactions. Stars like our sun use hydrogen as an energy source, and most of the stars in the sky are similar in structure. When the hydrogen runs out, the star swells up—becoming a red giant—turning to other fuels as energy sources. When all fuels are exhausted, the star dies, becoming a white-dwarf star, a neutron star, or perhaps a black hole. Almost all the stars in the sky, the ordinary ones and the peculiar ones, fit into this evolutionary scheme.

SOLID FACT

Stars are huge balls of hydrogen and helium gas.

Various stages of a newly forming star are known to exist: H I regions, dense molecular clouds, infrared sources, dust clouds, and the H II region that appears once a massive star forms.

Various classes of stars exist: main-sequence stars, red-giant stars, white-dwarf stars, and neutron stars.

Cepheid variables obey the period-luminosity relation, which makes them good distance indicators.

SECURE INTERPRETATION

The energy source of stars is nuclear fusion.

Stars evolve to the red-giant stage after the main-sequence stage.

White-dwarf stars and neutron stars evolve only by cooling.

Pulsars are neutron stars.

PROBABLE FACT

Cygnus X-1 is a black hole.

Solar activity comes from magnetic interactions on the solar surface.

WORKING MODELS

Star formation follows the sequence outlined in Chapter 10.

The planets formed from the solar nebula by chemical condensation.

Lightweight stars form white-dwarf stars, and middleweight stars become supernovae.

OPEN QUESTIONS

Where are the missing solar neutrinos?

What powers a supernova explosion?

★ PART FOUR ★

GALAXIES

Stars, like people, tend to be found in groups. The smallest of these stellar aggregations are the star clusters, clumps of 100 to 1,000,000 stars. These clusters are generally parts of larger galaxies, collections of 10^6 to 10^{12} stars. All the stars visible in the sky are part of our own Milky Way galaxy, which we see as a faint band of light stretching around the sky (Figure 13.1). Three other galaxies are visible to the unaided eye: the two Clouds of Magellan and the Andromeda galaxy. The Magellanic clouds are the two fuzzy patches in Figure 13.1; since they are near the South Celestial Pole, they are only visible from the tropics and the Southern Hemisphere of the earth.

The study of galaxies is difficult because galaxies are distant, faint objects. The first big question of astronomical research ("Where is it?") was only answered in the 1920s when the discovery of Cepheid variables in the Andromeda galaxy indicated that it was a gigantic star swarm similar to the Milky Way. Even now, distance estimates for galaxies are continually being revised. At present, most research on galaxies focuses on the second big question, "What is it?" Our own Milky Way is both the best and worst galaxy to study because interstellar dust limits our view of our immediate neighborhood. But while most researchers merely describe the nature of galaxies, classify, catalogue, and measure them, some tentative gropings toward the third big question, "How does it evolve?" are guiding our efforts. Some pieces of the evolutionary puzzle of galaxies are falling into place, and future observational efforts will be directed toward unraveling the past history of these giant star systems.

★ CHAPTER THIRTEEN ★

THE MILKY WAY

Next time you are far away from the glare of city lights, look up at the Milky Way. This faint, fuzzy band, stretching across the sky, has fascinated astronomers since the dawn of civilization. Most cultures have thought of this band as a road toward heaven or some mythical imperial court. The telescope demonstrated that this diffuse light was the combined light from uncountable numbers of stars, too faint to be seen individually.

Thanks to the interstellar dust, visible as great gashes in illustrations like Figure 13.1, it is quite difficult for us to discover just where we live within this huge stellar system. The difficulties of mapping the Milky Way are not insurmountable, and recent research has provided a fairly comprehensive picture of this galaxy that is our home.

★   MAIN IDEAS OF THIS CHAPTER   ★

We live about 10,000 pc from the center of a disk-shaped galaxy.

The disk of our spiral galaxy is broken into spiral arms.

The typical galactic star is a cool, dim, main-sequence star.

The stellar content of the galaxy can be divided into two populations of stars, the young Population I and the old Population II.

The galactic center is the location of high-energy activity not unlike the activity associated with supernovae.

The Milky Way rotates.

The Milky Way has evolved chemically, since supernova explosions increase the heavy-element abundance of galactic material.

★   13.1 SHAPE OF THE MILKY WAY   ★

Historically, the first big question of astronomical research, the question of the location of other objects in the universe, has had a considerable impact on the philosopher's view of the human condition. Before the Copernican revolution, we were seen as the center of the entire universe; the sun, planets, and stars were little lights in the sky that whirled around the earth. While Copernicus and his successors showed that the earth was but one of nine planets in the solar system, the insignificance of the location of the sun and its planets among the stars was not appreciated until the twentieth century. Were we somewhere near the center of a stellar grouping? Were the stars in the sky the only ones that existed? How large is the observable universe?

FIGURE 13.1   Chart of the Milky Way galaxy. This chart shows the entire sky, with the Milky Way running across it as a horizontal band. The two fuzzy objects at the lower right are the Magellanic clouds, the two galaxies nearest to us. (Martin and Tatiana Keskula, under the direction of Knut Lundmark, created this illustration. Composite painting of the Milky Way used by permission of Lund Observatory.)

☆   THE FLAT GALACTIC DISK   ☆

Look at Figure 13.1. The existence of a band of light around the sky, referred to as the Milky Way, shows that we live in a flat system of stars. The Milky Way band, seen in the sky, is the sum of the light from large numbers of very faint stars. The concentration of this light from faint stars in a band is the result of our position in this flat galaxy (Figure 13.2). As you look along the plane of the Galaxy, many stars are visible. An observer whose line of sight is directed upward, out of the galactic plane, will see few stars and no fuzzy band of light. More precise measurements of the distribution of stars indicate that our galaxy is several hundred parsecs thick.

This analysis only partially describes the shape of the Milky Way galaxy. How far does this flat galaxy extend in space? Are we in the middle or off to one side? Generations of astronomers, starting with William Herschel in the late eighteenth century, attacked this question by counting stars in various directions in the Milky Way. If stars were uniformly distributed within the Milky Way, the directions in which the largest number of stars were seen would be those in which the Milky Way extended furthest from us. This

333

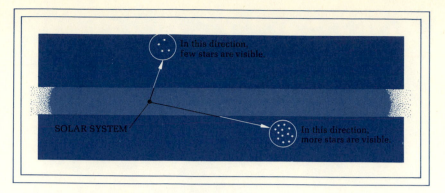

FIGURE 13.2 The Milky Way galaxy is flat. Star counts in different directions indicate that the stellar system we live in is far more extensive in the part of the sky perceived as the Milky Way.

technique is just an extension of the analysis depicted in Figure 13.2. Herschel and his successors did not notice any particular direction in which the faint stars of the Milky Way seemed most numerous, and so they naturally concluded that the Galaxy extended an equal distance in all directions. The Dutch astronomer J. C. Kapteyn, working in the early twentieth century, believed on the basis of his star counts that the visible Milky Way was a disk-shaped galaxy, about 2 or 3 kpc (1 kpc = 1000 pc) in radius, and the sun was somewhere near the center. The Copernican revolution may have thrown the earth out of the center of the universe, but until the early twentieth century we still believed that our solar system was at the center of the Galaxy.

What these early astronomers did not realize was the influence of interstellar dust on the visibility of the more distant parts of the Galaxy. Their 3-kpc disk was a reasonable map of the part of the Galaxy that is visible from the earth, but they did not realize that behind the dust clouds lay a huge, larger disk. Harlow Shapley's determination of the distances to the globular clusters extended our view beyond the small part of the *galactic disk* that we can see optically.

☆ **GLOBULAR CLUSTERS AND THE GALACTIC CENTER** ☆

The breakthrough in our understanding of the shape of the Milky Way galaxy came with the determination of the distances to star clusters. The two principal types of star clusters were described in Chapter 11 in connection with stellar evolution. *Open clusters*, sometimes called "galactic" clusters, are loose aggregations of relatively young blue stars. The Pleiades, shown in Color Photo 13, are an open cluster. *Globular clusters*, one of which is shown in Figure 11.1, are closely packed aggregations of stars that contain older stars with fewer heavy elements.

Star clusters are particularly useful in mapping the more distant portions of the Galaxy because their distances can be determined relatively easily. Most star clusters contain at least one or two variable stars, and the period-luminosity relation for these Cepheid variables can be used to determine the variable's luminosity and hence its distance, using the method described in section 11.4. The Hertzsprung-Russell diagram of a cluster can be used to pick out those stars that are on the main sequence, and the main-sequence stars in the cluster can be presumed to have the same luminosity as the main-sequence stars of similar color in our neighborhood. Box 13.1 provides a more detailed description of this second method of finding the distances to the star clusters.

The first hint that we live on the edge of our galaxy was provided by Harlow Shapley's determination of the distances of several globular clusters. When Shapley came to Mount Wilson in 1914, after receiving his doctoral degree from Princeton, he used the 60- and 100-inch telescopes at the Observatory to study the variable stars in clusters. The period-luminosity relation had recently been discovered by Henrietta Leavitt at Harvard, and Shapley wanted to explore this relation further by observing variable stars in globular clusters. Once an RR Lyrae variable, a short-period Cepheid with a period of half a day, has been found in a globular cluster, it is easy to determine the distance to the cluster, since all RR Lyrae variables are about 40 times as luminous as the sun in the visual band of the electromagnetic spectrum. Measure the brightness of the variable, and you can determine whether it is a bright near one or a dim far one. Box 13.2 provides a specific example. Shapley's distance measurements astounded the astronomical community, for he stated that these globular clusters were tens of kiloparsecs away. The system of globular clusters seemed to be 5 times the size

---

Box 13.1 Cluster Distances by Main-Sequence Fitting

The figure shows a color-magnitude diagram of both the stars in a star cluster (in color) and the absolute magnitudes of stars in the solar neighborhood (in black). The difference between the apparent and absolute magnitudes of a star is a measure of its distance. In this case, the distance modulus is $+5$. Thus, from equation B9.5,

$$m - M = -5 + 5 \log (\text{distance})$$

Letting $m - M = +5$, we have

$$+5 = -5 + 5 \log (\text{distance})$$
or $+10 = 5 \log (\text{distance})$

indicating that $\log (\text{distance}) = 2$ or that the cluster is 100 pc away. A similar calculation can be performed in the language of luminosity and flux.

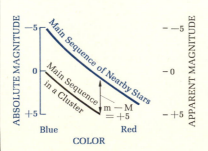

Box 13.2 Variable
Stars and the
Distance to M 13

The globular cluster M 13, a group of 10^5 stars, contains a number of RR Lyrae stars with apparent visual magnitude $+15.2$ and a period of 12 hours. Their absolute visual magnitude is $+0.8$. Applying the relationship between apparent magnitude, absolute magnitude, and distance, we have

$$m - M = 15.2 - 0.8 = 14.4$$
$$= -5 + 5 \log \text{(distance)}$$
$$\text{or } 19.4 = 5 \log \text{(distance)}$$

which gives $\log \text{(distance)} = 3.9$, or a distance of 7.9 kpc to M 13. This distance is something like three times the accepted size of the Milky Way before Shapley determined the distances to globular clusters like M 13.

While it is not done in practice, one can also use the language of luminosity and flux to see how the distance to a globular cluster like M 13 is measured. Photometry shows that an RR Lyrae variable with a period of half a day produces a flux of 1.9×10^{-12} ergs/cm²/sec in the visual part of the electromagnetic spectrum. The period-luminosity relation indicates that such a star has a visual luminosity 40 times the solar visual luminosity, or $40 \times 3.2 \times 10^{32}$ ergs/sec. The fundamental relationship between flux, luminosity, and distance (equation 9.1) can then be used:

$$\text{Flux} = \frac{\text{luminosity}}{4\pi(\text{distance})^2}$$

or

$$1.9 \times 10^{-12} = \frac{40 \times 3.2 \times 10^{32}}{4\pi(\text{distance})^2}$$

and this equation can be solved for the distance in the same way as the distance to Polaris was determined in Box 11.2.

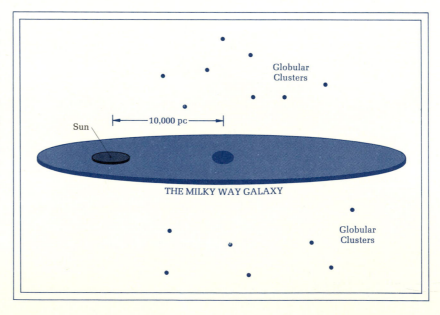

FIGURE 13.3   A schematic view of the Milky Way galaxy. The volume of space shown in Figure 13.2, including all the Milky Way stars visible to the eye, is shown in black. Until 1920, this region was thought to be the entire galaxy. Shapley's determination of the distances to the globular clusters (color dots) showed that we live about 10,000 pc from the core of our galaxy. Most of the galactic disk is invisible to us.

Globular
Clusters

10,000 pc

Sun

THE MILKY WAY GALAXY

Globular
Clusters

of the part of Milky Way galaxy that could be mapped by looking at individual stars (Figure 13.3). Many astronomers, comfortable with the small galaxy depicted in black in Figure 13.3, were unwilling to accept the fact that the Milky Way was 5 times as large as they had thought, and that once again our privileged position in the center of things was destroyed.

Figure 13.3 shows how the globular clusters in the Galaxy are distributed in space. While the stars of the Milky Way are concentrated in a disk, the old globular clusters form a spherical *halo* surrounding the galactic core. Current measurements of the distances to the globular clusters indicate that we are 10 kpc away from this core. Most of the galactic disk is invisible to us, hidden by interstellar dust.

☆　**MAPPING THE SPIRAL ARMS**　☆

Measurements of the distances to the globular clusters provide a rather crude map of our galaxy, schematically depicted in Figure 13.3, delineating the division of the galaxy into disk and halo regions. But these clusters in the galactic halo do not provide any clue to the distribution of matter in the disk. Is the disk a smooth, uniform one, or is matter within the disk organized into subunits? Mapping the disk is rather difficult, because we can only see a small portion of it. Distance measurements within the disk are complicated by the presence of interstellar dust, which absorbs light from distant stars and makes them appear fainter than they should at their distances. The dust thus makes stars seem farther away than they really are.

All that a stellar astronomer trying to map the galactic disk can do is plow ahead through the muck, compensate for the dust as well as possible, and try to map the Galaxy using a variety of tracers. The most reliable tracers, objects whose distances can readily be determined, are the open clusters. The distance of an open cluster can be measured by the main-sequence fitting method (Box 13.1). The distances of Cepheid variable stars can be measured through use of the period-luminosity relation. In some cases, spectral classification can place stars in the H-R diagram so well that their luminosities can be determined from their spectra. Once a stellar luminosity is known, a measurement of the brightness gives the distance. Use of these optical tracers provides the picture of the local part of the Galaxy described by the black dots in Figure 13.4.

Radio astronomers also try to map the Galaxy, since 21-cm radio waves pass unimpeded through the interstellar dust. Yet a radio astronomer looking at a cloud of neutral hydrogen gas (an H I region) has no direct way of determining how far away the H I cloud is. The inevitable astronomical question arises: Is it a small cloud close by or a huge distant one? Since our disk-shaped galaxy rotates, a number of radio astronomers have attempted to enlist galactic rotation as an aid to distance determinations, for the motion of an H I cloud can be partially measured by the Doppler shift. An assumption that the Galaxy is a circularly rotating disk can then provide some estimate of the distance to the cloud from an analysis of its motion. Unfortunately, the Galaxy is not a circularly rotating disk, but astronomers have to make do with what nature gives us and analyze the observations as well as

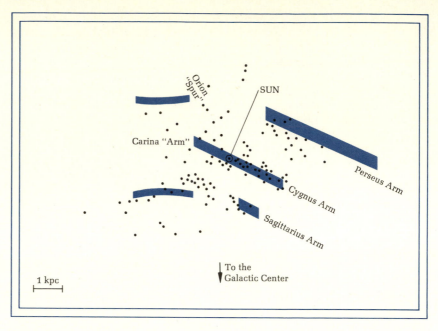

FIGURE 13.4 Spiral structure of the Milky Way in the vicinity of the sun. Black dots represent the locations of the optically observed objects. The color regions are the locations of spiral arms as derived from radio observations of the 21-cm line. While the observations are a little confused, it does seem evident that there are three well-defined spiral arms—the Sagittarius, Cygnus, and Perseus arms. Whether the Cygnus arm extends in the direction of Orion or Carina is uncertain; the figure shows the spiral feature in Carina as an arm, and the optical tracers in the direction of Orion as a spur. [Optical data from B. Bok and P. Bok. *The Milky Way,* 4th ed. (Cambridge, Mass.: Harvard, 1974), p. 210; radio data from G. Verschuur, "High-Velocity Neutral Hydrogen," *Annual Review of Astronomy and Astrophysics,* 13 (1975): 257–293.]

we can. The assumption of circular motion produces a map of the hydrogen clouds in our immediate neighborhood, shown in color in Figure 13.4. Further, radio astronomers have confirmed that the galactic disk has the radius of 15 kpc that Figure 13.3 shows.

Both the radio-astronomical studies and the studies of optical tracers indicate that in our neighborhood, at least, our galaxy contains several *spiral arms*. Photographs of other spiral galaxies (Color Photo 17, Figures 14.1, 14.3, and 14.5) indicate that the spiral structure is an expected feature of a galaxy like ours. These spiral arms are named for the constellations that lie in the same direction. Two spiral arms are clearly delineated: the Perseus arm, about 1 to 1½ kpc away from the sun toward the galactic rim, and the Sagittarius arm, an equal distance away in the direction of the galactic core. The arm in which the sun is located is not as well outlined, for there are spiral features that extend in the directions of Cygnus, Carina, and Orion. The Orion feature is not seen in the radio map. Some investigators believe that the Carina arm is part of the Sagittarius feature, and our spiral arm passes from Cygnus to Orion. Others believe that the Cygnus and Carina features form an arm and that the Orion feature is just a spur.

Another look at Figure 13.4 and some reflection on the last paragraph might lead a perceptive reader to conclude that we know next to nothing about the spiral structure of our own galaxy. It's not quite that bad, but our understanding is clearly limited by the murkiness of the interstellar dust. About 25 years ago, when the 21-cm radiation was first discovered, a number of astronomers hoped that the radio astronomers could push the dust aside and reveal the mysteries of galactic structure, in the same way that they have penetrated the interstellar clouds and are beginning to reveal the processes that lead to star formation (Chapter 10). Now, years after the 21-cm radiation was discovered, the hope of understanding spiral structure is still a hope, and the optimism of the early workers has been dampened. Science sometimes works that way; dramatic results come from unexpected directions while promising lines of research prove frustratingly unfruitful.

★ 13.2 CONTENT OF THE GALAXY ★

While studies of galactic structure cannot unravel the shape of the spiral arms in our disk, they can provide considerable insight on the type of objects that exist in our galaxy. Researchers wishing to investigate spiral structure have turned their attention to other galaxies, because the spiral structure of a distant galaxy such as Andromeda can be determined from a few photographs like Color Photo 17. Yet Andromeda is so far away that only the most luminous stars can be photographed individually, and none of them can be studied in detail. The stars in our own galaxy, though, are quite close to us. We start with our immediate stellar neighborhood, poetically described by Walter Baade as "the local swimming hole" of the Galaxy.

☆ CENSUS OF THE SOLAR NEIGHBORHOOD ☆

Table 13.1 lists the densities of various objects in the solar neighborhood. Evidently, the bright giant and supergiant stars that show up as the most striking stars in the sky are quite rare; the average galactic star is a cool, small, low-luminosity star with an M-type spectrum on the main sequence, or a white-dwarf star. There seem to be more and more stars on the main sequence as we follow it toward the dim bottom end. Is there no limit to the hordes of stellar glowworms? Recent and still controversial evidence indicates that the low-mass, dim stars start to thin out, becoming less numerous at 10^{-4} solar luminosities, roughly corresponding to a mass of 0.1 solar mass or roughly 100 times the mass of Jupiter. The rare red giants are far outnumbered by the main-sequence stars and white-dwarf stars. The number of unobservable nonpulsating neutron stars and invisible black holes can only be estimated; our present knowledge of the masses of the progenitors of these objects indicates that each is about one-fifth as abundant as the white-dwarf stars, leaving something like 1 billion of each in the entire Milky Way.

TABLE 13.1   CENSUS OF THE SOLAR NEIGHBORHOOD

TYPE OF OBJECT	NUMBER[a] PER 1000 pc³
Main-sequence stars:	
Spectral class O	3.5×10^{-5}
B	0.08
A	0.4
F	3.0
G	5.7
K	9.9
M	65
Giants, hotter than class G	0.03
Giants, spectral classes K, M	0.04
White-dwarf stars	30[b]
Interstellar clouds (H I regions)	0.08

[a] These numbers are calculated from the data in C. W. Allen, *Astrophysical Quantities*, 3rd ed. (London: Athlone Press, 1973).

[b] This number includes both visible and invisibly cool white-dwarf stars. The number of visible stars is from E. M. Sion and J. W. Liebert, "Space Motions and Luminosity Function of White Dwarfs," *Astrophysical Journal*, 213 (1977): 468–478; the number of invisible stars is based on H. L. Shipman (paper in preparation).

Table 13.1 describes a reasonably representative sample of the types of stars found in the galactic disk. Many halo stars are found in the globular clusters, and study of these objects reveals some significant differences from the census of Table 13.1. The stars at the upper end of the main sequence—the hot, luminous O, B, and A stars—are missing from globular clusters. No gas has ever been found in one. Red-giant stars are somewhat more abundant in globular clusters, but there are no extremely luminous giant stars like Betelgeuse in them. Stars seem to be divided into two *populations*, one characteristic of the disk and the other characteristic of the halo.

☆   **STELLAR POPULATIONS**   ☆

The existence of two distinct tribes of stars, the somewhat heterogeneous group found in the disk and the group of the halo that contains no hot main-sequence stars, was discovered by California astronomer Walter Baade during World War II. The blackout of Los Angeles created by wartime conditions eliminated the light pollution that has all but destroyed observing conditions at Mount Wilson. Baade, on the mountain at that time, was able to secure some superb photographs of the Andromeda galaxy because he did not have to work through the haze of city lights. Analysis of his photographs of Andromeda and of other elliptical galaxies indicated that the stars in Andromeda were divided into two groups, which he called *Population I* and *Population II*.

These populations correspond to the disk and halo populations of our own galaxy, and Table 13.2 summarizes the differences between these two groupings of stars. More recent research indicates that these two populations represent two extremes. The solar neighborhood, for example, does not contain as many blue main-sequence stars as an extreme Population I group like the Pleiades star cluster, but it still contains more gas, more hot main-sequence stars, and more heavy elements than an extreme Population II object like the globular cluster M 13.

The distinction between the two types of stellar populations turned out to be important to the use of Cepheid variable stars as distance indicators. Disk Population I stars contain ten times as many heavy elements (metals) as the halo Population II stars; as a result, the period-luminosity relations for these two groups of stars are different. Early workers determined the distance to the Andromeda galaxy by measuring the periods of its Cepheids. They used the period-luminosity relation for Population I stars in the spiral arms of the Milky Way galaxy to determine the luminosities of Population II variables in Andromeda. Consequently, erroneous luminosity values were assigned to the Andromeda stars. The distance measurements, made by comparing the luminosities of the Andromeda stars to their brightnesses, were wrong. Proper comparison of the variables places Andromeda at its presently accepted distance of 680 kpc, making it similar in size and shape to our galaxy.

Color Photo 17, showing the Andromeda galaxy, illustrates the distinction between the two types of stellar populations. The nucleus of Andromeda contains stars with Population II characteristics. The yellow color of the nucleus in the photograph comes from the abundance of yellow and red stars in a Population II group. The spiral arms are much bluer than the nucleus, indicating that they contain luminous, blue main-sequence stars. Population II stars are found in places where all the stars were formed tens of billions of years ago, for only the faint red stars still survive on the main sequence. Population I stars are younger, for the hot blue stars are still around. The spiral arms contain the gas clouds that are necessary for star formation, and so they are a logical place to find the young Population I stars.

TABLE 13.2 STELLAR POPULATIONS

CHARACTERISTIC	POPULATION I (Disk and spiral arms)	POPULATION II (Halo)
Most luminous stars	Blue giants	Red giants
Heavy-element (metal) content	High	Low
Cluster types	Open	Globular
Gas content	High (4–5%)	Low (practically zero)
Galactic orbits	Circular	Eccentric

★ 13.3 THE GALACTIC CORE ★

The nucleus of our galaxy is hidden from optical astronomers by the familiar problem of galactic-structure studies, interstellar dust. The small black area in Figure 13.3 shows the limit of our optical view. Radio and infrared astronomers can penetrate the smoky veil, but their observations are more difficult to interpret because our understanding of the kinds of objects that emit radio and infrared radiation is far more limited. Fortunately, the core of the Andromeda galaxy is similar to our own galactic core, and a composite model of the central regions of both of these galaxies can be constructed from optical observations of Andromeda and observations of our own core in other parts of the electromagnetic spectrum.

☆ STARS AND GAS CLOUDS ☆

Someone approaching the galactic nucleus would see the stars become much more numerous than they are out in the galactic fringe where our sun is located. At the center, the star density is some 10^5 to 10^6 stars/pc^3, in contrast to the density near the sun of about 0.1 stars/pc^3. About 10^8 stars are crowded into the nuclear region, which is about 8 pc across. Apparently many of these stars are red giants, for they emit strongly at a wavelength of 2.2 μ, radiation that can penetrate the interstellar dust and be observed by infrared astronomers. It is only the cool stars that can emit so strongly in the infrared. Recent, tentative investigations indicate that these stars are probably Population II, metal-poor stars.

In a ring 300 pc from the very center of the Galaxy, we find a number of molecular clouds. The largest of these, named Sagittarius B2, is one of the best hunting grounds for interstellar molecules. Many of the molecular types mentioned in Chapter 10 have been discovered in this set of dense clouds. Little is known about these clouds except for their existence, since they are optically invisible.

Many of the other types of objects associated with molecular clouds are found at the galactic center. Recall that the Orion complex described in Chapter 10 contained molecular clouds, dust, protostars, and an H II region that we see as the Orion nebula. Optical radiation from H II regions cannot be seen, of course, but radio radiation similar to that produced by Orion is visible. The H II regions in the galactic core seem to be associated with the ring of molecular clouds. A distinct surprise was the discovery that most of the radiation coming from the galactic core was emitted at a wavelength of 100 μ. This part of the electromagnetic spectrum is one of the most difficult to observe, because the atmosphere is opaque to this radiation, which is emitted by everything on the earth—telescopes, detectors, the atmosphere, the balloon you fly the experiment on. Looking for a source of 100-μ radiation in the sky is similar to trying to find a tiny flashlight when looking through the flame of a household furnace. The current interpretation is that this 100-μ radiation comes from heated dust; a few solar masses of dust heated to a temperature of a few hundred degrees by the stars of the nucleus will produce this radiation. At shorter wavelengths in the infrared spec-

trum, a few point sources of radiation have been found; these may be protostars or they may be small dust clouds surrounding massive stars that have
already been formed.

Figure 13.5 shows, in a very tentative and schematic way, the different
kinds of gas clouds at the galactic core that produce radiation associated
with low-energy astrophysical activity. The great diversity of radiation
sources may be confusing at first; look back at Figure 10.8 and notice that all
the types of objects seen in the Orion nebula are also found at the galactic
center. The galactic core is far larger than Orion, of course, but the same
types of objects and the same forms of radiation are there. Figure 13.5 represents, to some extent, a scientific flight of fancy, because the spatial relation-
ships among these different components of the galactic core have not yet
been completely unraveled. The objects depicted in the drawing do exist,
but their locations relative to each other may in fact be different from those
indicated in the illustration.

☆ **HIGH-ENERGY ELECTRONS** ☆

The preceding section and Figure 13.5 would seem to indicate that the galactic core is a rather large analog of the Orion nebula. Yet we have not exhausted our analysis of the core, for more energetic events seem to be taking
place there as well. The clouds of ionized hydrogen (the H II regions) and

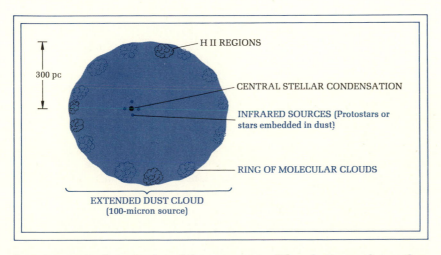

FIGURE 13.5 A schematic view of the components of the galactic core that produce
thermal radiation, radiation not associated with high-speed electrons. In black are
those objects that would produce optical radiation if we could see them: the stars at
the center and the H II regions, which appear to be associated with the ring of molecular clouds. The molecular clouds (color) are detected by radio astronomers. The
entire nuclear region contains a diffuse dust cloud that emits radiation at a wavelength of 100 μ.

the molecular clouds emit only a small fraction of the total radiation emanating from the galactic core.

The galactic core was the first radio source to be discovered, in the pioneering days of radio astronomy, and it is the strongest source of radio radiation in the sky apart from the sun. Most of the radio radiation from the galactic center is continuous and polarized, showing the two hallmarks of synchrotron radiation. This synchrotron source is named Sagittarius A. (In the early days of radio astronomy, a radio source was named for the constellation near it in space, and was lettered A, B, and so on depending on its brightness.) Close observations of Sagittarius A show that the synchrotron source is composed of a number of small knots of synchrotron radiation, each about 1 pc across. One small cloud of high-speed electrons is smaller than 10^{-3} pc (200 AU).

The existence of synchrotron radiation from the galactic core indicates that something more than the quiescent, low-energy process of star formation is occurring there. Synchrotron radiation, also produced in the Crab nebula, comes from electrons accelerated to speeds approaching the speed of light. Some kind of high-energy, explosive phenomenon must be producing these high-speed electrons. While the energy emitted by this synchrotron source is only 5×10^{38} ergs/sec, far less than 1 percent of the energy emitted by the stars at the core, it cannot come from ordinary stellar events. The galactic center is also an x-ray source, and x rays tend to come from places in the universe where high-energy, explosive events are going on. In the galactic core, cosmic violence is taking place on a huge scale; the synchrotron radiation from Sagittarius A is 20 times as energetic as the synchrotron radiation from the Crab nebula. Other galaxies are more powerful still.

The difference between the galactic core and the local stellar neighborhood can be summarized by imagining what it would be like to live on a planet in that part of the Galaxy. The stellar environment would be far more exciting than our relatively placid one. The night sky would be ablaze with light, with a million stars in the sky brighter than Sirius, the brightest star in our sky. The nearest and brightest star might be as bright as the full moon. The star that such a planet was circling would be one of many celestial luminaries. If this star were very near the core, a few bright infrared sources would be only a few parsecs away. But life would be difficult on such a planet. The low-energy events involved in starbirth present no problem for a living organism, but the high-energy electrons would look like deadly cosmic rays to an inhabitant of such a planet. In this respect, such a planet would be a little like Jupiter's satellite Io, the one that circles in the heart of the Jovian belt of high-speed electrons.

Thus the Milky Way contains three distinct regions: the *spiral arms*, where the sun circles, containing gas, dust, and Population I stars; the *halo*, containing Population II stars and globular clusters; and the *nucleus*, containing a wide variety of objects, including some clouds of gas normally associated with newly forming stars and sources of high-speed electrons. These different types of objects must be moving around each other, so that the force of gravity does not pull all of them into the galactic core. A complete description of the Milky Way galaxy must include a description of the motions of the different objects in it.

The motions of stars in the solar neighborhood can give a clue to the overall motion of the Galaxy. A star's motion through space is observed in two different ways, each of which can provide some information about the star's motion in three dimensions (Figure 13.6). Spectroscopy, through the Doppler shift (Chapter 9), can measure the star's *radial velocity* toward or away from us. Measuring the star's *transverse velocity*, or its velocity perpendicular to the line of sight, is a little more difficult. Photographs of a star's position, taken many years apart, can show a tiny displacement of the star's position in the sky called the star's *proper motion*. But the proper motion is an angular change in position, and it is impossible to tell directly whether the star is a nearby one moving at a low speed or a distant speedy one. A star can have a large proper motion either because it is close or because it has a high velocity (Figure 13.7). Mathematically, this relation among velocity, proper motion, and distance is

$$\text{Proper motion} = \frac{\text{transverse velocity}}{4.74 \times \text{distance}}$$ (13.1)

where the proper motion is measured in seconds of arc per year, the distance in parsecs, and the transverse velocity in kilometers per second. The proper motion can be large either because the transverse velocity is large or because the distance is small. The proper motions of stars are very small, imperceptible to the naked eye over the course of a human lifetime.

Study of the motions of nearby stars can provide us with a clue to the motion of the entire Milky Way galaxy. When we and the stars near us rotate around the galactic center, stars nearer the core orbit faster than stars further

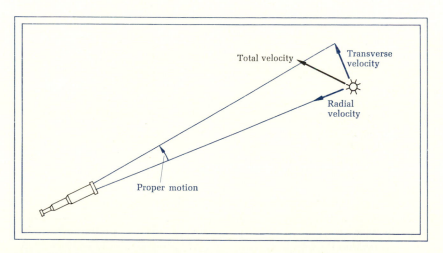

FIGURE 13.6   A star's total velocity through space is observationally measured in two ways. Doppler shifts can determine radial velocity, and photographing the star's position at different times measures the proper motion. Transverse velocity can only be determined if we know the star's distance.

FIGURE 13.7 A star can
have a high proper motion
either because it is nearby
or because it has a high
velocity.

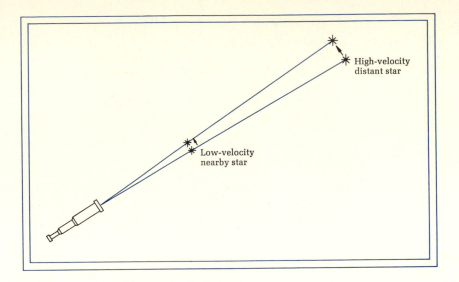

away, just as planets near the sun zip around rapidly and planets far from
the sun are quite sluggish. Figure 13.8 shows the motions of stars in their galactic orbits. Stars located in the direction of the galactic center (from our
point of view) are overtaking us, while stars far away from the galactic
center are being overtaken by the sun and move "backward." The motions of
these stars relative to the sun can be measured by measuring radial velocities and proper motions. Proper motions show stars on the outside sliding
back, and stars on the inside moving ahead. The pattern of radial velocities
is more complex, but the rotation of the Galaxy shows up quite clearly.

Detailed, quantitative analysis of the motions illustrated in Figure 13.8
can provide some information on the dynamics of the Milky Way. The sun
and its attendant planets are rotating around the galactic nucleus at a speed
of 250 km/sec, traveling in the general direction of Cygnus. Since the galactic center is 10 kpc away, it takes us 250 million years to make one circuit
of the Galaxy. This time period is sometimes called the "cosmic year," and
its length reflects the slow pace of galactic rotation. One cosmic year ago,
dinosaurs roamed the face of the earth, and the evolution of the human race
was far, far in the future. Two cosmic years ago, only a few multicellular
organisms existed in the ocean, and the land was barren. Human beings
have existed for only 0.01 cosmic years, and civilizations first existed only
forty-millionths of a cosmic year ago. The Milky Way itself is about 80
cosmic years old.

The nearby stars do not travel around the Galaxy in a uniform swarm like
airplanes flying in formation. Each star travels on its own individual orbit,
slightly different from the rest. We can measure the sun's motion relative to
the stars around it by looking for systematic trends in stellar motions, since
we observe a general tendency for stars to move toward us if we look in the
direction in space toward which we move. The sun carries the solar system
toward a point in the constellation Hercules, sometimes called the "solar
apex," at a speed of 20 km/sec, far slower than the speed of galactic rotation.

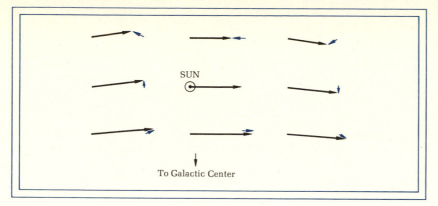

FIGURE 13.8 When the sun and the stars near it rotate around the galactic center, the sun appears to overtake the stars farther away from the galactic center, and stars nearer the galactic center move faster than the sun. The black arrows show the total motion of each star, and the color arrows show the motion of each star relative to the sun.

Radio astronomers can also analyze stellar motions by observing Doppler shifts of the 21-cm line. Radio astronomers, however, have no direct way of determining the distance to a particular hydrogen cloud, and so their analyses cannot be as sophisticated as those of the optical astronomers, even though their radio telescopes can see clear across the Galaxy while optical astronomers must work in the solar neighborhood. They have attempted to map the Galaxy by using galactic rotation as a clue to the location of individual hydrogen clouds, with limited success. However, they have been able to measure the rotational velocities of various parts of the Galaxy.

Some high-velocity clouds have been discovered by radio astronomers trying to map the Galaxy. Many of these clouds are located in the outer parts of the Galaxy and seem to be falling into the disk from the halo. The origin of these clouds is unknown. In the direction of the galactic center, the innermost spiral arms appear to be expanding away from the center at speeds of 50 to 150 km/sec. The current theoretical interpretation of this phenomenon is that an explosion in the galactic center produced a shock wave that propelled these arms outward. Once again, the galactic core is the source of explosive events that contrast with the languid motions of galactic rotation.

★ 13.5 STRUCTURAL EVOLUTION ★
OF THE MILKY WAY

The preceding sections have treated the second big question of astronomy as it applies to the Milky Way galaxy. "What is it?" The many uncertainties in our description of our own galaxy might call for some caution in approaching the third big question, "How does the Milky Way evolve?" However, a look at galactic evolution can perhaps highlight the important features of our spiral galaxy, separating the major results of evolutionary processes from minor details. Why is the Milky Way a flat spiral galaxy? Why are there three separate regions—spiral arms, halo, and nucleus? What maintains the spiral arms? Why are metal-poor Population II stars in the halo and metal-rich Population I stars in the disk? These are some of the questions that must be answered if we are to understand the events that shaped the Milky Way.

☆ FORMATION OF THE DISK, ☆
HALO, AND NUCLEUS

The separation of the Milky Way into its three components took place when the Galaxy was formed. Currently, theorists view the formation of the Milky Way as an event quite similar, dynamically, to the formation of the solar system. In the beginning, there was a huge cloud containing a few billion solar masses of hydrogen and helium gas. This rotating gas cloud collapsed to form a rotating disk, just as the solar nebula formed a disk that eventually made the planets. Some of the gas in this cloud did not collapse to make a disk, for some presently obscure reason, but was left behind as the galactic halo. One of the key features of the structure of our galaxy, the distinction between disk and halo, is thus understood in a fairly general way.

The different origins of the galactic disk and halo are preserved in the motions of these two classes of objects. Halo objects, like the globular clusters, do not share our rapid rotation around the galactic center. Globular clusters orbit the galactic nucleus on long, cometlike elliptical orbits, spending most of their time out in the halo and only occasionally dipping into the nuclear regions.

The collapse of any gas cloud under the influence of gravity, be it the small cloud that formed the solar system or the huge cloud that formed the Milky Way, should be accompanied by the formation of a central condensation, according to theory. In the solar system, this condensation contains most of the mass of the system, and is known as the sun. The center of the Milky Way contains only a small fraction of the total mass of the Galaxy, but its distinction from the rest of the galactic disk is again understood in a fairly simple way as a result of the processes that took place when the Galaxy was formed.

☆ BLACK HOLES IN GALAXIES AND ☆
STAR CLUSTERS

If a galaxy or a smaller grouping of stars, a star cluster, were to develop a sufficiently massive central condensation, this large mass could not remain for long as a stable ball of gas. A central condensation containing even a small fraction of the mass of the system must eventually become a black hole, for no pressure can keep it from collapsing indefinitely. Consider how such a mass might form in a globular cluster, for example. Two stars colliding in the nucleus of a globular cluster would coalesce, forming one object with the combined mass of the two stars. Computer calculations indicate that after about 1 million years, stars in a very closely packed central condensation would collide sufficiently frequently to form a massive central object containing several thousand solar masses of material. Such an object would eventually become a black hole.

This story would be just another story and have no place in a textbook were it not for some recent discoveries that indicate that black holes might exist at the centers of globular clusters. Four globular clusters have been identified as x-ray sources. It is not yet clear whether the x-ray sources must be located at the cores of these globular clusters, and it is possible that the x-ray sources are close binary systems that just happen to be located in the globular clusters. A growing number of astronomers suspect that a large black hole, the residue of many stellar collisions, lies at the core of each of these four globular clusters. Such a hole could emit x rays when gas in the cluster fell toward it, forming an accretion disk. It is too early to tell whether this particular black-hole bandwagon is heading toward the truth or following a path heading toward the limbo of fruitless speculation. Black holes are strange, fascinating objects that tend to stir up passions not usually seen in the cold rationality of scientific debate, and the case for black holes in globular clusters is not yet firmly established.

Some investigators attribute the high-energy activity in galactic cores to phenomena connected with a black hole in a galactic nucleus. This interpretation was supported at one time by the alleged detection of gravitational radiation from the galactic center. Such gravitational radiation, seen as a small fluctuation in the gravitational field observed at the earth, might well have been produced when a star fell into a central black hole. Subsequent attempts to confirm the detection of this gravitational radiation have not been successful; as a result there is nothing to prove that there must be a black hole in the galactic core.

☆ **SPIRAL STRUCTURE** ☆

A very generalized, theoretical picture of the formation of our galaxy can therefore explain, in a very vague, nonquantitative way, the broad division of the Milky Way galaxy into halo, disk, and nucleus. However, this simple picture cannot explain the existence of spiral structure in the galactic disk. The presence of spiral structure is, in fact, somewhat puzzling, considering the rate at which stars orbit the Galaxy. With the stars nearest the center moving faster than the stars on the outside, a spiral arm would be stretched by galactic rotation. The stars in the inner part of the arm would outpace those in the outer part of the arm, and after one or two rotations of the Galaxy the arm would be stretched so far as to be unrecognizable as a separate feature. Yet the Milky Way does contain spiral arms, and our sun orbits the center once every 250 million years. Why do the spiral arms persist?

In the 1960s, this question was answered by C. C. Lin and his collaborators at MIT with the *density-wave theory*. They abandoned the picture of a spiral arm as one collection of stars that must remain together for eternity. In the density-wave theory, a spiral arm is just a pattern that can persist even while individual stars move in and out of the pattern, just as ocean waves can travel many miles without actually moving water molecules that distance. Gravitational forces tend to hold the spiral pattern together. A spiral arm can be thought of as a trough moving through the Galaxy, and the individual objects that fill that particular trough can be different at different times.

Figure 13.9 illustrates the persistence of a spiral arm with different objects in it. The left and right panels of this figure are "snapshots" of the Milky

Way taken 30 million years (one-eighth of a cosmic year) apart. In the interval between the two pictures, the Galaxy has made one-eighth of a revolution. In the first picture, two spiral arms, A and B, are visible. They are traced out by numbered stars, with single-digit numbers for arm A (stars 1–8) and double-digit numbers for arm B (stars 11–18). The spiral arms are still there 3×10^7 years later, but the objects in them are different. While the stars have made one-eighth of a revolution, the spiral-arm patterns have moved somewhat more slowly. Arm A, for example, has lagged behind the rotation, so that stars 2, 4, 6, and 8 have moved ahead and are now in a different arm, arm C. Stars 12, 14, 16, and 18, which used to be part of arm B, have moved ahead and have been caught in the gravitational trough that forms arm A. Again, this spiral arm has persisted as a pattern, even though the individual objects in it are not the same.

The same distinction between the motion of a wave pattern and the motion of individual objects appears in all waves. Water molecules do not travel all the way across the Pacific Ocean to end up on a beach in California and help make that state the paradise of surfing. Watch ocean waves as they come in, and you will notice that each individual element of water just sloshes back and forth when the pattern of the waves moves across the ocean.

The density-wave theory is more than just a nice idea. Computer calculations show that in fact model galaxies will produce spiral patterns that do resemble the spiral patterns in real galaxies. Our understanding of galactic structure is sufficiently primitive so that precise matching between theory and reality is sketchy at best, but the agreement is reassuring. For example, the theory states that spiral arms should be about 3 kpc apart, and examination of the Milky Way and other galaxies shows that the spiral arms are in fact about that distance apart.

Thus the theoretical ingredients for explaining the shape of the Milky Way are all present. The density-wave theory can explain the persistence of the spiral pattern, and a generalized description of the formation of the Galaxy can explain the distinction among halo, disk, and nucleus. One important distinction is yet to be explained. Different populations of stars exist in the disk and the halo. Why does the halo contain old, metal-poor stars, and why is the disk the site of the formation of new stars? The structural evolu-

FIGURE 13.9 Two "snapshots" of a galaxy like the Milky Way illustrate, schematically, how a spiral arm persists as a density wave. Individual stars, numbered in the illustration, can move from one arm to another. The text describes the processes involved.

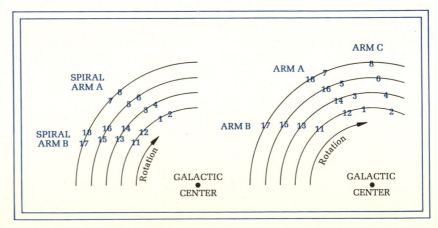

tion of the Galaxy has been explained, and we now seek a model for its chemical evolution.

★ 13.6 STELLAR POPULATIONS ★ AND NUCLEOSYNTHESIS

The chemical evolution of the Galaxy is governed by the processing of galactic material through the interiors of stars, which change hydrogen to helium and fuse helium to form heavier elements. Figure 13.10 summarizes the chemical processing that occurs. The primordial gas that formed the Galaxy was comprised of hydrogen and helium, and it contained none of the heavier elements that we refer to as metals. This gas formed stars, which then followed the usual path of stellar evolution. About half the mass that formed stars then became locked up in stellar remnants like white-dwarf stars, neutron stars, and black holes. The rest of the gas, some of which had undergone nuclear processing in the inner regions of massive stars, was returned to interstellar space when supernova explosions spewed the guts of stars across the Galaxy. The material ejected in supernova explosions has a higher abundance of heavy elements than the average parcel of galactic material, and so it is supernova explosions that produce a gradual enrichment of the metal content of galactic gas.

Gas can be processed through stellar interiors and supernova explosions many times. The processed gas then forms second-generation stars, some of which explode, eject matter, and form third-generation stars. The cosmic recycling machine takes matter, makes stars of it, and returns some matter to the interstellar medium, where stars can form again. This recycling process is inefficient, because in every trip around the cycle about one-third of the mass of a star becomes trapped in some kind of a stellar remnant—white-dwarf star, neutron star, or black hole.

A number of bits of observational data confirm the theoretical picture of chemical evolution shown in Figure 13.10. In 1958, Geoffrey and Margaret Burbidge, William Fowler, and Fred Hoyle calculated the distribution of individual elements that would be produced in a supernova explosion. They asked what kinds of heavy elements would be fabricated in the high-

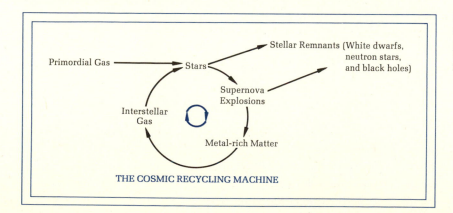

FIGURE 13.10 Chemical evolution of the Galaxy causes the interstellar gas to become progressively richer in heavy elements (metals).

temperature environment of a stellar interior. Comparison of their calculations with the observed distribution of heavy elements indicated that supernovae could, in fact, produce different elements in the right proportions. Work along these lines continues, as investigators seek to show that the observed abundances of different elements match up with the expected production rates in supernovae.

This theoretical scheme for the chemical evolution of the Galaxy can explain the different metal abundances of stars in the galactic halo and in the galactic disk. Halo stars were formed early in the evolution of the Galaxy; at the time they were formed, galactic reprocessing had not worked for long and the galactic gas contained few heavy elements. The halo now contains little gas, and no new stars are being formed there. The Population II stars found in the halo thus show low metal abundances. A surprising fact is that we have not discovered any primordial stars, remnants of the original gas that formed the Milky Way, which should contain no heavy elements at all.

The disk stars, in contrast to the halo stars, are located in regions where star formation continues and cosmic recycling proceeds. With each trip through the bowels of a supernova, a parcel of galactic gas is enriched in heavy elements.

The heavy elements such as carbon and oxygen, so essential for life, would not exist in our galaxy were it not for the processing of galactic gas in the high-temperature environment of a supernova core. This book is made of material that was once part of a titanic stellar explosion. It is less likely that life will be found on planets of Population II stars, because the heavy elements that form biologically significant chemicals are only one-tenth as abundant there as they are on our own planet. The chemical distinction between disk and halo is explained by the chemical-evolution scheme depicted in Figure 13.10.

SUMMARY

The Milky Way is a thin disk of stars broken into spiral arms, with its center about 10 kpc from the sun, in the direction of the constellation Sagittarius. There is a strong distinction between the stars in the disk and the stars in the halo regions of the Galaxy—between the disk Population I and halo Population II stars. We are only just beginning to understand the nucleus of the Galaxy; in addition to stars, it contains a source of high-energy electrons that produces synchrotron radiation. The rotating disk contains spiral arms. The process of star formation and the supernova explosions in the galactic disk continuously enrich the disk material with elements heavier than hydrogen and helium.

This chapter has covered:

The use of Cepheid and RR Lyrae variables in the determination of our location in the Milky Way;

The differences between Population I and Population II stars;

A description of the galactic core;

The motion and rotation of the Milky Way galaxy;

A theory describing how the structural features of our galaxy evolved; and

The role of supernovae in forming elements heavier than hydrogen and helium.

KEY CONCEPTS

Density-wave theory	Nucleus	Proper motion
Galactic disk	Open clusters	Radial velocity
Globular clusters	Population I	Spiral arms
Halo	Population II	Transverse velocity

REVIEW QUESTIONS

1. How do we know that the Milky Way is a flat galaxy?

2. Why can't we see the galactic core optically?

3. Why were star clusters such an important tool for mapping the Milky Way?

4. Summarize the differences between Population I and Population II stars.

5. In what respects does the galactic nucleus resemble the Orion nebula?

6. In what respects does the galactic nucleus resemble the Crab nebula?

7. Suppose that the Milky Way galaxy did not rotate, that all objects in it circled around the center in arbitrary orbits. In what respects would the proper motions and radial velocities of stars differ?

8. In what respects is a spiral arm similar to an ocean wave?

9. Living things are made of organic molecules, in which carbon is the basis of the molecular structure. Would it be as easy for life to form on planets surrounding a Population II star as it would be for life to form on planets around a Population I star?

10. A certain fraction of the mass of the Galaxy is locked up in stellar remnants, white-dwarf stars, neutron stars, and black holes. Will this fraction increase or decrease with time? Describe the probable appearance of the solar neighborhood 30 billion years from now.

FURTHER READING

Bok, B. J., and P. F. Bok. *The Milky Way.* 4th ed. Cambridge, Mass.: Harvard University, 1974.

Sanders, R. H., and G. T. Wrixon. "The Center of the Galaxy." *Scientific American* (April 1974): 67–78.

★ CHAPTER FOURTEEN ★

GALAXIES AND QUASARS

Our own Milky Way galaxy is an insignificantly small part of the universe, for there are some 10^9 to 10^{10} galaxies like it within the range of large ground-based telescopes. But only the nearest galaxies like Messier 101 (Figure 14.1) are near enough to investigate in detail.

The shape of a galaxy is its most obvious distinguishing characteristic. Other differences among galaxies, such as variations in color and stellar content, are correlated with galactic shapes. Some galaxies do more than just emit starlight; they produce hordes of high-speed electrons from high-energy events in their cores. These active galaxies are the most luminous objects in the universe, and they include the quasars, galaxies that have such explosively active cores that the rest of the galaxy is invisible.

★ MAIN IDEAS OF THIS CHAPTER ★

Galaxies can be separated by shape into spirals, ellipticals, and irregulars.

All but the nearest galaxies travel away from us because of the expansion of the universe.

The differences among galaxies reflect the rotation rates of the galaxies and the time when their stars formed.

High-energy processes in the centers of some galaxies produce synchrotron radiation.

★ 14.1 CLASSIFICATION BY SHAPE ★

☆ THE FIRST STEP IN AN OBSERVATIONAL ☆
SCIENCE

Someone seeking to understand the diversity of a number of objects in the natural world will try to classify them before trying to understand them. An astronomer studying galaxies in the universe is confronted with a bewildering diversity of objects, demonstrated by the large number of different types of galaxies in the photographs in this chapter. The first step in bringing order from chaos is sorting the 10^9 visible galaxies into a smaller number of categories. The problem of understanding the differences among a billion galaxies is thus reduced to the problem of understanding the differences among a dozen or so classes of galaxies. This primitive stage of research in an observational science is marked by the publication of catalogues, inventories of particular parts of the universe.

Different investigators search randomly for a unifying principle, trying different classification schemes in the hope of finding one that illuminates the underlying physical cause of the differences among galaxies. A theorist can provide explanation for the diversity of classes in a good classification scheme, and once some concrete theory can be provided, the science

FIGURE 14.1 The spiral galaxy Messier 101. (Hale Observatories)

matures. The questions asked by researchers become more focused. The classification scheme is tested in the hope that the different classes of galaxies really are different in some fundamental way. We may, just may, be at this transitional stage in our study of galaxies. The shape-classification scheme introduced in this section is the one most widely used, and the vague shadows of some theoretical basis for it are beginning to emerge.

The development of a scientific field from classification to solid theoretical model is evident in two subfields of astronomy treated earlier in this book, the fields of celestial mechanics (Part 1) and stellar astronomy (Part 3).

Celestial mechanics, the study of the motion of the planets, is a very mature science. The classification of the moving lights in the sky, the division of these objects into stars and planets, was made by the ancients. The Copernican revolution, culminating in the Newtonian theory of gravitation, provided a concrete theoretical explanation for the different behavior of the two classes of objects, and subsequent work has refined our understanding to the point where the match between theory and observation is extremely precise, good to 1 part in 10^9 or so.

Stellar astronomers of the nineteenth and early twentieth centuries were approximately where astronomers studying galaxies are now, making large catalogues of objects and analyzing a few of them as closely as possible from a largely observational slant. The key theoretical step was the explanation of the spectral sequence by M. N. Saha and C. Payne-Gaposhkin in the 1920s. They discovered that the different spectral classes are a consequence of the differing temperatures of stars. The theory of stellar evolution that was developed in the 1950s and 1960s completed our understanding of the diversity of stellar spectra, since it demonstrated why stars populate different regions of the H-R diagram.

We have no H-R diagram for galaxies, since no one has penetrated to the roots of galactic diversity. The classification of galaxies by shape, almost universally adopted in current catalogues, is hopefully related to the rotation of the galaxies. Yet the theoretical explanations of why differently rotating galaxies have different shapes, of why the ones with spiral arms contain recently formed stars while the ones with no spiral arms contain old stars, are only vague. Classification by shape has the advantage of requiring relatively little telescope time; as a result, many, many galaxies have been classified by now.

☆ CLASSES OF GALAXIES ☆

The three broad categories of galactic shapes are illustrated in Figures 14.1 through 14.7. *Elliptical galaxies*, about 30 percent of all galaxies, show a relatively uniform distribution of stars, since their brightness decreases smoothly from center to edge. The two companions of the Andromeda galaxy, shown in Color Photo 17, are typical elliptical galaxies, and a more detailed picture of one of them is shown in Figure 14.2. *Spirals* form a second category, the largest, which contains about 45 percent of all galaxies. The distinguishing feature of these galaxies is their spiral arms, evident in Figures 14.1 and 14.3. The remaining galaxies are classified as *irregulars*, so named because they were at first thought not to fall into any meaningful shape category. Most irregulars, called Irregular I (or sometimes Im), have a generally linear shape like the Large Magellanic Cloud (Color Photo 16) and represent a genuine galaxy category. (Some investigators classify the Large Magellanic Cloud as a very loose one-armed spiral.) The name Irregular II is now reserved for the miscellaneous bin that any classification scheme must have, the class of galaxies that have shapes showing no regularity whatsoever.

FIGURE 14.2 The small elliptical galaxy NGC 205. This galaxy is one of the companions to the Andromeda galaxy (Color Photo 17). (Hale Observatories)

FIGURE 14.3 The spiral galaxy Messier 51. The knot of luminous material at the end of one of the spiral arms is another galaxy. This spiral galaxy is unusual in that the spiral pattern has been generated by the close encounter of these two galaxies. (Hale Observatories)

The elliptical and spiral classifications can be subdivided further. Look back at Color Photo 17; one of the companions of the Andromeda galaxy is round, and the other one is noticeably flattened. The flatness of an elliptical galaxy provides another criterion for classification. Elliptical galaxies are subclassified from E0 (circular) to E7 (the flattest); the number following the designation E is $10(a - b)/a$, where a and b are the long and short dimensions of the galaxy (Figure 14.4). These classifications partially reflect the rotation rate of the galaxy, but they are of somewhat limited usefulness. A very flat, E7 galaxy would be classified E0 if we were seeing it face on, and so the classification of ellipticals partially reflects their orientation in space. Since the orientation of a galaxy in space is not a property that should affect its evolution, the classification of ellipticals from E0 through E7 is only a partial measurement of their rotation rates.

Spirals show a wider variety of shapes than ellipticals do. Figures 14.5 and 14.6 illustrate the variety of spiral shapes that have been observed. All investigators agree on the basic division between the ordinary spirals, designated S, and the barred spirals, designated SB, that have a bar in the middle. Further division into subcategories designated a, b, and c, illustrated in Figures 14.5 and 14.6, is based on three criteria. Running from a to c, the size of the nuclear bulge decreases, the spiral arms become more loosely wound, and the distribution of material within a spiral arm becomes more patchy. Yet another distinction, often made only with barred spirals, is between those whose spiral arms protrude from a ring (the r galaxies) and those whose arms come from the end of the bar (designated s).

Galaxy researchers have added a plethora of subscripts to the basic set of ordinary/barred, a/b/c, and ring/no-ring categories outlined above. For the most part, these new categories represent refinements of the basic system. The usefulness of any particular refinement must be demonstrated. For example, Sidney van den Bergh of Canada's David Dunlap Observatory recognized that the extent and development of the spiral arms was correlated with

FIGURE 14.4 The class of elliptical galaxies is subdivided according to the flatness of the galaxy, from E0 to E7.

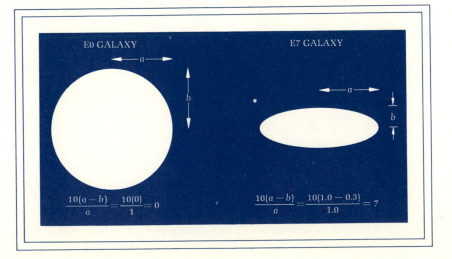

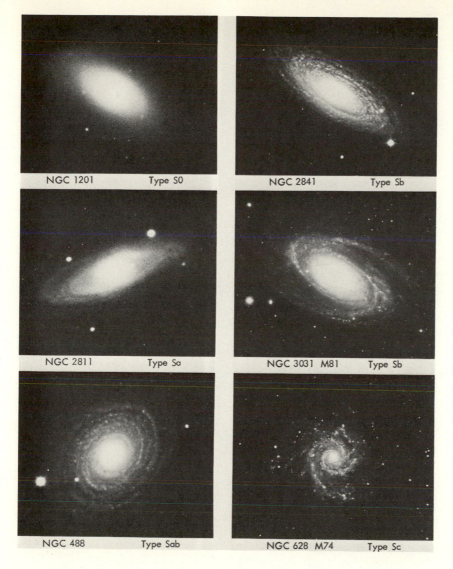

FIGURE 14.5 Ordinary spiral galaxies. (Hale Observatories)

NGC 1201 — Type S0	NGC 2841 — Type Sb
NGC 2811 — Type Sa	NGC 3031 M81 — Type Sb
NGC 488 — Type Sab	NGC 628 M74 — Type Sc

the luminosity of the galaxy. NGC 2841, in Figure 14.5, for example, is in luminosity class I with well-developed arms; NGC 2811, in the same figure, has shorter arms and is in luminosity class II-III. Van den Bergh's classifications turned out to be useful because they were related to an intrinsic property of the galaxy: its luminosity.

Most galaxies are named by catalogue numbers. Two lists of these fuzzy objects appear most often in galaxy names: the *New General Catalogue* (NGC), a list compiled by the Irish astronomer J. L. E. Dreyer in the late nineteenth century, and the list of Charles Messier (M). The designation IC refers to the *Index Catalogue*, an extension of the NGC, and galaxies that are radio

FIGURE 14.6 Barred spiral
galaxies. (Hale
Observatories)

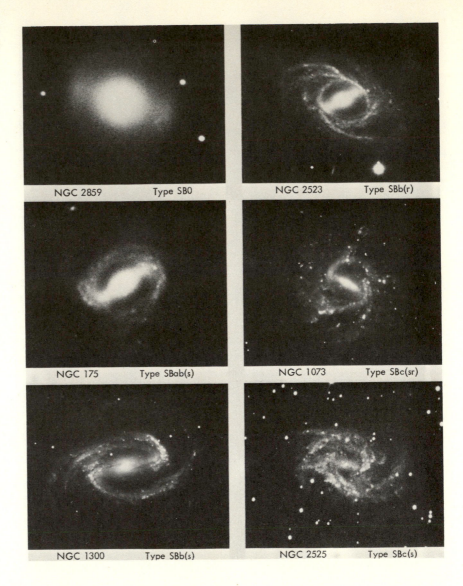

NGC 2859   Type SB0

NGC 2523   Type SBb(r)

NGC 175   Type SBab(s)

NGC 1073   Type SBc(sr)

NGC 1300   Type SBb(s)

NGC 2525   Type SBc(s)

sources are sometimes given 3C numbers according to their positions in the *Third Cambridge Catalogue* of radio sources.

Figure 14.7 summarizes the appearances of galaxies of different shape classifications. This classification scheme, originally suggested by Edwin Hubble in the 1930s, is very useful. Measurements of the color of the total light from spiral and irregular galaxies demonstrated a color progression from red to blue running from the E's and the S0's, the transitions between spirals and ellipticals, to the irregulars. This color sequence suggests that there is some physical property underlying the purely empirical galaxy classification scheme, for the shape of a galaxy is apparently related to some-

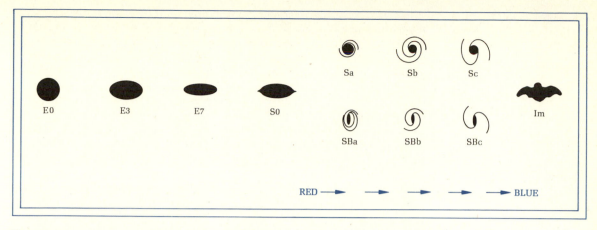

FIGURE 14.7 A summary of the galaxy classification scheme. Among spirals, the trend from class a to class c corresponds to decreasing nuclear size, more open spiral arms, and more patchy spiral arms. This trend follows the trend in colors of galaxies from red to blue.

thing else. But why is there such a diversity of galactic shapes? This question can only be answered if we have a better idea of what galaxies are.

★ 14.2 DISTANCES OF GALAXIES AND ★
THE EXPANDING UNIVERSE

Galaxies can be classified with no knowledge whatsoever of their fundamental nature. They look like little fuzzy patches in the telescope eyepiece. An understanding of the fundamental type of object we are dealing with is dependent on a measurement of the distances to galaxies. Furthermore, a precise value for the distance is needed in order to estimate the luminosity of a galaxy, the total amount of energy that it puts out every second. The distance measurement is an answer to the first big scientific question, "Where is it?" It is worth going back into history and examining the fundamental revelation of the extent of the universe that was brought about by the first measurement of the distances to galaxies.

In the early twentieth century, the prevailing view in the astronomical community was that spiral galaxies were luminous whirlpools of interstellar gas, located within the Milky Way, a few hundred parsecs from the earth. A few astronomers turned the large telescopes at Mount Wilson and Lick Observatory onto these objects, and the beginnings of the classification scheme described in the last section were devised. Yet the fundamental nature of these objects remained obscure. Recall, too, that in the early twentieth century, the boundaries of the Milky Way were thought to be 2 or 3 kpc away. The universe was thought to be a comparatively small place.

Just as the Cepheid variables and their period-luminosity relation were the cornerstone of Harlow Shapley's determination of the size of the Milky Way galaxy and our location in it, so were they the key to the determination of the distances of other galaxies. Shapley himself thought that these other galaxies were patches of interstellar gas, in keeping with the established view of the time. Another astronomer working at the Mount Wilson Observatory, Edwin P. Hubble, was able to photograph individual stars in other galaxies and measure their distances. The photograph of NGC 205 (Figure 14.2) shows individual stars in that companion to Andromeda. In 1923, he discovered the first Cepheid variable in the main galaxy. He immediately used such variables to measure the distance to the Andromeda galaxy and demonstrated that it is a huge collection of stars, comparable to our Milky Way galaxy. Hubble's original estimate of the distance to Andromeda was half the currently accepted value, because he did not know about the differences between the two types of stellar populations. Until this distinction was recognized by Walter Baade in the 1940s, the Milky Way was thought to be two to three times larger than Andromeda. But the essential step had been made, for the Andromeda galaxy was shown to be the same general type of object as the Milky Way.

Cepheid variables determine the distance to a galaxy like Andromeda because their luminosity can be measured by use of the period-luminosity relation. This process of distance measurement follows the same steps as the measurement of the distance to the globular cluster M 13 discussed in Box 13.2. A series of photographs of a galaxy shows a Cepheid variable regularly increasing and decreasing in brightness. The accumulation of a sufficient number of photographs taken at different times allows the astronomer to measure the period of the star, and the use of the period-luminosity relation of Figure 11.10 fixes the luminosity of the Cepheid. Since the brightness of the Cepheid is determined by its luminosity and distance, and the luminosity and brightness are both known, the distance can be readily determined.

However, most galaxies are so far away that Cepheid variables are too faint to photograph. Another method for determining the distances to these galaxies uses the expansion of the universe as a distance indicator. Hubble, continuing his work on galaxies with his colleague Milton Humason, began to obtain spectra of these objects, and he discovered that, surprisingly, all but the nearest galaxies were receding from us, showing a Doppler shift toward the red end of the spectrum. Hubble and Humason, through a series of measurements, were able to show that the velocity of recession, measured by the redshift, increases in proportion to the distance of a galaxy (Figure 14.8). The more distant the galaxy, the faster it recedes from us. The mathematical expression of this relation between the velocity of recession and the distance is called *Hubble's law:*

$$\text{Velocity} = 50 \times \text{distance} \qquad (14.1)$$

where the velocity is measured in kilometers per second and the distance is measured in megaparsecs.

The constant in Hubble's law, 50 km/sec/mpc, is called the *Hubble constant* in honor of this pioneer of galaxy research. It is measured by determining the distance to a fairly distant galaxy by other methods, an arduous task. This text provisionally adopts the value of 50 for this number; its pre-

RELATION BETWEEN RED-SHIFT AND DISTANCE

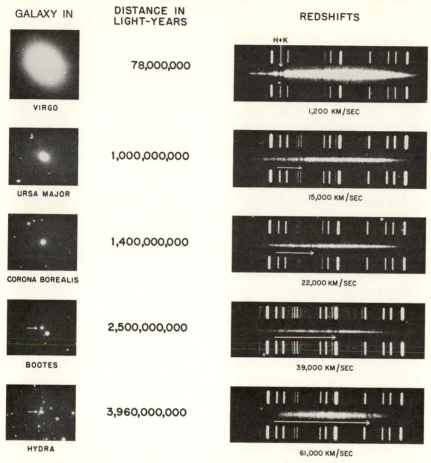

GALAXY IN	DISTANCE IN LIGHT-YEARS	REDSHIFTS
VIRGO	78,000,000	1,200 KM/SEC
URSA MAJOR	1,000,000,000	15,000 KM/SEC
CORONA BOREALIS	1,400,000,000	22,000 KM/SEC
BOOTES	2,500,000,000	39,000 KM/SEC
HYDRA	3,960,000,000	61,000 KM/SEC

Red-shifts are expressed as velocities, $c\, d\lambda/\lambda$. Arrows indicate shift for calcium lines H and K. One light-year equals about 9.5 trillion kilometers, or 9.5×10^{12} kilometers.

Distances are based on an expansion rate of 50 km/sec per million parsecs.

FIGURE 14.8 Increasingly distant galaxies show greater and greater redshifts in their spectra, indicating that they are moving away from us with increasing velocities. (Hale Observatories)

cise value is not firmly established, and the current value of 50 is ten times smaller than the number Hubble originally published. Hubble's law does not hold for nearby galaxies, which are in orbit around the Milky Way and have substantial velocities of their own superposed on the general expansion of the universe. But for a distant galaxy, Hubble's law provides a simple way to measure the distance. Obtain a spectrum of the galaxy, measure the redshift, determine a velocity from this Doppler shift, plug it into Hubble's law, and the distance pops out.

Hubble's law indicates that the universe expands, since all the galaxies in it recede from us. The increase of recession velocity with distance might seem to indicate that the Milky Way is the center of the expansion, but this simple interpretation is wrong. The distance between every pair of galaxies in the universe increases according to Hubble's law because it is the entire universe that expands. In this respect, the universe is like a giant jungle gym made out of telescoping pipe, where the pipe lengthens with time and anyone in this jungle gym will see all other parts of the jungle gym, and all other objects in it, move farther and farther away. Another popular analogy is the raisin-cake analogy, in which the universe is likened to a raisin cake that expands when it is baked. A rising raisin cake increases in size and, as the dough expands, every raisin moves farther and farther away from every other raisin (Figure 14.9), just as every galaxy in the expanding universe recedes from every other galaxy. Imagine yourself as a tiny being sitting on any of the raisins in Figure 14.9, and you will note that all the other raisins move away from you with time.

The recognition of galaxies as star systems comparable in size to our own Milky Way reinforced the astronomical perception of the insignificance of our own planet. Here we are, one of nine planets around one of 10^{11} stars in the Milky Way. But the Milky Way itself is one of about 10^9 galaxies visible to modern telescopes, making our sun only one of 10^{20} stars in the visible universe. Such a view is a long way from the perspective of pre-Copernican days, only six centuries in the past, when it was thought that the earth was the center of the universe and that stars and planets were nearby lights in the sky, wheeling around us to challenge the human mind. Astronomers through the ages have taken up the challenge of trying to understand just what the luminous objects in the sky are and have recognized our own small place in the cosmic scheme.

★ 14.3 PROPERTIES OF GALAXIES ★

The measurement of the distances to galaxies thus establishes their basic nature as giant collections of stars, and Hubble's law enables us to measure the distance to any galaxy with some precision. The next stage in extraga-

FIGURE 14.9 The raisin cake analogy to the expanding universe. Every raisin in the dough moves away from every other raisin as the dough rises, so that any individual raisin can be perceived as the center of the expansion. The only way one can define the center of the dough is by reference to the edge of the dough, not by reference to the changing position of the raisins as the dough expands.

lactic research is the stage under way now, the stage of measuring the properties of galaxies in the hope of understanding what lies behind the Hubble classification scheme. We ask what kinds of stars are present in any individual galaxy, how many of them there are, and what evolutionary stage they are in. What else exists in a galaxy besides the stars in it?

Perhaps the easiest galactic property to measure is the total luminosity of a galaxy. Here the measurement proceeds in exactly the same way as the measurement of stellar luminosity. Photometry determines the brightness of any galaxy; the relationship between brightness, luminosity, and distance allows the luminosity to be calculated easily, once the distance is known from Hubble's law or from the recognition of Cepheid variables in the galaxy. The color of a galaxy can also be measured using photometric techniques. Table 14.1 lists the colors and luminosities of various galaxies.

In order to determine the number of stars in a galaxy, we must determine the total amount of mass that it contains. The only way to measure the mass of a galaxy is by analyzing its influence on the galaxy's rotation. The speed with which individual stars orbit around a galactic nucleus depends on their distance from the center, a measurable quantity, and on the mass of the galaxy. The more massive the center of attraction, the faster any orbiting object must move in order to avoid falling into the middle. Measurements of

TABLE 14.1 PROPERTIES OF GALAXIES

CLASS	NAME	B-V COLOR[b]	LUMINOSITY (Solar luminosities)	GAS MASS (Solar masses)	TOTAL MASS (Solar masses)	MASS RATIOS Gas/Total	M/L
E	NGC 4486 (M 87)	0.97	8×10^{10}	Near 0	4×10^{12}	~0	44
E	Leo II	0.91	6×10^5	Near 0	1×10^7	~0	15 (est.)
E5	NGC 205	0.67	3×10^8	Near 0	1×10^{10}	~0	30
S0	NGC 3115	0.84	5×10^9	3×10^8	7×10^{10}	0.005	14
Sa	NGC 681	0.59	2×10^{10}	Small	2×10^{10}	~0	1.1
Sb	NGC 224 (M 31)	0.62	7×10^{10}	8×10^9	2×10^{11}	0.04	2.8
Sb/Sc	Milky Way	?	?	1×10^{10}	2×10^{11}	0.05	?
Sc	NGC 5457 (M 101)	0.21	2×10^{10}	9×10^9	2×10^{11}	0.06	7.6
SBm	LMC[a]	0.30	3×10^9	5×10^8	1×10^{10}	0.05	3.2
Im	SMC[a]	0.24	7×10^8	5×10^8	2×10^9	0.32	2.2

[a] LMC and SMC are the Large and Small Magellanic Clouds.

[b] The B-V colors have been corrected for dust in our galaxy and in the subject galaxy and for redshift.

Source: E. M. Burbidge and G. R. Burbidge, "The Masses of Galaxies," and M. S. Roberts, "Radio Observations of Neutral Hydrogen in Galaxies," in A. Sandage, M. Sandage, and J. Kristian (eds.), *Galaxies and the Universe* (Chicago: University of Chicago Press, 1976); G. de Vaucouleurs and A. de Vaucouleurs, *Reference Catalogue of Bright Galaxies* (Austin: University of Texas Press, 1964). Photographs: NGC 205, Figure 14.2; NGC 224, Color Photo 17; Milky Way, Figure 13.1; NGC 5457, Figure 14.1; LMC, Color Photo 16. The positions of the galaxies in the Local Group—Leo II, NGC 205, M 31, the Milky Way, and the two Magellanic clouds—are indicated in Figure 14.10.

Doppler shifts can produce a *rotation curve* for a galaxy, a determination of the rotational speed of the whirling stars at different distances from the center. A first estimate of the mass of the galaxy can then be made by simply applying Kepler's third law (Chapter 9), the relationship among the mass, orbital period, and orbital size of any two orbiting point masses. But the mass of a galaxy is not concentrated in a point at the nucleus, and more precise estimates of its mass are made by fitting a model of the mass distribution to the rotation curve. Radio or optical astronomers can determine rotation curves of galaxies by observing Doppler shifts. Radio measurements can also provide an estimate of the total mass of neutral hydrogen in the galaxy, a quantity listed in Table 14.1.

The sample galaxies listed in Table 14.1 demonstrate that elliptical galaxies can vary greatly in size but do not vary too much in their other properties. Masses range from 4×10^{12} solar masses for a titan like Messier 87 (also known as NGC 4486) to 10^7 solar masses for Leo II, a dwarf companion of the Milky Way. Recent research indicates that even smaller elliptical galaxies may exist, and there may well be a continuous distribution of objects ranging from the giant ellipticals like M 87 down to the globular clusters. None of the ellipticals listed in the table has any detectable abundance of gas, and there is only one elliptical that has a detectable amount of neutral hydrogen. In this elliptical, NGC 4472, 0.0002 of the total mass is neutral hydrogen gas. Elliptical galaxies are all rather red, and the more luminous ones tend to be redder.

The spiral galaxies listed in Table 14.1 tend to be more uniformly distributed in mass than the ellipticals, but their other properties are more varied. Masses of spirals range from 10^{10} to a few times 10^{11} solar masses, and so our own Milky Way is a spiral of average mass (2×10^{11} solar masses). Yet the gas contents of spirals range from near zero for the elliptical-like S0's to about 0.06 for the Sc's, and a corresponding trend in color ranges from the relatively red S0's (with high B-V values) to the far bluer Sc's. The irregulars lie at the extreme end of the spiral sequence as far as color and gas content are concerned; they have far more gas than even the most gas-rich spirals.

The color of a galaxy is presumably related to its stellar content, for the light from the normal galaxies listed in Table 14.1 is the combined light of their stars. The *mass-to-light (M/L) ratio*, obtained by dividing the total mass of the galaxy by its total luminosity, is another number that indicates the types of stars that exist in a particular galaxy. Any given star is characterized by a value of this ratio, normally expressed in solar units. A faint, dim dwarf star, with a mass of $\frac{1}{3}$ solar mass, has a luminosity of 0.01 suns and an M/L ratio of $\frac{1}{3} \div 0.01 = 33$. A blue, hot, massive main-sequence star, with a mass of 7 solar masses and a luminosity of 2800 times the solar luminosity, has a lower M/L ratio of $7 \div 2800 = 0.0025$. The M/L ratio of a whole galaxy is the composite M/L ratio of its individual stars, and a galaxy with a low M/L ratio has more hot, blue luminous stars than a galaxy with a high M/L ratio.

The data on M/L ratios and color are correlated, and they show that the Hubble classification sequence of galaxies corresponds to a variation in stellar type. At the top of the table are the ellipticals, with high M/L ratios and red colors. Attempts to calculate a mix of star types that produces the correct color and M/L ratio show that elliptical galaxies contain mostly old stars, stars similar to the Population II stars in the galactic halo. (Metal abundances do vary among ellipticals, though.) Among spiral galaxies, there is a progression from Sa through Sc, with the galaxies becoming progressively bluer and the M/L ratios becoming lower as more and more massive blue main-sequence stars are brought into the mix. At the blue extreme are the irregulars, with very blue colors and low M/L ratios.

Thus there does seem to be a trend of galactic properties with shape. The vertical arrangement of Table 14.1 parallels the Hubble classification, and even with this limited sample of galaxies the direction of the variation is evident. While it is not true that indicators such as the M/L ratio decrease smoothly along the sequence, the general direction is clear. More blue, luminous, massive main-sequence stars and more gas are contained in the stellar population of a galaxy as one progresses along the sequence from ellipticals to irregulars. There is no obvious correlation of these properties with galactic mass. While most research on galaxies seeks to refine this trend, theorists are beginning to ask "Why?" Why does this trend exist? Can we explain the differences between ellipticals on the one hand and spirals on the other? Some tentative gropings toward a possible scenario for the evolution of different types of galaxies are emerging from the data.

★ 14.4 SPECULATIONS ON ★
GALACTIC EVOLUTION

Our picture of the evolution of other galaxies builds on the sketchy evolutionary history of the Milky Way outlined in Chapter 13. To recapitulate: The Milky Way began as a rotating cloud of gas. When this rotating cloud collapsed gravitationally, a disk formed. Gas that did not become incorporated into the galactic disk formed the halo, consisting of primeval stars with low metal abundances. At this time, only the low-mass halo stars are left since all the high-mass stars have evolved. In the disk, the recycling of matter through stars, supernova remnants, and the interstellar medium results in a steady stream of star formation and the continuous enrichment of the galactic gas.

The extension of this picture to other types of galaxies would then seem clear. An elliptical galaxy forms from a protogalactic cloud that is spinning slowly, if at all. No disk is formed, and star formation occurs in an initial burst at the beginning of galactic evolution. Subsequently, the overall shape of the elliptical galaxy changes little. The M/L ratio increases, and the galaxy becomes redder while high-mass stars peel off the upper end of the main sequence, never to be replaced by newly formed stars. At the present time, the elliptical galaxy consists of stars less massive than the sun, contributing relatively little light and giving both a red color and a high M/L ratio, in agreement with observations.

The irregulars represent the other pole of galactic evolution. In this scenario, they begin as rapidly rotating clouds of gas, and no halo or nucleus forms from the collapsing cloud. Star formation is continuous, and there is an ample supply of gas to make new stars at any time; this is indicated by the high gas mass/total mass ratios in Table 14.1. The prevalence of recently formed stars allows blue, hot, luminous main-sequence stars to be observed even now, providing a blue color and a high luminosity (and hence a low M/L ratio), in agreement with observations once again. The recycling of stars into more stars continues actively. Presumably, the spirals represent an intermediate case, ranging from the elliptical-like S0's and Sa's to the irregular-like Sc's.

There is little precise, quantitative support for this working hypothesis, but there are reasons for believing it is valid enough to buttress or shoot down by gathering more data. Detailed models of the population of stars in an elliptical galaxy are in good agreement with observations of the total stellar radiation from these types of galaxies. The large gas masses of irregular galaxies, combined with their blue color (Color Photo 16), support the idea that star formation continues in these galaxies. Time will tell whether the last two columns of Table 14.1 contain the forerunner of a galaxian equivalent of the H-R diagram.

This working hypothesis for galactic evolution is, like stellar-evolution theory, a form of scientific knowledge. Yet these two fields are in far different stages of maturity, and consequently the two theories should be given different degrees of credence. A stellar theorist can use a computer program to "cook" a model ball of 2×10^{33} g of gas through the processes of stellar evolution. Comparison of this model star with the sun, and of other model stars with other real stars, produce faith in the model. No such quantitative matching of theory and observation exists for galaxies. In fact, the theory itself is scattered through the scientific literature, usually in those final sections of scientific papers where investigators tend to speculate a little. The ideas presented in this section are an amalgamation of the theories that have been presented in various places, and all they do so far is explain the general trends of color, M/L ratio, and gas content along the Hubble sequence of shapes of galaxies.

★ 14.5 CLUSTERS OF GALAXIES ★

Just as stars tend to group together in star clusters, galaxies tend to be associated with each other in groups or clusters. Our own Milky Way is part of a small group of galaxies called the Local Group. Figure 14.10 illustrates the Local Group as it would be seen from a vantage point directly above the plane of the Milky Way. There are three large spiral galaxies in the Local Group, M 31, M 33, and the Milky Way. Maffei 1, marked with a question mark, is a large elliptical galaxy seen through dust in the plane of our own galaxy, and its distance is uncertain. It may or may not be a member of our own group of galaxies. The remainder of the galaxies in the Local Group are smaller, with a few irregulars and many, many small ellipticals. Figure 14.11 illustrates a small group of galaxies similar to the Local Group. M 81 and M 82 are a rewarding sight through a small telescope.

The Local Group is a typical small group of galaxies, dominated by galaxies with high gas content, spirals and irregulars. Most groups of galaxies with fewer than 1000 members tend to have spirals as their brightest members. The rich, regular clusters, with 1000 members or more, tend to be

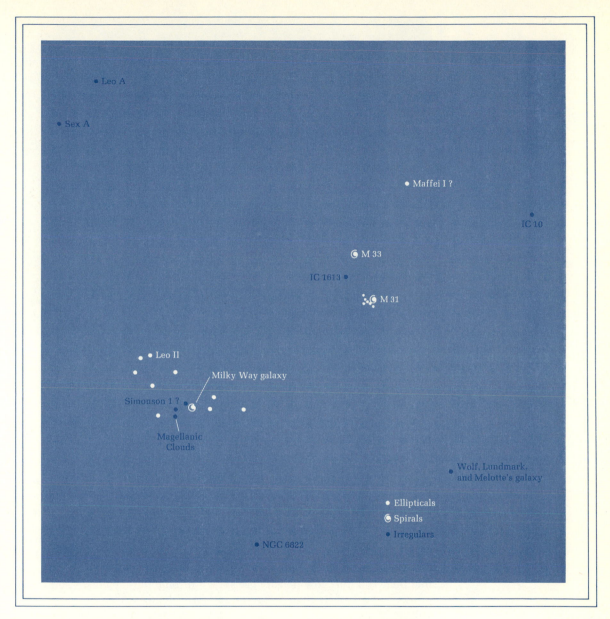

FIGURE 14.10   The Local Group of galaxies. The distances of Simonson 1 and Maffei 1 are uncertain; Simonson 1 may be a distant spiral arm of our galaxy, and the giant elliptical galaxy Maffei 1 may be too distant to be a member of the Local Group. [Locations from S. C. Simonson, *Astrophysical Journal Letters*, 201 (1975): L103; H. Spinrad et al., *Astrophysical Journal Letters*, 163 (1971): L25; and G. de Vaucouleurs, in A. Sandage, M. Sandage, and J. Kristian, eds., *Galaxies and the Universe* (Chicago: University of Chicago Press, 1976), pp. 557–597.]

FIGURE 14.11   The M 81 group of galaxies, a group similar to the Local Group. The large spiral galaxy is M 81, a galaxy not unlike the Milky Way. The irregular galaxy M 82 is at the left, and two small elliptical galaxies in this group are also visible in the field. (Hale Observatories)

spherical in shape and have giant ellipticals as their most prominent members. The nearest of these rich clusters is in Virgo, northwest of the bright star Spica, and is a rewarding sight in a telescope with a 12-inch or larger mirror.

Rich clusters of galaxies probably contain more than just galaxies. Several of these clusters have been discovered as x-ray sources by the Uhuru satellite in its survey of the x-ray sky. These x rays come from the entire cluster, and so they are not coming from, say, one big black hole in one of the more massive cluster galaxies. The current interpretation of this x-ray emission is that there is a thin, low-density gas spread throughout the cluster, emitting x rays because of its high temperature of 10^6 degrees. This uniform inter-cluster medium contains less mass than the galaxies do.

The motions of galaxies in large clusters indicate that there is some missing mass in these clusters, some mass that we have not yet found. Measurements of the motions of individual galaxies in the Coma cluster of galaxies (Figure 14.12) indicate that these galaxies are orbiting around the cluster center as one would expect. However, the speeds of these galaxies are surprisingly high, so high that, at their present rate of speed, they are traveling too fast to be gravitationally bound to the Coma cluster if the galaxies in the cluster have normal masses. There must be some extra mass in the cluster to

FIGURE 14.12 The center portion of the Coma cluster of galaxies, a rich cluster dominated by giant ellipticals. (Hale Observatories)

hold the galaxies to it if the cluster is not to fly apart. The amount of extra mass needed varies from one cluster to another, but it ranges from five to ten times the mass of the galaxies in the cluster. Where is this missing mass? No one has yet discovered any light coming from it. Perhaps these galaxies in clusters, like Coma, are surrounded by huge halos of low-mass, invisibly dim stars, as some investigators have postulated. The puzzle remains.

★ 14.6 ACTIVE GALAXIES ★

Most galaxies emit almost all their energy in the form of starlight, and thus their structure and evolution are determined by the evolution and motion of the stars in them. Yet the slow, stately, low-energy process of stellar evolution and motion is not the only event that occurs in galaxies. Observations show that the core of our own Milky Way contains clouds of high-energy electrons, producing synchrotron radiation when they spiral around magnetic field lines. Huge clouds of dust near the galactic center emit large quantities of infrared radiation.

Messier 82, shown in Figure 14.13, is an example of a galaxy that shows more nonstellar radiation than our own Milky Way. Figure 14.13 is a photograph of this galaxy in the light of a hydrogen emission line; it shows a number of filaments of gas extending away from the galactic nucleus. M 82 is a very powerful infrared source, emitting 2×10^{44} ergs/sec of infrared

radiation, about 100 times as much energy as the Milky Way emits in the infrared, and about 5 times as much energy as the Milky Way emits in the form of starlight. Numerous other active galaxies have very high infrared luminosities. It is not yet clear whether the infrared radiation is simply heat from hot dust or is from some more exotic process. At one time M 82 was thought to be exploding. This innocuous-appearing galaxy, not seeming at all unusual in the optical photograph of Figure 14.11, contains a number of surprises, and its true nature is not yet understood. It is nevertheless a form of galaxy that emits most of its energy outside the visible part of the electromagnetic spectrum, where starlight is emitted.

☆ RADIO GALAXIES ☆

Galaxies that emit something more than just plain starlight generally emit this extra radiation in some other part of the electromagnetic spectrum. Synchrotron radiation is perhaps the best-studied form of high-energy emission from active galaxies. Because radio radiation can penetrate the earth's atmosphere, the synchrotron emission from radio galaxies can be studied from the ground in a fairly routine way. In order to be called a *radio galaxy*, a galaxy must produce at least 10^{42} ergs/sec in the radio part of the spectrum. (Our Milky Way emits 4×10^{43} ergs/sec of radiation optically, 40 times what a minimally active radio galaxy emits.) This radio radiation indicates that high-energy electrons are being produced somewhere in the radio galaxy by a violent, explosive process. The normal events of stellar evolution do not produce large quantities of high-speed electrons.

Many radio galaxies emit part of their radio radiation from very compact clouds, only a few light months across in one extreme case. These tiny sources can only be resolved by pairs of radio telescopes observing the source simultaneously from opposite sides of the earth. Signals from the two telescopes are brought together and combined by computer, and these simultaneous observations provide some of the information that we would get from a radio telescope the size of the earth. These pairs of radio telescopes can distinguish radio sources as small as 3×10^{-4} sec of arc, equivalent to the size of a dime located 6000 km away. Strangely, some of these compact sources exhibit rapid angular motions that seem to indicate great speeds. If the galaxies containing them are at the distance indicated by their redshifts, these sources are apparently moving at several times the speed of light. While there are physical processes that can produce the illusion of such a rapid expansion, it is also possible that the incomplete picture provided by these pairs of radio telescopes is the source of the illusion.

In a typical radio galaxy, the radio emission comes from two giant clouds of high-speed electrons, one on either side of the parent galaxy. Figure 14.14 shows Cygnus A, the brightest radio galaxy. The contour lines show the two clouds of radio emission, each about 17 kpc across and some 50 kpc away from the central galaxy. In the core of the central galaxy are two tiny clouds of radio emission, only a parsec or so across, aligned in the same direction in space as the two giant clouds visible in Figure 14.14. The same central object must have produced both the large and small clouds.

Some radio galaxies are still larger than Cygnus A. The largest object in the known universe is a radio galaxy, 3C 236. This object contains a central galaxy surrounded by two enormous clouds of radio emission, stretching 6 Mpc from one end to the other. Either of these clouds could engulf the entire Local Group of galaxies.

☆ **BRIGHT NUCLEI** ☆

The seat of the activity in most active galaxies is the galactic nucleus. The first suspicion that galactic nuclei played a key role in galactic activity was voiced by Carl Seyfert in 1943, who recognized that a small number of galaxies had bright, almost starlike nuclei. The spectra of these nuclei showed emission lines, and such galaxies have been named *Seyfert galaxies*. Many radio galaxies have Seyfert nuclei.

The presence of emission lines in the spectrum of any astronomical object indicates the presence of an intense source of ultraviolet radiation. Emission lines come from low-density gas, produced by the recombination of gas atoms that have been split apart or ionized by high-energy ultraviolet photons. The energy produced by emission lines in Seyfert galaxies is so great that no reasonable collection of stars could produce enough ultraviolet radiation. The cores of Seyfert galaxies thus contain a strong source of high-speed electrons, a source that can produce the required amount of

FIGURE 14.14 Cygnus A, a strong radio galaxy. The contour lines show the location of the radio-emitting clouds, and the photograph (taken by Walter Baade with the Hale 5-m reflector) shows the galaxy itself. (From A. Moffet, "Strong Nonthermal Radio Emission from Galaxies," in A. Sandage, M. Sandage, and J. Kristian, eds., *Galaxies and the Universe*, published by the University of Chicago Press. © Copyright 1975 by the University of Chicago. All rights reserved.)

ultraviolet radiation by the synchrotron process. In this respect, a Seyfert galaxy is similar to the Crab nebula, where high-speed electrons produce ultraviolet radiation by the synchrotron process. The ultraviolet radiation then produces the emission lines from gas filaments.

Although the Seyferts may be similar to the Crab in the type of processes that occur, the energies involved are far greater than the energies normally associated with galaxies. The most luminous Seyfert galaxy is 3C 120, a galaxy with a relatively modest radio luminosity of 5×10^{42} ergs/sec, but a nuclear luminosity of 3×10^{44} ergs/sec. Its infrared luminosity is 3×10^{46} ergs/sec. 3C 120 is almost 1000 times as powerful as the Milky Way, and it emits almost as much energy in emission lines alone as the Milky Way emits in the form of starlight. The bulk of this high luminosity is traceable to the production of high-speed electrons.

Some active galaxies have such bright nuclei that the outlying galaxy is barely visible. A recently discovered type of active galaxy, called a *BL Lac object* after the first one discovered, has a nucleus that produces so much optical synchrotron radiation that no other radiation from the galactic nucleus can be seen. The nature of these objects was quite mysterious until recently, and controversies about them have not yet subsided. With no emission lines from filaments or absorption lines from stars, it was impossible to determine the recession velocity of these objects, and their distance and fundamental nature were unknown. But in the last 2 years, spectra of the stars in the galaxies surrounding these bright nuclei have been photographed, and redshifts have been measured. BL Lac itself is as luminous as the most luminous Seyfert galaxy, 3C 120. The BL Lac objects and the luminous Seyfert galaxies like 3C 120 are the most active galaxies known. It is still possible, though unlikely, that the stellar spectra photographed around the BL Lac objects are some kind of illusion, and that they are not hyperactive galaxies but some still more surprising type of object.

★ 14.7 QUASARS ★

In some active galaxies, the radiation from the nucleus so swamps the image of the outlying galaxy that the only thing visible on a photograph is the starlike image of the nucleus. These starlike objects were first named "quasistellar" objects when they were discovered in 1963, and the name was subsequently shortened to *quasar*. In a photograph, a quasar looks just like a star (Figure 14.15), and before the spectra of these faint objects were obtained, astronomers thought they were ordinary galactic stars.

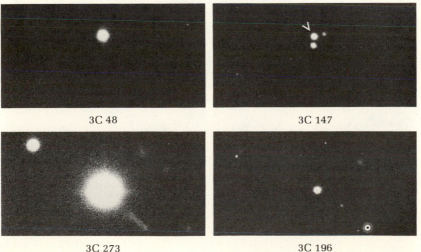

3C 48

3C 147

3C 273

3C 196

FIGURE 14.15 Photographs of four quasars, showing that these objects look like stars. Only one, 3C 273, shows anything unusual, a protruding jet. (Hale Observatories)

But it is the spectra of the quasars that show their true nature as the most luminous distant objects in the known universe. Plotted as a graph in Figure 14.16, the spectrum of the nearest and brightest quasar, 3C 273, shows that most of the radiation from this object is concentrated in emission lines. In this respect, a quasar spectrum resembles that of the Orion nebula.

The surprise in the quasar spectrum is the location of these emission lines at wavelengths shifted far to the red of the wavelengths the photons had when they left the quasar. This redshift is shown by an arrow in Figure 14.16; in 3C 273, the photons have wavelengths 16 percent longer than the wavelengths they had when they left the quasar. (The redshift z of a quasar is the wavelength shift $\Delta\lambda$ expressed as a fraction of the original wavelength λ; thus $z = \Delta\lambda/\lambda$.) Ordinary stars show far smaller Doppler shifts (an angstrom or so) due to their radial velocities. This redshift shows that 3C 273 recedes from us with a speed of 47,000 km/sec, about 16 percent of the speed of light. Quasars with even higher redshifts exist; a quasar with one of the highest redshifts is OQ 172, with a redshift z of 3.53. This quasar recedes from us at 91 percent of the speed of light, and its photons have been traveling toward us for almost $15\frac{1}{2}$ billion years. (Note that the standard Doppler formula cannot be used to compute velocities from such large redshifts; see Box 14.1 if you are interested in how to do it.)

Most astronomers believe that the enormous redshifts in quasars are due to the expansion of the universe, but the implied distances to the quasars then become very, very large. 3C 273 is, for example, 900 mpc away. Quasars

FIGURE 14.16   The spectrum of the quasar 3C 273, shown as a graph of light intensity versus wavelength. The peaks indicate that most of the light energy is concentrated at specific wavelengths in emission lines; compare this spectrum with the spectrum of the Orion nebula (Figure 10.5). The spectrum has been shifted toward the red, as demonstrated by the contrast between the observed wavelengths (black) and the emitted wavelengths (color). (Adapted from J. A. Baldwin, *Astrophysical Journal Letters,* 196, L91–L93, published by the University of Chicago Press. Copyright © 1975 by The American Astronomical Society. All rights reserved.)

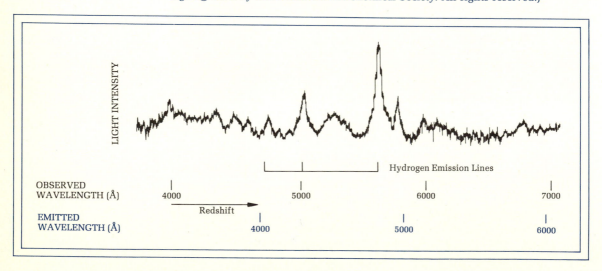

BOX 14.1 LARGE REDSHIFTS AND VELOCITIES

The usual Doppler-shift formula, equation 9.6, cannot be used to calculate velocities when redshifts are large. Instead, the relativistic formula must be used. Many people find it hard to believe that redshifts greater than 1 can exist, but such redshifts can easily be achieved when the velocity of a receding object approaches c. For readers interested in what the relativistic Doppler formula looks like, it is

$$z = \left(1 - \frac{v^2}{c^2}\right)^{-1/2} \left(1 + \frac{v}{c}\right) - 1$$

$$= \frac{\Delta\lambda}{\lambda}$$

For such large speeds, Hubble's law is no longer precisely correct, and other formulations must be used.

---

must then have very high luminosities to be visible at such distances. Take the quasar 3C 147, illustrated in Figure 14.15, as an example. This quasar is of a brightness comparable to those of the stars in the field, but the quasar is about 10^7 times as far away as those stars, because it has a redshift of 0.545. It can only be as bright as the stars if it is 10^{14} times as luminous as those stars. Precise calculations of its brightness and distance show that this quasar is 2×10^{12} times as luminous as the sun, or 200 times as luminous as the Milky Way galaxy. Such high luminosities, which can range up to 10,000 times the luminosity of the Milky Way, are not expected as part of the normal process of galactic evolution.

Quasars share many of the properties of active galaxies, but the energies involved are 10 to 100 times greater. The nearest quasar, 3C 273, is an intense infrared source, emitting most of its energy in this part of the spectrum, as the nucleus of the Milky Way does. It is a radio source not unlike the radio galaxies, since its radio emission is seen to come from both large and small clouds of high-speed electrons. Its emission-line spectrum resembles that of many Seyfert galaxies.

Careful observation and analysis of quasars have resulted in a fairly complete model for these objects. Somewhere in the center is an energy source, of unknown constitution, that generates roughly 10^{46} ergs/sec of energy in the form of high-speed electrons. This emission rate is 2×10^{12} times the luminosity of the sun. These high-speed electrons then spiral around magnetic fields, producing synchrotron radiation in many parts of the electromagnetic spectrum. Ultraviolet synchrotron radiation ionizes gas atoms in filaments in the galaxy, producing emission lines in the same way that the Orion nebula produces emission lines from stellar ultraviolet photons. Somehow a great deal of infrared radiation is produced in 3C 273, probably from heated dust grains, but possibly from the synchrotron process. There are some cooler clouds, further away from the nucleus of the quasar, that absorb photons from the quasar spectrum, producing absorption lines. This nice model produces all the right types of radiation: continuous radiation from radio to x-ray wavelengths, large quantities of infrared emission, emission lines, and absorption lines.

Yet the quasar model has one serious omission. The nature of the central energy source is unknown. Suggestions as to possible quasar powerhouses, ranging from the reasonable to the ridiculous, have appeared in the astronomical literature. Is it a huge black hole eating matter in the heart of a large galaxy, shooting high-speed electrons out of its surroundings by some poorly understood process? Is it a giant pulsar, producing high-speed electrons in the same (unknown) way as the pulsar in the Crab nebula does? The trouble is that 10^{46} ergs/sec is a great deal of energy to ask from a model quasar power source. Even 1000 times as many stars as there are in the Milky Way would not do the job, and stars emit low-energy photons of starlight rather than high-speed electrons. There are other, more technical problems with the quasar model, but the chief weakness is that no one has yet come up with a convincing model for the central energy source.

Some astronomers are so dissatisfied with the quasar model that they seek to relieve the energy problem by supposing that the quasars are not as far away as their redshifts indicate. Bringing the quasars closer reduces the energy requirements but presents another problem: Where do the redshifts come from? No known physical effect can account for the redshift in a possible quasar model, and so if the quasars are indeed closer than their redshifts indicate, some new laws of physics are involved. Most of us like to work within the known physical laws rather than postulate new ones, and the majority of the astronomical community supports the quasar interpretation of the last few pages—that quasars are very luminous active galaxies. Yet a minority view persists, and there is evidence supporting their position. I have dealt with this controversy at some length in another book.[1] The study of quasars may show that they are objects in which new physical laws operate, or it will more likely show that they are the most luminous objects in the universe. In either case they are fascinating.

SUMMARY

Most galaxies are huge collections of stars, looking like fuzzy wisps in the telescope. Their shapes seem correlated with their evolution, since the elliptical galaxies made all their stars at one time, and star formation goes on continuously in the spiral arms of spirals and in the ellipticals. Some galaxies do more than just produce starlight, since some central energy source in their nucleus produces great clouds of high-speed electrons. Synchrotron radiation generated by these high-speed electrons can be observed, particularly by radio astronomers. It is most likely that quasars are extremely distant and extremely luminous active galaxies, the most luminous objects in the universe.

This chapter has covered:

The separation of galaxies by shape into spirals, ellipticals, and irregulars;

The differences in the content of galaxies with different shapes;

[1] Harry L. Shipman, *Black Holes, Quasars, and the Universe* (Boston: Houghton Mifflin, 1976), chap. 12. See also G. Field, H. Arp, and J. Bahcall, *The Redshift Controversy* (New York: Benjamin, 1973).

The expansion of the universe and Hubble's law;

A tentative, speculative explanation for the differences among galaxies;

The high-energy processes that occur in Seyfert and radio galaxies; and

The nature of quasars.

KEY CONCEPTS

Elliptical galaxies	Irregular galaxies	Rotation curve
Hubble constant	M/L ratio	Seyfert galaxy
Hubble's law	Radio galaxy	Spiral galaxies

REVIEW QUESTIONS

1. Review the differences among spiral, elliptical, and irregular galaxies, testing your understanding by examining the photographs in this chapter.

2. Why is classification the first step in an observational science?

3. Why doesn't Hubble's law imply that we live at the center of the universe?

4. Which would have a higher M/L ratio, an open cluster or a globular cluster?

5. Would you expect to find an open cluster in an elliptical galaxy?

6. How does the evolution of the halo of a spiral galaxy and the evolution of elliptical galaxies differ from the evolution of the spiral arms of spirals and the evolution of irregulars?

7. Review the similarities and differences among the Local Group (Figure 14.10), the M 81 group (Figure 14.11), and the Coma cluster (Figure 14.12).

8. Review the different manifestations of high-energy activity in galaxies.

9. Describe the similarities among quasars, the Crab nebula, and the Orion nebula.

FURTHER READING

Hubble, E. *The Realm of the Nebulae.* New Haven, Conn.: Yale University Press, 1936 (also available from Dover Publications, New York).

Sandage, A. *The Hubble Atlas of Galaxies.* Washington, D.C.: Carnegie Institution, 1961.

———, M. Sandage, and J. Kristian, eds. *Galaxies and the Universe.* Stars and Stellar Systems series, vol. 9. Chicago: University of Chicago Press, 1975.

Shipman, Harry L. *Black Holes, Quasars, and the Universe.* Boston: Houghton Mifflin, 1976.

★ CHAPTER FIFTEEN ★

EVOLUTION OF THE UNIVERSE

Where is it all going? Every astronomer since the first cave dweller who kept track of the lunar phases on a piece of bone has asked this question. For all the intensive observation of the sky, century after century, the slow pace of the evolution of the universe has stymied those who wish to see the universe change significantly in a few human lifetimes. In the 1960s, the theories of cosmology, proposed in various forms since the dawn of civilization, at last were confronted with the real world of observations. The current evidence vindicates the theoretical schemes for the evolution of the universe that were based on the gravitational theory of Albert Einstein (Figure 15.1). The universe began in a primeval explosion, the big bang, about 20 billion years ago, and it has expanded and evolved ever since.

★   MAIN IDEAS OF THIS CHAPTER   ★

The principal difference between the cosmological theories accepted today and those of a century ago is that the contemporary view sees the universe as an evolving one.

A variety of clues indicate that the universe began expanding 20 billion years ago.

The existence of a primeval explosion is confirmed by the observation of radiation from that explosion.

Galaxies formed about 1 billion years after the big bang.

The expanding universe may or may not expand forever. While the evidence at hand seems to indicate that the expansion of the universe will never stop, this conclusion is not completely certain.

★   15.1   THE EVOLUTIONARY UNIVERSE   ★

The study of the structure and evolution of the universe, the field of *cosmology*, is the oldest branch of astronomy. Reasonably concrete models of the workings of the universe have been part of philosophical literature since the time of the ancient Greeks, and the mythologies of even the earliest cultures contain some tale of how the universe was formed, how it is made up, and how it evolves. These ideas were, however, based on the philosophical and religious views of the time, not on any hard evidence, and could not be called scientific theories in any sense.

   Over the centuries, one trend in cosmology has been the continuous expansion of what is perceived as the universe. In their early myths, the Mediterranean cultures indicated the end of the universe to be somewhere near the Mediterranean Sea, since they did not understand the true size of the earth. Before Copernicus, the planets were believed to be tiny lights in the sky, and the entire universe was not thought to be too much larger than the

FIGURE 15.1 The theory of gravitation developed by Albert Einstein is the basis of modern cosmology. (Yerkes Observatory photograph)

earth itself. Copernicus demonstrated the insignificance of our planet, showing that the earth was only one of nine planets in the solar system. The distances to the stars were first measured in the nineteenth century, and the universe became yet larger. The most recent leap in the ever-expanding human perception of the universe, the discovery of the true size of the Milky Way galaxy and the determination of the distances to other galaxies, occurred in the early twentieth century (Chapters 13 and 14).

Once the size and composition of the universe is roughly known, its evolution becomes the focus of cosmological study. In the geocentric, pre-Copernican viewpoint, the evolution of the universe was regarded as something that mere humans could never comprehend. There was a strong distinction between the earth, where matter behaved predictably and the interactions and motions of objects could be at least described and partially understood, and the incomprehensible heavens. The sun and planets were beyond the grasp of the human mind. Ptolemy and his followers found the motions of the planets so complex that they only hoped to describe them sufficiently well so that astrologers could concoct horoscopes. For a long time, the Bible was regarded as the only guide to what lies beyond the earth.

The Copernican revolution demolished the heavenly wheels of Ptolemy, and replaced the geocentric universe with a cold piece of clockwork. Newtonian gravity described the motion of the planets quite accurately; they were no longer incomprehensible celestial lights. Once started, the motions of the planets in their orbits would continue onward to eternity. The stars were just a backdrop to the action on center stage, the orbiting planets. Mathematicians developed increasingly elaborate techniques for refining the calculation of planetary orbits to more and more decimal places. Gravity keeps the planets orbiting forever, and the universe will not evolve.

The everlasting, clockwork universe, rather compatible with the rationalist philosophy prevailing in the eighteenth and nineteenth centuries, did not survive the upheavals that occurred in physics in the early part of the twentieth century. Albert Einstein began the revolution by forging a new theory of gravitation, a new description of the attraction of massive bodies. The application of his theory to the solar system resulted in only miniscule changes in the theoretical paths of the planets, changes that were useful in confirming his theory but that did not affect the fundamental view of the evolution of the planetary system.

A number of theorists soon applied Einstein's gravitational theory to the entire universe. The Russian Alexander Friedmann and the Belgian clergyman Abbe Georges Lemaitre both were able to prove that, if Einstein's equations were correct, the universe must evolve, either expanding or contracting. Because the idea of an evolving universe was totally inconsistent with pre-twentieth-century cosmological theory and with the prevailing ideas of Western philosophy, Einstein modified his theory of gravity, introducing an extra physical force that could balance gravity at very large distances and keep a static universe from collapsing. The theoretical work had not proved that the universe must expand.

In retrospect, Einstein's introduction of this extra force was a mistake. In the crucial decade of the 1920s, when Friedmann and Lemaitre were exploring the cosmological consequences of the Einstein equations, a number of American astronomers were measuring the redshifts of galaxies, which were now known to be the most distant objects in the universe. Data for the most distant galaxies were obtained by Edwin Hubble and Milton Humason, working with the 100-inch telescope at Mount Wilson, then the largest telescope in the world. Hubble announced his discovery of an expanding universe in 1929. Arthur S. Eddington, in 1930, provided the crucial connecting link, for he realized that the expanding model universes of Friedmann and Lemaitre produced a theoretical relationship between redshift and distance that corresponded precisely with Hubble's law. This discovery made the first direct connection between theoretical and observational descriptions of the evolution of the entire universe. Cosmology had become a science, now that theory and reality could be compared.

Lemaitre described the origin of the expanding universe as a *primeval atom*, a dense glob of matter. The explosion of this primeval atom created the motion that is now observed as the expansion of the universe and described by Hubble's law. This cosmological scheme further evolved in the 1940s and 1950s, when George Gamow, his co-workers, and Chushiro Hayashi explored the events that occurred in the primeval explosion. Because of the central importance of the explosion of the primeval atom, this cosmological theory has become known as the *big bang theory.*

A big bang universe is the epitome of an evolving one, for the initial explosion marks a definite beginning, and the expansion of the universe leads to an eventual end. The idea of an endlessly cycling clockwork universe is totally inconsistent with the big bang picture. Figure 15.2 depicts the evolutionary feature of the big bang universe. The "film" starts with a definite beginning—the big bang—and the appearance of the universe is vastly different at different times.

There is, however, a nonevolutionary explanation of the redshifts described by Hubble's law. In 1948, the British astronomers Hermann Bondi, Thomas Gold, and Fred Hoyle proposed that the universe expanded because the continuous creation of new matter kept pushing it outward. Their theory presumes that the universe does not evolve on a large scale, in great contrast to the big bang picture. Two "snapshots" of a small part of the universe will always show the same number of galaxies (Figure 15.3). Twenty billion years ago, a frame in the film showed three galaxies. Although they have moved out of the picture by now because of the expansion of the universe, the presence of three galaxies in the frame is preserved, because two new galaxies, shown in color, have been created. This nonevolutionary cosmology is called the *steady state theory.*

An important consequence of the steady state theory is the continuous creation of matter in empty space. If the galaxies in the universe are to separate from each other, and yet if the universe is not to evolve, new galaxies

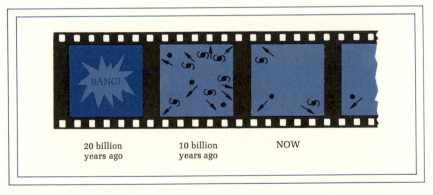

FIGURE 15.2 A "movie" illustrating the big bang theory. The film has a definite beginning, 20 billion years ago, with the explosion of the primeval atom. The expansion of the universe keeps pushing the galaxies apart, so snapshots of the universe taken at different times will show fewer and fewer galaxies in a given region of space.

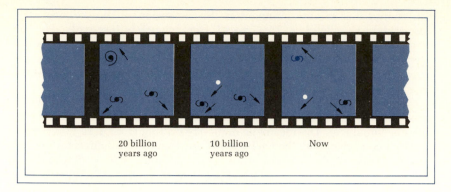

20 billion   10 billion   Now
years ago   years ago

FIGURE 15.3   A "movie" illustrating the steady state theory. Although galaxies move away from each other, in agreement with Hubble's law, the number of galaxies in any particular part of space, or any frame of the movie, is kept constant because new galaxies (color) are continuously created. This universe has no definite beginning; the film fragment shown here is just part of an infinitely long piece of film, showing similar frames infinitely far into both the past and future.

must be born to take the place of the old galaxies that leave the picture. Continuous creation violates the commonly accepted law—familiar to those who have had a chemistry course—that matter can neither be created nor destroyed. The creation of matter proceeds so slowly that it could not be observed in a laboratory; if creation were uniform through the universe, 1 g of new matter would be created each year in a sphere 2.5 AU on a side. Thus, while continuous creation violates the law of conservation of matter, it does not violate it on an experimentally detectable scale.

In the 1950s, the big bang and steady state theories competed as equals for support from the astronomical community. Both explained the one piece of observational evidence relating to cosmology, Hubble's law. The cosmological debates were philosophical in tone; one could only decide which theory was philosophically preferable—one in which the universe began at a definite time or one in which the universe existed forever. In the 1960s, more fragments of cosmological evidence accumulated, and as the pile of evidence mounted, the steady state theory became more and more difficult to defend. Most astronomers believe that the conventional big bang theory is the correct one. The remaining sections of this chapter describe some of the details of the big bang picture, showing how this theory is consistent with the evidence.

★   15.2 AGE OF THE UNIVERSE   ★

The big bang theory of the universe begins at a definite time, when the primeval atom exploded 20 billion years ago (Figure 15.2). In contrast, the steady state universe did not begin at any definite time; Figure 15.3 is only a short fragment of a film, showing similar frames extending infinitely into both the past and the future. This distinction between the two theories is testable. The age of very old objects in the universe equals the age of the entire universe in the big bang model. In the steady state theory, any particular object can be young or old, and no such concordance of ages should exist.

The age of the entire universe is determined by its expansion rate, described by Hubble's law. To determine an expansion age for the universe from Hubble's law, start by assuming that the expansion has always progressed at the same rate. Take a particular galaxy in the Coma cluster, presently receding from us at a speed of 7000 km/sec and located 140 Mpc from us. Its speed indicates that every billion years it will move 7 Mpc further away. At its present speed, it would take 20 billion years to go from a location right next to us, in the primeval atom, to its present location. Thus Hubble's law indicates that the Universe is 20 billion years old, if the expansion rate has remained constant. A similar calculation of the age of the universe can be performed for any galaxy (Box 15.1). The age of the universe, as given by Hubble's law, is called the *Hubble time,* the age provided by the assumption that the expansion rate is always the same. If the universe expanded more rapidly in the past, as it would if the gravitational attraction among objects in the universe had been slowing their speeds, the universe would be younger than one Hubble time, perhaps as young as half a Hubble time. With the presently accepted value of 50 km/sec/Mpc for the Hubble constant, the Hubble time is 20 billion years.

Box 15.1   The Hubble Time

The Hubble time can be calculated from the velocity and distance of any galaxy. Suppose a galaxy is a distance D from us. Hubble's law states that that galaxy will recede from us with a velocity

$$V = HD$$

That galaxy and the Milky Way were at the same point in space when the primeval atom exploded, at the time of the big bang T years ago. Assuming the expansion of the universe is constant, and using the formula distance = rate × time, we set

$$D = VT = (HD)T$$

or, canceling D from both sides of the equation,

$$HT = 1$$

or

$$H = \frac{1}{T}$$

If $H = 50$ km/sec/Mpc, to calculate $1/T$ we need to cope with the units that H is expressed in:

$$\frac{1}{T} = \frac{50 \text{ km}}{\text{sec Mpc}}$$

$$\times \frac{1 \text{ Mpc}}{3.1 \times 10^{19} \text{ km}}$$

$$\times \frac{3.1 \times 10^{16} \text{ sec}}{1 \text{ billion years}}$$

$$= \frac{1}{20 \text{ billion years}}$$

and so $T = 20$ billion years.

A number of strands of evidence point to some event occurring 10 to 20 billion years ago as the origin of many of the objects in the universe. The ages of stars can be measured from stellar interior theory (Chapter 11). Observe a globular cluster, and determine the temperature or color of the most massive stars that are still on the main sequence. Look at a group of theoretical stellar evolutionary tracks, and ask how long those stars would remain on the main sequence. This time span is the age of the cluster, because stars on the upper end of the main sequence are those that are just ending their main-sequence stage, about to peel off the main sequence and head toward the red-giant part of the H-R diagram. Allowing for uncertainties in the stellar interior theory and in the determination of the temperatures of globular-cluster stars, the globular clusters in our galaxy are between 10 and 15 billion years old. These globular clusters were formed shortly after the Milky Way galaxy formed, according to our present view of galactic evolution, and so their age should be quite close to the age of the Galaxy.

The elements themselves also provide an indication of the age of the universe. When supernova explosions produce new heavy elements, some radioactive elements are produced along with stable, nonradioactive ones. A nucleus of a radioactive element will eventually decay, transforming itself into a different element. Since the relative amounts of radioactive and nonradioactive elements fabricated in the cores of supernovae can be estimated, it is possible to determine the average age of the material made by supernovae by determining how much of the radioactive material is still around. Samples of material from meteorites are used in this analysis, since meteorites have not been processed by planetary evolution (Chapter 7).

The age of elements measured in this way is some sort of average time interval that has passed since a typical parcel of galactic material passed through the bowels of a supernova. Some refinements of the analysis are possible, for the amount of short-lived radioactive elements present now can indicate the rate of supernova activity just prior to the formation of the solar system. Studies of radioactive elements provide a variety of numbers, depending on what element is considered, but the most reliable radioactive chronometer sets the age of the universe at between 11 and 23 billion years.

Thus, taken as a whole, three strands of evidence indicate that globular clusters, the elements, and the universe all formed between 10 and 20 billion years ago. None of these measurements is particularly precise, but their coincidence is gratifying because they are all independent. None of these measurements depends on the others, and any error in one would not affect the others. Thus the evidence favors the theory that all the matter in the universe originated at some definite time, 20 billion years ago, in a tremendous primeval explosion, the big bang.

★ 15.3 THE BIG BANG ★
AND ITS RELICS

Perhaps the best way to verify the big bang theory is to find relics of this primeval explosion that marked the origin of the universe. Yet a big bang relic cannot be recognized unless we have at least a vague idea of what it is supposed to look like. Fossils, big bang or otherwise, do not come with neat little labels on them. How are they recognized? Once again, the interplay between the model world and the real world appears. A big bang model is either proved or shot down by a comparison of the model and the real world as revealed by observations. The model states what big bang relics should be seen, and observation shows that they are seen.

☆   THE CHAOTIC BEGINNING   ☆

How did the big bang start? To find out, we follow the expanding universe backward in time. The galaxies become closer, closer, and closer together, eventually merging into one cosmic blob. We follow the expansion still further backward in time, and we see a process that is identical to the collapse of matter at the singularity inside a black hole, with the arrow of time reversed. Theory states that the expansion of the universe began from a state of zero volume and infinite density, the cosmological singularity, just as the collapse of a star into a black hole ends in a singularity of zero volume. The word "singularity" is just a smoke screen to hide our absolute bafflement about the beginnings of the big bang from the unsuspecting reader; the idea of a singularity with infinite density, zero volume, infinite gravitational forces, and infinite temperature is patently absurd. Clearly there is something that we do not know about the beginnings of the universe.

In recent years a potential method of determining the degree of violence in the early part of the expansion from this cosmological singularity has been suggested. The British theoretical astrophysicist Stephen Hawking realized that an early turbulent universe would produce hordes of tiny, tiny black holes, far smaller than the black holes produced by collapsing stars. Subsequent calculations showed that such miniature black holes would do something larger ones could not do—evaporate in a burst of gamma rays when quantum mechanical pressures forced their matter outside the event horizon. No one has yet seen these gamma rays, but their detection would possibly confirm a turbulent early universe, and their absence would mean that the initial expansion was quite smooth and uniform.

Fortunately, our ignorance of the very early stages of the primeval explosion does not affect our model of what happens later. Whatever happened at the very beginning, all the turbulence and chaos of the first few moments of cosmic evolution soon dissipated, and the universe then expanded and cooled smoothly. In the smooth primeval expansion, hordes of subnuclear particles bounced off each other, but the temperature and density were so high that every reaction was soon balanced by a counterreaction. In this first microsecond, all particles were in equilibrium with each other, and we need not follow these collisions in detail in order to understand what happened. The universe expanded and cooled; after 1 microsecond ticked off on the cosmic clock that started running at the hypothetical moment of zero volume and infinite density, the temperature dropped to 10^{13} K. At this point in time, the hordes of subnuclear particles disappeared, leaving high-energy gamma rays bouncing around off each other, off electrons, and off the nuclear particles, protons and neutrons.

This picture of the early stages of the universe is an optimistic one, because it brushes all the uncertainties surrounding the cosmological singularity under the rug. Most astrophysicists believe that it is at least true enough to provide a good starting point for asking, "What happened in this big bang that we can now see?" Those of us who share that viewpoint believe that the early events only matter insofar as they produce something that can be observed. Several physicists point out, quite correctly, that we may be building a big bang model on quicksand, because the early universe may well have been more chaotic than the previous paragraph indicates. The limitations of astronomy, an observational science, are again evident. It might be nice to go into the laboratory and test various big bang models, the chaotic and the uniform ones, but no one can make a laboratory replica of a miniature big bang and watch the universe begin under carefully controlled conditions. All we can do is hope that our simplified model is correct, follow it onward, and check it by observation.

The constituents of the early big bang universe at, say, a time of 1 second (or a temperature of 10^{10} K) are photons, protons, electrons, and neutrons. Atoms, which consist of protons, neutrons, and electrons bound together, are not present, since any that were around in the beginning would not survive the enormously high temperatures of the very early big bang. Yet as the big bang progressed, temperatures became cool enough so that a colliding neutron and proton could stick together without being disrupted by a passing particle. One neutron and one proton form a kind of hydrogen nucleus, called *deuterium* or heavy hydrogen to contrast it with the usual hydrogen nucleus, a bare proton.

The building up of elements did not stop with deuterium. A deuterium nucleus might collide with a proton and form a helium-3 nucleus, with two protons and one neutron; or it might add a neutron and form tritium, with two neutrons and a proton. Either nucleus could then add a neutron or proton to form a very stable helium nucleus, with two protons and two neutrons. Elements heavier than helium are difficult to make in a big bang model, for no stable nucleus of five particles exists, and a helium nucleus can only be built further by colliding with a rare deuterium or tritium nucleus.

The synthesis of helium in the early stages of the hot big bang was complete by the time the universe was 20 minutes old. Depending on the density of the early universe, some deuterium might be left over. Deuterium would have more difficulty surviving in a dense universe, because a deuterium nucleus would have a high probability of running into a neutron or proton before the nuclear building stage of the big bang was over. But the amount of helium that is produced depends very little on the details, for almost all the neutrons present in the 1-second-old universe managed to pick up two protons and another neutron and become helium nuclei. Detailed calculations, following the various nuclear reactions in quantitative terms, indicate that, for a wide range of initial conditions, about one-quarter of the mass of the universe was transformed from hydrogen to helium in the early big bang.

Helium synthesis in the big bang is an important milestone since it provides a potential observational check on the correctness of the model. Matter in the universe has not undergone any wholesale nuclear processing since the big bang. Supernovae have added small quantities of heavier elements (metals) to the mix, but if one-quarter of the mass of the universe was transformed into helium in the early moments of the explosion, the matter we now see in the universe should have at least one-quarter of its mass in the form of helium. A variety of techniques have been used to determine the helium abundance in a variety of objects, and so far the verdict is encouraging. There are no objects that anyone has seen that started their lives with significantly less than one-quarter of their mass in the form of helium, and the helium abundances of a wide variety of objects are rather close to the 25 percent figure predicted by the big bang theory. Of course, the helium could have been made somewhere else, but this other helium factory would have had to be omnipresent; even the matter of other galaxies is 25 percent helium.

The helium test of the big bang model is one that the theoretical scheme can fail quite easily; but passing the test does not mean that the big bang model is unambiguously correct. Helium is not difficult to make in nuclear reactions; our sun makes helium all the time. The difficult task is spewing helium out into the interstellar medium from stellar cores. While it is difficult to concoct a model of galactic evolution that allows stars to make one-quarter of the universe into helium but not to make large quantities of other heavy elements (which are not seen), this theory could be wrong: One-quarter of the universe could have been made into helium in stellar interiors. Is there no more direct test of the big bang model?

☆ **THE MICROWAVE BACKGROUND** ☆

We follow the big bang onward, starting from the end of helium synthesis, 20 minutes after the beginning. The expansion goes on, the gas comprised of 75 percent hydrogen and 25 percent helium cools, and nothing significant happens for a million years. During this first million years of expansion, the gas was ionized. Photons zipped around, bouncing off the electrically charged protons, helium nuclei, and electrons, but nothing changed significantly until the million-year mark.

At the million-year point, the universe had cooled to a temperature of 3000 K, and electrons and protons were able to recombine and form atoms without being immediately ionized. At this time, collisions between electrons and helium nuclei produced hydrogen and helium atoms, making the gas electrically neutral. An important consequence of the electrical neutrality of the gas was the transparency of the gas to photons. Photons no longer bounced off electrons, protons, and helium nuclei because such particles no longer existed. These photons could then travel unimpeded through the universe, eventually reaching our radio telescopes. The redshift of a photon traveling since the universe was a million years old is immense, and the 3000-K optical photons that existed then have been redshifted into the microwave and radio parts of the electromagnetic spectrum. Detailed calculations by Ralph Alpher, George Gamow, and Robert Herman, completed in the late 1940s, showed that these photons would have a black-body spectrum that radio astronomers might be able to see. Again, the big bang model offers an opportunity for confirmation by observation. Observation of these photons would be a confirmation of the later stages of the primeval fireball.

When Gamow and others noted that observation of the primeval fireball was possible in principle, radio-astronomical technology was quite primitive. Such a faint signal could not possibly be distinguished from the other sources of background noise with the equipment then available, and Gamow's prediction of background radiation originating from the big bang languished in the journals for a long, long time. This radiation from the primeval fireball was only discovered when technological improvements of the 1960s increased the sensitivity of radio telescopes.

The discovery of this *microwave background* radiation, named for its location in the electromagnetic spectrum, occurred in 1965. The steady state theory was then dealt a blow from which it has not recovered. Two scientists working at Bell Telephone Laboratories, Arno Penzias and Robert Wilson, were trying to isolate all the sources of background static in a radio receiving system. They were trying to track down just what it was that produced the background hiss similar to that you can readily hear on a radio set that is not tuned to any particular station. About 100 miles from Penzias and Wilson's laboratory, a team of Princeton scientists was looking for the microwave background that had been predicted by Gamow almost 20 years before. Penzias and Wilson discovered a source of static that they could not identify, and when the two groups made contact it became evident that Penzias and Wilson had found the microwave background radiation.

How could one be sure that this tiny bit of static detected by the radio telescopes at Princeton and at Bell Labs was in fact the primeval fireball? To be sure, the amount of radiation observed was quite close to the amount that theory stated should be detected. Skeptics answered, however, that this radiation might come from a cloud of distant radio galaxies, radiating at just the right level of intensity, that was too far away to be separated into individual galaxies. While the primeval fireball radiation—the microwave background—had been discovered, its nature remained to be confirmed.

The primeval fireball radiation can be distinguished from a whole collection of distant radio galaxies in two ways. The big bang radiation should be equally intense in all directions, because the universe was uniform at the time it was emitted. Any part of the million-year-old universe was an equally intense photon emitter. A collection of distant radio galaxies, examined closely, would look lumpy rather than smooth. Careful examination of the microwave background radiation shows that it is smooth, as a big bang relic should be. Furthermore, the theory predicts that the big bang relic should have a black-body spectrum, which the microwave background does (Figure 15.4). The homogeneity and spectral shape of the background radiation identify it as the product of the primeval fireball.

This confirmation of the existence of a titanic explosion 20 billion years ago makes the steady state theory very difficult to defend. The observations in Figure 15.4 are literally observations of the million-year-old universe; they show that the universe was rather hot then, totally unlike the present universe. The steady state theory will admit no such change, for its universe 20 billion years ago should have been a collection of galaxies, not a hot ra-

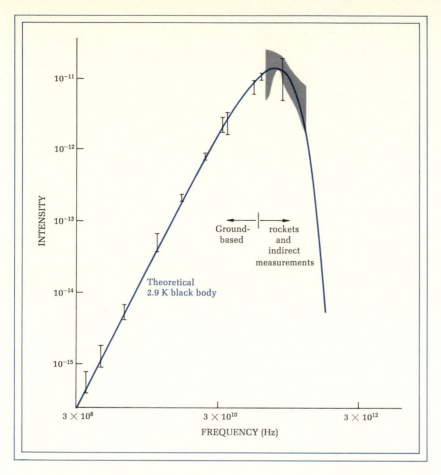

FIGURE 15.4   Observations (black) and theoretical expectations (color) of the micro-wave background. The vertical lines show measurements at one frequency, and the length of the line is the estimated error of the measurement. The shaded area is a recent rocket measurement at several frequencies. [Rocket measurement by D. P. Woody, J. C. Mather, N. S. Nishioka, and P. L. Richards, "Measurements of the Spectrum of the Submillimeter Cosmic Background," *Physical Review Letters*, 34 (1975): 1036–1039. Other data from P. Thaddeus, "The Short-Wavelength Spectrum of the Microwave Background," *Annual Review of Astronomy and Astrophysics*, 10 (1972): 305–334.]

diating gas. Of course, it is possible to salvage the steady state theory by finding another source for the microwave background radiation. A cloud of grains, heated to a temperature of 3 K and spread throughout the universe, would produce a microwave background just like that produced by the big bang. But how are such grains produced? It turns out that they have to be long skinny needles if they are not to evaporate too fast. Such a salvage scheme for the steady state theory is contrived, according to most astronomers. Surely it would be a remarkable coincidence for these mysterious and magical grains to be at just the right temperature to mimic the primeval fireball. Of course, it is not absolutely impossible that the microwave background comes from some such source. It is just sufficiently unlikely enough that most of us believe that the existence of the microwave background confirms the big bang model.

<h2 style="text-align:center">★ 15.4 GALAXY FORMATION: THE ★
QUASAR ERA</h2>

After the million-year mark, the universe continued to expand and cool. The next event of interest was the formation of structure in the universe. Observations of the microwave background show that the universe was remarkably homogeneous 1 million years after the beginning of the explosion; no protogalaxies existed at that time. Yet much of the matter in the universe is now condensed into galaxies, some of which congregate together in great clusters like the Coma and Virgo clusters of galaxies. How did these clusters form?

The process of galaxy formation is the biggest gap in the contemporary big bang model. The theoretical issues to be addressed are the same issues that confront theorists trying to understand star formation (Chapter 10). How do condensations form within an initially uniform gas cloud? Galaxy-formation theorists, unlike those studying the formation of stars, have almost no observations to work with. Was there any turbulence in the early universe? What was the role of magnetic fields? When did condensations first start to form? So far, no one has come up with any compelling theoretical scheme for the origin of galaxies.

With all these theoretical uncertainties, can the problem of galaxy formation be attacked at all at the present time, or must it wait until a new generation of telescopes has penetrated the early days of the universe? A few isolated facts can be deduced from present-day observations. These facts do allow some constraints to be placed on possible models for galaxy formation.

One point of attack is the question, "How is the matter in the universe distributed?" Should a theorist attempt to make large clusters of galaxies, or individual large galaxies like our own, or should one try to fill the universe with small elliptical galaxies and globular clusters? Princeton's P. James E. Peebles has posed this question by taking a number of catalogues of galaxies and combining all the data, asking how galaxies tend to clump together. The answer is, in some respects, frustrating for a galaxy-formation theorist, because there is no preferential size for lumps in the universe. Galaxies are distributed in units ranging all the way from single isolated galaxies to the giant rich clusters like Virgo and Coma. There may even be large superclusters, clusters of clusters of galaxies. Yet Peebles has shown how these different size units are distributed, indicating that there are far more small groups than large ones. His ratios of different cluster sizes may yet provide a bench mark for theorists to use in testing possible mechanisms of galaxy formation.

Another approach to the galaxy-formation problem is the question, "When did galaxies form?" Looking out into the universe, we are also looking backward in time. Light reaching us from the sun left the solar surface about 8 minutes ago; light from the nearest star, Alpha Centauri, left it 4.3 years ago; and light from the Andromeda galaxy left it 2.2 million years ago. The photons that created the color image of this galaxy, Color Photo 17, left it when the human race was beginning to become intelligent. As we probe deeper and deeper into the far reaches of the universe, we are probing backward in time, looking at earlier and earlier frames of the cosmic movie.

But even millions of years are only a small time interval on the cosmic scale. Twenty billion years have passed since the big bang. The most distant galaxy known is a twenty-second-magnitude radio galaxy, 3C 123, at the outer limits of observability. In 1975, the redshift of this galaxy was measured to be 0.637, placing it at a distance of 2300 Mpc, or 8 billion light years. Light that left this galaxy before the solar system was formed is only now reaching our telescopes.

To probe further out into space, further back in time, we need to observe more luminous objects. Ordinary galaxies like our own Milky Way and even 3C 123 are too faint to be observed at such large distances. The quasars, far more luminous than the galaxies are, can be seen at far greater distances. The current holder of the record for the most distant quasar known is the radio source OQ 172, with a redshift of 3.53 (OQ = Ohio State Radio Survey, list Q). Light left this quasar 15 billion years ago, about 5 billion years after the big bang. Quasars, the most powerful searchlights of the universe, allow us to probe the past because they are visible at large distances.

There are enough quasars in the universe so that a count of their numbers can indicate when quasars were most abundant in the universe. Most quasars have redshifts near 2, and light from an object with a redshift of 2 has been traveling for about 13 billion years. Thus quasars were most common in the universe at this time, 7 billion years after the big bang, and they are still seen in considerable quantities about 10 billion years after the beginning. This early epoch of the universe can then be called the quasar era. Few quasars exist now. Were such quasars in existence, they would be easily seen because of their high luminosity. But was there a beginning to the quasar era? Was there a time when quasars did not exist? There is a shortage of quasars with high redshifts, redshifts substantially greater than 2, but this shortage may be due to our inability to locate and discover them.

Because quasars are still somewhat mysterious, it is difficult to interpret exactly what was going on in the quasar era. Assuming the conventional wisdom, that quasars are very energetic active galaxies, the existence of quasars 7 billion years after the big bang indicates that galaxy formation must have occurred before that time. But are quasars galaxies that formed as quasars, or are they galaxies that had existed for a few billion years before becoming hyperactive? In either case, if the redshifts of quasars are due to the expansion of the universe, as the conventional interpretation assumes, their existence indicates that structure in the universe existed 7 billion years after the big bang.

★ 15.5 THE FUTURE ★

☆ TWO POSSIBLE UNIVERSES ☆

After galaxies formed, not much happened to the evolution of the universe on a cosmic scale. The expansion continued, with galaxies flying apart according to Hubble's law. Individual galaxies, of course, continued to evolve in their own ways. Elliptical galaxies lost their luminous blue stars, since these massive main-sequence stars soon lost their energy resources of central hydrogen fuel. Now, some 20 billion years after the beginning, and 10 to 20 billion years after galaxy formation, only the yellow and red main-sequence stars are left in these elliptical galaxies. As time passes, these stars, too, will burn out. In the future an elliptical galaxy will be rather dim. An observer on a planet surrounding a star in an elliptical galaxy will, some 40 billion years from now, see nothing but a group of dim red dwarfs in the sky. The nearest of these will barely be distinguishable with eyes as sensitive as ours. A few red giants may be visible. While new stars do form in spirals, and the sky of an observer in a spiral galaxy would contain some luminous, young new stars, the trend of evolution in spiral galaxies is not unlike the trend in ellipticals. When stars evolve, an increasing fraction of the mass of these galaxies is locked up in the stellar graveyard of white-dwarf stars, neutron stars, and black holes.

Yet these events in the formation of individual galaxies have only an aesthetic impact on the evolution of the entire universe. The only remaining question about the evolution of its structure is whether the expansion will last forever. The mad rush of galaxies away from each other, the result of the primeval explosion, is continually slowed by gravity. Every galaxy in the universe pulls every other galaxy toward it, slowing the expansion. Whether the expansion can be slowed to a stop depends on the nearness of galaxies to each other and on their masses, on the density of the universe. The denser the universe is, the greater the gravitational force among galaxies is, and the more the expansion rate will be slowed. If the mean density of the universe is greater than the *critical density* of 4.7×10^{-30} g/cm^3, or 1 particle/m^3, gravitation will be strong enough to slow the expansion to a stop. If the density of matter in the universe is less than the critical density, the expansion will continue forever.

A universe in which the expansion is eventually stopped is a *closed universe*, because its evolution eventually ends. While individual galaxies fade away as their luminous stars die, the expansion of the universe does end and turn into a contraction (Figure 15.5, top). The first three frames of the movie depicting the evolution of the universe are the same for both open and closed universes, resembling Figure 15.2. But the expansion of the closed universe stops, and the next frames show the big-bang movie run backward. Galaxies now rush toward each other, eventually coalescing. The temperature of the coalesced gas of galaxies then increases, and matter becomes ionized by the high temperature. Compression continues. The collapse of a closed universe is identical with the collapse of a star to form a black hole, but on a far larger scale.

What happens next? We face the same dilemma that we faced in describing the black hole. The laws of physics that we know cannot describe matter

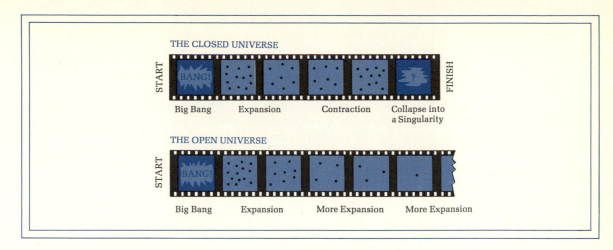

FIGURE 15.5 Simulated "movies" depicting the evolution of a closed and an open universe. The closed universe (top) has a definite beginning—the big bang—and a definite end when all the matter in the universe collapses in the same way that the matter of a star collapses to form a black hole. The open universe (bottom) expands forever and has no definite end.

near the singularity that lies at the end of the evolution of the closed universe, when matter is irresistibly compressed by gravity. Einstein's equations predict that all the mass will coalesce into a point of zero volume and infinite density, the cosmological singularity. We cannot sweep the cosmological singularity under the rug in the way that cosmic censorship swept the singularity in a black hole beneath the event horizon, since we are inside the event horizon of a closed universe and our Milky Way will, tens or hundreds of billions of years from now, become part of that cosmological singularity. Thus, the future of a closed universe beyond the contraction, beyond the end of the movie in Figure 15.5, is mysterious.

It is even possible that the universe will re-expand from the contracted state depicted in the final frame of the closed-universe movie. Some people speculate that the eventual fate of a cosmological singularity is re-expansion, and such a picture would correspond to a series of film segments like the top film of Figure 15.5 laid end to end. The pulsating universe would last forever, going through endless cycles of big bang, expansion, contraction, collapse to a singularity, big bang again, and so forth. But these extrapolations beyond the singularities that mark the beginning and end of a closed universe are just speculations presently removed from the domain of testable scientific theory.

The future evolution of an *open universe*, where the expansion continues forever, is far better understood, and is depicted in the film in the bottom of Figure 15.5. Galaxies never stop moving away from each other, and fewer and fewer galaxies are to be found in any one slice of space. It is only groups and clusters of galaxies, similar to our Local Group, that do not share in the overall expansion because their gravitational force binds them together. The dots in the film thus represent groups and clusters of galaxies. The galaxies, too, will become dimmer and dimmer, while the brighter stars in them run out of fuel and die. In another 5 billion years, our sun will die; in 50 billion years, perhaps half the mass of the galaxy will be tied up in stellar remnants. The sky will be far dimmer then, and it will become yet dimmer as time goes on. The evolution of an open universe resembles the ticking of a clock that is wound up and left to run down. The open universe does not die, it only fades away.

☆ THE MASS DENSITY ☆
OF THE UNIVERSE

What kind of a universe do we live in? Is it the top or bottom film in Figure 15.5 that correctly describes the future evolution of the universe? While all of us harbor some aesthetic prejudices in favor of one scenario or another, observational facts rather than philosophical prejudices must control the decision. The evidence is not yet complete, but the trend of the discoveries of the last several years has been in the direction of an open universe, one where the expansion never stops.

The future of the universe is governed by its density. One way to test the future of the universe is to add up all the mass in the universe and see if it can be filled to a density exceeding the critical density, the density needed to close the universe. The mass concentrated in galaxies is measured by counting the number of galaxies in a given volume of space, measuring their luminosities, and assuming a mass-to-light (M/L) ratio for each of the galaxies counted. The M/L ratio is needed because one can only measure the total amount of light energy emitted by the galaxies in, say, a cubic megaparsec of the universe. The uncertainties are considerable. Even allowing for possible sources of error, the mass in the form of galaxies as we know them is only a few percent of the critical density. Galaxies cannot close the universe.

However, galaxies do not contain all the mass in the universe. Analyses of the velocities of galaxies in clusters (section 14.5) indicate that most of the mass in these clusters of galaxies is hidden. Yet this hidden mass in clusters of galaxies is not sufficient to close the universe, even if all galaxies have such extra mass associated with them. How else can the mass needed to close the universe be hidden? A variety of researchers have explored a number of possibilities. The only two viable forms for the hidden mass are gas and small condensed objects. A diffuse intergalactic gas, uniformly spread through space, could just barely close the universe and remain unobserved. Anything smaller than a star and bigger than a dust grain would be invisible, and a collection of such objects could easily close the universe, although it is difficult to imagine where they might come from. Black holes too could close the universe and remain invisible.

Thus, measurements of the present mass density of the universe tend to argue that the density of the universe is less than the critical density, and the expansion will continue forever. However, the verdict is uncertain because some mass could be hidden. Another approach to the cosmological problem involves measuring the past mass density of the universe by an indirect technique. The first 20 minutes of the big bang saw the buildup of elements from hydrogen to helium. The first stage in this building process was the formation of deuterium, or heavy hydrogen, from the collision of a proton and a neutron. A dense universe would consume more of the deuterium, since further nuclear processing would easily transform the deuterium to helium. In a low-density universe, a deuterium nucleus would have a better chance of survival, because it would have a greater chance of not colliding with other particles before the end of the nuclear-processing stage. A measurement of the present abundance of deuterium in the universe can thus provide some insight into the density of the universe at the time of nuclear processing.

Figure 15.6 illustrates the way in which the test is applied. The graph in black shows the theoretical calculations of how much deuterium should be left over in universes with different present densities. Deuterium was observed in interstellar space by the Copernicus satellite when the absorption of starlight by deuterium was detected and measured. Analysis of the measurements shows that the deuterium abundance is about 1.5×10^{-5} of the hydrogen abundance. If this deuterium is all left over from the big bang, the present density of the universe is about 20 percent of the critical density, the universe is open, and the expansion will continue forever.

The deuterium test of the future of the universe is an imperfect test. If the deuterium observed by the Copernicus satellite is not a big bang relic, the present abundance of deuterium has no cosmological significance. In such a case, a closed universe could produce a small fraction of the deuterium presently observed, and the rest could be made somewhere else. But where? Theorists are now prospecting for possible deuterium-building events. So far, a variety of attempts to make deuterium without making obnoxiously large quantities of lithium, beryllium, and boron have not succeeded: The deuterium factories make quantities of these other elements that considerably exceed their observed abundances. Future work, however, may show that the deuterium need not be a big bang relic. At this time, the deuterium test provides an equivocal, tentative verdict in favor of an open universe.

☆ CHANGES IN THE EXPANSION RATE ☆

Since direct and indirect measurements of the density of the universe provide only tentative answers to the cosmological question, we seek other tests to probe the future of the universe. One consequence of a closed universe is a continual change in the expansion rate. The expansion of the universe must be slowing down if it is to stop eventually. It is not possible to look into

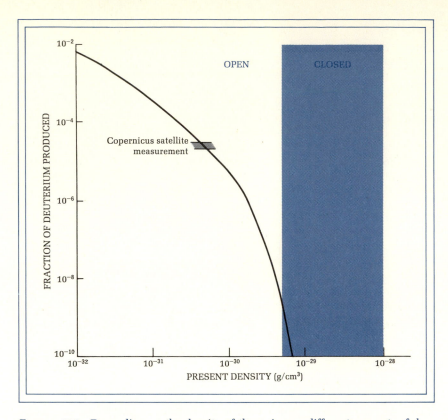

FIGURE 15.6 Depending on the density of the universe, different amounts of deuterium were produced in the big bang. A satellite measurement of the abundance of interstellar deuterium indicates that the universe is open if this helium was made in the big bang. [Deuterium abundance curve from R. V. Wagoner, "Big Bang Nucleosynthesis Revisited," *Astrophysical Journal,* 179 (1973): 349.]

the future and determine future changes in the expansion rate, for a change in the Hubble constant would be imperceptible for billions of years even if it could be measured with an accuracy of a few percent. But we can look into the past and see if the expansion rate has changed. If the change is sufficient, the future evolution of the expansion rate can be predicted, and we can see whether the expansion will stop. Once again, there are direct and indirect ways to determine whether the expansion rate has changed in the past.

The direct method of measuring the deceleration of the expanding universe is the oldest cosmological test, a test that has been explored since the 200-inch telescope was completed in the late 1940s. The test, in principle, is fairly easy to make. Measure the distance to a distant elliptical galaxy, billions of light years away in space and in time, without using the redshift of this galaxy. Once the distance is measured, you know how far into the past you are looking. A measurement of the galaxy's redshift will then measure the expansion rate of the universe at that time. If the expansion rate is changing, that galaxy will have a higher redshift than it would if the expansion rate were uniform (Figure 15.7).

In practice, the implementation of this test is a bit sticky, because it is difficult to measure the distance to a distant elliptical galaxy without using its redshift. All you can see in faraway clusters of galaxies is the image of the galaxy itself. You must then assume that the luminosity of, say, the brightest elliptical galaxy in a distant rich cluster is the same as the luminosity of a nearby corresponding galaxy in a rich cluster. This nearby galaxy is near enough so that changes in the expansion rate have a negligible effect on a distance measurement made from Hubble's law, near enough to be on the part of the curves in Figure 15.7 where the open- and closed-universe lines coincide. The distance of the faraway galaxy can now be measured by comparing its brightness to the nearby one and using the usual relation among brightness, luminosity, and distance.

A straightforward application of this test, made on the basis of the accumulation of 20 years of observations and published in 1971, showed that the expansion rate of the universe was changing fast enough for the universe to be closed, contradicting the results of the two tests mentioned earlier. But this test involves possible problems. The elliptical galaxies used in the test

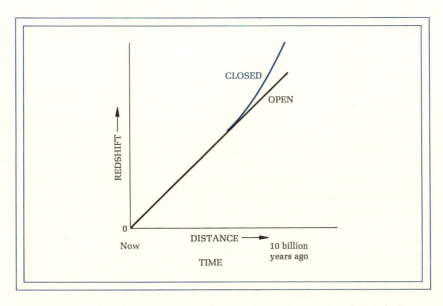

FIGURE 15.7 The relationship between redshift and distance can show whether or not the expansion rate of the universe has been changing. Light left progressively distant galaxies longer and longer ago. If the expansion rate of the universe was greater in the past, the redshifts of these galaxies would be larger (color) than it would be if the expansion rate did not change (black).

are just at the limits of visibility, and it is possible, even probable, that the more luminous of the faint, distant galaxies were observed, biasing the test results. One tends to pick the most luminous elliptical galaxies to put on the observing list because the less luminous ones are invisibly dim. Further, the straightforward use of this test presumes that elliptical galaxies had the same luminosity 10 billion years ago as they do now. If this assumption is invalid, the test will produce erroneous results. There is no clear consensus in the astronomical community on the verdict of the redshift test; it may well tell us more about the evolution of the luminosity of giant elliptical galaxies than about cosmology.

☆ COSMIC AGES ☆

An indirect method of measuring the change in the expansion rate may yet prove to be more successful than the direct method described above. The age of the universe presumed in this book, 20 billion years, is equal to the Hubble time. If the expansion rate has changed, it then was faster in the past, and the big bang occurred something less than 20 billion years ago. Faster-moving galaxies would have moved as far away from each other in less time. Thus, if we can measure or guess the actual age of the universe and compare it to the Hubble time, we can determine whether the expansion rate has changed. If the universe is to be closed, it must be younger than two-thirds of the Hubble time. With a Hubble constant of 50 km/sec/Mpc and a Hubble time of 20 billion years, the universe must be younger than 13 billion years if it is to be closed.

The actual age of the universe is difficult to measure. No primeval observer left a readily visible clock around, to tick off the seconds that have passed since creation. We can, however, measure the ages of elements and stars (section 15.1). Most of the observations indicate ages greater than 13 billion years, again tending toward an open universe. But the uncertainties in these determinations are sufficiently large so that a definite conclusion cannot be made. This test is, perhaps, one for the future.

Someone who views science as a cosmic computer, always ready with unequivocal answers, may well be disappointed with the material in this section. While the evidence shows a clear trend toward an open universe, the verdict of each test is clouded by the complexity of the universe. Not one of these tests is a clean one; the effects of cosmic evolution are entangled with lesser events such as the evolution of the luminosity of elliptical galaxies or the possible production of deuterium. The noncosmological processes must be understood before the tests can be applied reliably. Progress in recent years has been encouraging, for all the tests described here have either been implemented for the first time or significantly improved since 1970. When will we know what will happen to the universe? It would be foolish to guess when we will untangle the complex web of cosmic evolution to the point at which one or more of these tests probing the future evolution of the universe can give a definitive answer.

Virtually all astronomers accept the big bang theory for the past evolution of the universe. About 20 billion years ago, all the mass of the universe was concentrated in the primeval atom. The high temperatures in this cosmic egg caused it to explode, and we have observed several relics of that explosion. The expansion of the universe is the result of the big bang. More direct relics are the helium atoms that compose one-quarter of the mass of the universe and, more importantly, the microwave background radiation. The next stage of cosmic evolution, galaxy formation, is not understood. Also cloudy is the future of the universe. While current cosmological tests appear to indicate that it will continue to expand forever, none of the tests is simple and unequivocal.

This chapter has covered:

The difference between the evolving cosmological models accepted today and the endless clockwork models of the nineteenth century and before;

The outlines of cosmic evolution according to the big bang and steady state theories;

Estimates of the age of the universe by various methods;

The observations of the microwave background, and their significance as evidence for the correctness of the big bang theory;

What little we know about galaxy formation; and

The verdicts of four tests probing the future evolution of the universe.

KEY CONCEPTS

Big bang theory
Closed universe
Cosmology

Critical density
Deuterium
Hubble time
Microwave background

Open universe
Primeval atom
Steady state theory

REVIEW QUESTIONS

1. What is the essential difference between the big bang and steady state theories of cosmology?

2. Briefly describe three methods of measuring the age of the universe.

3. Justify the statement that the three methods of measuring the age of the universe described in section 15.2 are independent of each other.

4. In what respects does the beginning of the big bang resemble the collapse of a star to form a black hole?

5. Why does the observation of the microwave background show that the steady state theory contradicts observation?

6. What are the two observational facts that bear on the problem of galaxy formation?

7. Summarize the four probes of the future of the universe and their results.

8. Briefly enumerate the uncertainties and complications in one of the four cosmological tests.

FURTHER READING

Gingerich, O., ed. *New Frontiers in Astronomy.* San Francisco: Freeman, 1976.

Hawking, S. W. "The Quantum Mechanics of Black Holes." *Scientific American* (January 1977): 34–49. A description of the primordial black holes that might have been made in a chaotic big bang.

John, Laurie, ed. *Cosmology Now.* London: BBC Publications, 1974.

Sciama, D. W. *Modern Cosmology.* Cambridge: Cambridge University Press, 1971.

The study of galaxies is part of the field of cosmology, the study of the structure and evolution of the entire universe. We can only observe the universe by analyzing its contents, and galaxies are the only objects that are visible at cosmologically interesting distances. Once we understand them better, the enigmatic quasars may allow us to plumb the depths of the far-off universe.

There are few firm theories in the field of galaxies and cosmology, because observational data pertaining to the properties and distances of these complex aggregates of stars are extremely difficult to obtain. It is important to distinguish between concrete theories supported by evidence, such as the big bang model of the expanding universe, and working hypotheses like the theory of evolution of galaxies.

OBSERVATIONAL FACT

Two types of stellar populations exist in the Milky Way galaxy.

Elliptical, spiral, and irregular galaxies differ in shape, gas content, and stellar content.

Quasars and active galaxies emit synchrotron radiation.

The microwave background exists and almost certainly has a black-body spectrum.

The universe expands according to Hubble's law.

CONCRETE THEORY

Quasars are galaxies with central sources of high-speed electrons.

Spiral structure is caused by density waves.

The chemical evolution of the Milky Way galaxy is caused by the gradual enrichment of the galactic gas by supernovae.

The universe began with a big bang.

WORKING HYPOTHESIS

The differences between stellar populations and different types of galaxies are somehow caused by the difference in galactic rotation and gas content.

The universe is open according to current evidence.

★ PART FIVE ★

LIFE

The evolution of astronomical bodies can be understood without invoking the presence of life on them. The universe is so large, and living beings so small, that the activities of living organisms can scarcely affect the appearance of even a small part of the universe—a planet. On the cosmic scale, life is unimportant.

But here we are. One species of living organisms on our planet has evolved to the point where intelligence and curiosity have led to an understanding of the evolutionary processes that shape the entire universe and many of the objects in it. Are there other environments where intelligent life has also developed? This question was, until a few years ago, the kind of question that could be answered only by science fiction writers who had no need to reconcile their stories with facts. But in the last 20 years, advances in astronomy and biology have enabled exobiologists to begin approaching this question scientifically, breaking it down into more specific questions that research can answer. Exobiology, the study of life in extraterrestrial environments, is an infant science. Some people argue that it is not yet a science, for no example of an exobiological system, a life form outside the earth, is known to exist. Yet there is a pervasive sense of optimism among exobiologists that we are not alone in the universe, that the processes that led to the development of intelligent life on the earth are natural processes that should occur elsewhere. A few astronomers are exploring ways in which we might communicate with other life forms.

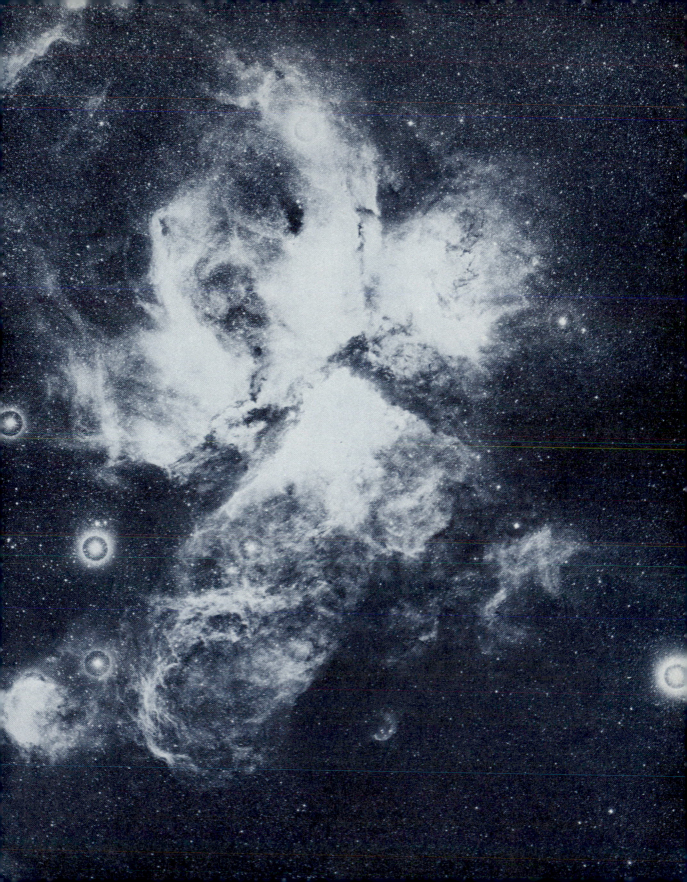

★ CHAPTER SIXTEEN ★

LIFE IN THE UNIVERSE

Is there life elsewhere in the universe? How could we communicate with it? A scientific approach to these questions requires their decomposition into more specific questions that researchers can answer. We can do more than appeal to the huge number of stars in the Milky Way (Figure 16.1), simply stating (or hoping) that at least a few will have environments similar to our own and that life will evolve there. Recent advances in astronomy and biology permit a more serious assessment of the likelihood and locations of other possible life sites. Which stars are good suns, and how many are there? Identification of good suns can guide the search for signals from extraterrestrial civilizations.

★ **MAIN IDEAS OF THIS CHAPTER** ★

The subdivision of the question of the existence of extraterrestrial life delineates specific questions that biologists, astronomers, and social anthropologists can answer.

The astronomical evidence indicates that many good life sites exist.

The biological evidence indicates that the emergence of life on a good life site is probable but not necessarily inevitable.

The principal uncertainty in determining the abundance of extraterrestrial civilizations is a sociological one, the lifetime of such civilizations.

Communication with extraterrestrial civilizations is best done through the transmission of radio signals.

★ **16.1 THE FRAMEWORK** ★

How many civilizations are there in our galaxy? The future of research in interstellar communication hinges on an optimistic answer to this question. If life is not abundant in the universe, lifetimes would be wasted searching for life around barren suns with radio telescopes. But if life is abundant, the nearest stars might harbor communicative civilizations, and an attempt to detect signals from other civilizations might be a worthwhile endeavor.

An estimate of the number of civilizations begins with the presumption that such a culture will exist on a planet in orbit around some star in the Galaxy, using the luminosity of that star as the energy source that drives its biological processes. In principle, all of the 2×10^{11} stars in the Galaxy could support life. In practice, some fraction of those stars will actually harbor a living, communicative civilization. We must estimate that fraction to determine the number of such cultures that we could contact.

A number of conditions must be met if a communicative civilization is to exist near a star. The star must have a planetary system. At least one planet in the system must be in an orbit that maintains a viable environment on the

FIGURE 16.1 Closeup of a region of the Milky Way galaxy in the direction of the constellation Sagittarius. Most of the stars in this photograph are good life sites. With so many stars in the Milky Way galaxy, it would be extremely paradoxical if our sun were the only one with living beings around it. (Lick Observatory photograph)

planetary surface. Life forms must actually evolve on the planet. If we are to discover a life form outside the solar system in the near future, we must be able to communicate with it. Therefore, we require an intelligent species that has developed the technology needed for interstellar communication, and that species must exist on that planet now.

The probability that a particular star has communicative life around it is the joint probability that all the conditions are met. This probability is the product of each of the individual probabilities. If we seek a star with an earthlike planet, and there is a probability of $\frac{1}{2}$ that the star has planets and a probability of $\frac{1}{2}$ that any planetary system contains an earthlike planet, then the probability that the star has a planetary system *and* that the system contains an earthlike planet is $\frac{1}{2} \times \frac{1}{2}$ or $\frac{1}{4}$.

Thus we can decompose the probability that a star has a communicative civilization near it, the fraction of all stars with communicative civilizations, into a series of probabilities, each one of which refers to some condition necessary for the existence of a communicative civilization. Several such decompositions exist. The conventional one, the conventional expression for the probability P that a communicative civilization exists around a given star, is

$$P = f_P n_L f_L f_I f_C (L/2 \times 10^{10} \text{ years})$$ (16.1)

where the various individual probabilities or fractions, the f's, have the following meanings:

f_P: We presume that a star, to harbor a civilization, must have planets. f_P is the fraction of all stars with planets.

n_L: Not just any planet will suffice; a planet must be capable of supporting life if it is to support a culture. n_L is the average number of good planets per planetary system, called n because there may be more than one planet in an individual system and we must count all of them.

f_L: Does life evolve, given a suitable environment? f_L is the fraction of potential life-supporting planets on which life does, in fact, evolve.

f_I, f_C: Amoebas can't build radio telescopes to communicate with us, and unless they exist on Mars we will have a hard time finding them. A communicative culture must have developed both intelligence and technology. f_I is the fraction of ecosystems in which an intelligent species evolves, and f_C is the fraction of intelligent species interested in and capable of interstellar communication.

$L/2 \times 10^{10}$ years: Communication over interstellar distances requires an active respondent, for interstellar archaeology is presently impossible. We cannot communicate with a dead civilization or a civilization not yet in existence; we require an active, flourishing civilization at the other end of the communication link. The probability that a civilization will now be in its communicative stage is equal to the duration of its communicative stage L divided by the total lifetime of its parent star, here taken to be roughly equal to the age of the Milky Way galaxy or 2×10^{10} years.

The total number of communicative civilizations in the Galaxy is the total number of stars multiplied by the probability P that any one of them has a civilization around it. Taking 2×10^{11} as the number N_* of stars in the galaxy, and putting the whole thing together, we have for the number N of civilizations in the Milky Way,

$$N = N_* P$$

or
$$N = N_* f_P n_L f_L f_I f_C \left(\frac{L}{2 \times 10^{10} \text{ years}} \right)$$ (16.2)

or
$$N = 10 f_P n_L f_L f_I f_C L$$

This equation, expressing the number of communicative civilizations in the Milky Way, was initially presented by Frank Drake of Cornell University at the first scientific conference on extraterrestrial life, and it is often referred to as Drake's equation, even though there are others who can claim to have thought of it independently. It provides a framework for asking how many other civilizations there are in the Milky Way, because each of the f's can be numerically estimated.

This simple equation does a magnificent job of separating a significant but poorly posed question, "How many other cultures are out there?" into a series of separate questions, some of which at least can be intelligently addressed. We need no longer rely on an assumption that some of the stars in the Milky Way must be like our sun and examine the foundations on which that faith can rest. Three disciplines contribute to the assignment of numbers to the terms in Drake's equation. Astronomers can estimate the terms dealing with planets, f_P and n_L; biologists can estimate f_L and specify what sorts of environments should be included in n_L; and the last three factors depend on our knowledge of social evolution, contributed by the disciplines of sociology and anthropology. The next several sections of this chapter will assign a numerical value to each of these factors.

★ 16.2 DISCOVERING OTHER ★ PLANETARY SYSTEMS

Do most stars have planetary systems? It is difficult, but not impossible, to imagine life forms existing in the cold near-vacuum of interstellar space. A planet provides a convenient base from which a living organism can draw raw materials as it uses the energy coming from its sun. We seek to evaluate f_P. How many stars have planets?

Some indirect evidence indicates that most stars have planets, that f_P is fairly close to 1. The story of star formation, described in Chapter 10, indicates that the formation of planets is a natural, common occurrence. Support for this idea comes from the slow rotation rate of cool F-, G-, K-, and M-type main-sequence stars, stars like the sun, as compared with the rapid rotation of their hotter counterparts. One of the strange features of the solar system is the concentration of its angular (rotational) momentum in the orbital motions of the planets. If the orbital motion of the planets were given to the sun, the sun would spin far faster, as fast as the hotter stars on the main sequence. Therefore, it is at least plausible to argue that the slow rotation rate of cool main-sequence stars, most of the stars in the Galaxy, is caused by the presence of planetary systems.

Yet more direct evidence would be desirable, for there are many potential flaws in the arguments of the previous paragraph. The discovery of just one planet around a nearby star would convince the most hidebound skeptic that planetary systems are not a rarity. However, planets are miniscule cosmic bodies, emitting no light of their own. The reflected light from planets around distant stars would be swamped by the light of their suns, just as planets in our own system are invisible when they are located right next to the sun in the sky. Indirect methods of planetary searching are the only ones that offer promise.

Planets around other stars can be discovered by their gravitational influence on the motion of their suns. In our solar system, most of the mass is concentrated in the sun and Jupiter. These two objects orbit around their common center of mass, located close to but not at the center of the sun (Figure 16.2). Someone carefully observing the sun from a great distance would notice a slight irregularity or *perturbation* in its path through space, caused by the motion of the sun and Jupiter around each other. The mass of Jupiter is just sufficient to cause the sun to deviate from the straight-line path it would follow were it barren of planets. Other planets like Saturn influence the sun's path in similar ways.

The detection of planets by the observation of gravitational perturbations of the motion of the parent stars is a difficult observational task. Consider the problems involved in detecting Jupiter from the nearest star Alpha Centauri as an example. The presence of Jupiter would cause the sun's path to deviate from a straight line by 0.003 sec of arc, 0.3 percent of the size of the images of stars on a typical photographic plate, or the size of a dime located 500 km away! Measuring such a tiny deviation is almost, but not quite, beyond the abilities of contemporary astronomy.

Over the years, a few specialists in the measurement of stellar positions (astrometrists) have suspected irregularities in the motions of several stars in binary systems. Someone measuring a particular binary star in an effort to measure that star's mass (Chapter 9) noticed a possible irregular trend in the motion of one of the components of the binary that just might be caused by the presence of a third body, a planet, in the system. The research paper on the binary star carefully noted the presence of a suspected planetary perturbation. In none of these cases has the perturbation been confirmed by subsequent research. Occasionally the author of the original paper was quite

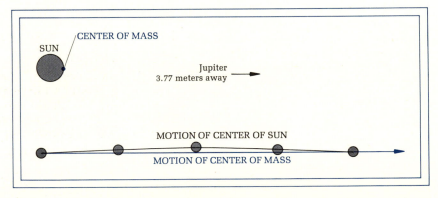

FIGURE 16.2   The center of mass of the sun-Jupiter binary system is located just outside the surface of the sun. The center of mass of the system moves uniformly through space, and the sun deviates from uniform motion because of the presence of Jupiter. Saturn has a similar but smaller effect on the sun's motion.

upset to see this 30-year-old suspicion, shown by later research to be just a suspicion and not reality, appear on a list of "planetary suspects" in the secondary literature of textbooks and magazine articles, authored by nonastrometrists. These planetary suspects have proliferated like rabbits in spite of their ephemeral existence. The most celebrated case involved an alleged planet around Barnard's star, the second nearest star to the sun. Some theorists debated whether there were one, two, or even three planets around this object. Sadly, the observations indicating that there were any planets at all turned out to be suspect, for the largest perturbation in the motion of the star occurred when the telescope used to make the photographs was refurbished.[1]

Yet the fact that no planetary discoveries have been confirmed does not mean that the planets are not there; it only indicates that we have not found them yet. Modern observational techniques have progressed to the point where an intensive effort to detect perturbations in the motions of nearby stars caused by planets could discover a planet with a mass equal to the mass of Jupiter around most of the nearby stars, if such planets existed. The outcome of such a search could answer the question, "Are planets common?"

Thus there are some grounds for a cautiously optimistic view that planetary systems are common in the universe, that f_P is near 1. All that we know about star and planetary formation is consistent with this attitude. The natural sequence of events leading to the formation of our own planetary system and the slow rotation of cool main-sequence stars like the sun support the prevalence of planetary systems. Unfortunately one cannot appeal to the actual discovery of extrasolar planetary systems to support this optimistic view.

★ 16.3 GOOD SUNS ★

The mere existence of a planetary system around a star does not necessarily mean that that system contains a good life site. We seek both an average value of the number of life sites per planetary system and some idea of which types of stars are most likely to contain them. Such information not only will help estimate the abundance of communicative civilizations in the Milky Way, but also will direct the search for extraterrestrial signals to the *good suns,* to those where life is most likely.

The primary factor governing the ability of a planet to support living organisms is the surface temperature of the planet. This temperature is determined largely but not entirely by the amount of radiation a planet receives from its sun. A planetary surface will be roasting hot if the radiation blast is too intense, and freezing cold if the stellar radiation is too feeble. A star with a given luminosity has an *ecosphere* around it, a volume of space within

[1] For a thorough discussion of these planetary suspects, see George Gatewood, "On the Astrometric Detection of Planetary Systems," *Icarus* 27 (1976): 1–25.

which a planet must orbit if it is to be neither too hot nor too cold. The definition of ecosphere is not precise, because other factors like albedo and the strength of the greenhouse effect (Chapter 5) also influence a planet's temperature. However, it is generally possible to broadly define the habitable zone around any given star. The ecospheres of several nearby stars of differing luminosities are sketched in Figure 16.3.

While the ecospheres of hot stars are far from the star and the ecospheres of cool, low-luminosity stars are very close to the star, this distinction alone does not define the selection of good suns. Planets in our own solar system tend to be spaced farther apart in the outer solar system. For the four stars shown in Figure 16.3, there are roughly one to two planets in each ecosphere, counting the asteroids as a planet. It might seem as though any star would be a good sun. But the hot and cool ones present serious limitations.

A star significantly hotter than Procyon, with an O-, B-, or A-type spectrum, would not remain on the main sequence long enough to permit the development of intelligent life before it became a red giant and fried any life on its planets. The evolution of intelligent life took $4\frac{1}{2}$ billion years on the earth, and there is no reason to believe that it will proceed at any drastically different pace elsewhere. It would thus seem unlikely that life could evolve to the intelligent stage on a planet of an A-type star like Sirius, with its main-sequence lifetime of 1 billion years. Sirius will swell up, become a red giant, and burn off any incipient algae that might be trying to evolve further. While quantitative, precise estimates are difficult (how short a main-sequence lifetime is acceptable?), a requirement of 2 billion years of main-sequence evolution for the development of life rules out the O, B, and A stars as good suns. Red giants, too, can be ruled out as life sites for the same reason, for they cannot provide a stable climate for 2 billion years.

The extremely cool, low-luminosity, M-type main-sequence stars have tiny ecospheres, for planets must huddle within a few hundredths of an astronomical unit of the central star to be warm enough to provide the liquid water that is necessary for our form of life. No such close planets exist in our system. Even if such planets were to exist around an M star, tidal forces would make them uninhabitable by stopping their rotation, in the same way that tidal forces cause the moon to show the same side to the earth at all times. The atmosphere of a nonrotating planet would freeze on the dark side, making the planet uninhabitable. Again, quantitative estimates of just how close a rotating planet could be are difficult. Most investigators agree that a distance of 0.2 AU is far too small to allow for rotation, and so the low-luminosity M-type stars at the bottom end of the main sequence (and the white-dwarf stars) are ruled out as good suns.

One other limitation on the number of good suns may also play a role in determining the average number of habitable planets around stars. Astronomers do not agree whether binary star systems are good possible life sites. The presence of a second star close to a potential sun limits the kind of orbits that a planet can follow. It must orbit either one star or the other, or orbit at a great distance from the pair, lest gravitational forces kick it out of the system eventually. Will planets evolve in a binary system, or will the second star sweep up all the gas that would be needed to make planets? Considering that our own solar system is almost a binary (if Jupiter were 50

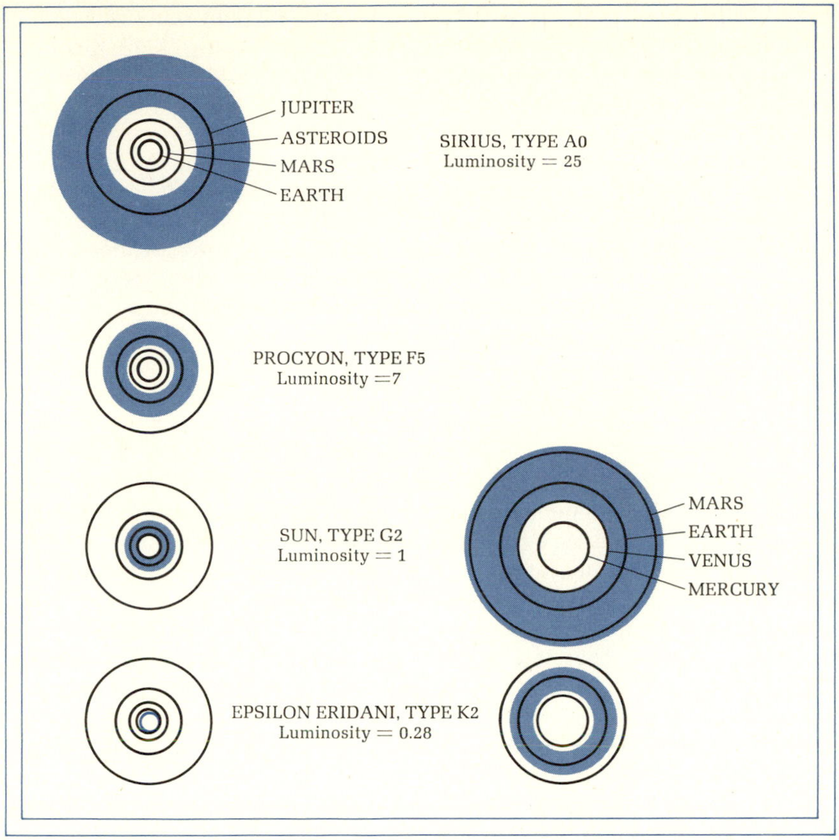
JUPITER
ASTEROIDS
MARS
EARTH
SIRIUS, TYPE A0
Luminosity = 25
PROCYON, TYPE F5
Luminosity =7
MARS
EARTH
VENUS
MERCURY
SUN, TYPE G2
Luminosity = 1
EPSILON ERIDANI, TYPE K2
Luminosity = 0.28

FIGURE 16.3 Depending on the luminosity of the central star, a planet like the earth could circle it at various distances and still remain habitable. The color zones depict the *ecosphere* of a star, the region within which a habitable planet's orbit will lie. Shown in black are the orbits of various planets in our solar system, with an average asteroid treated as a planet. Roughly one or two planets are found in each ecosphere.

times its actual size, it would be called a star), I believe it is unreasonable to rule out binary systems.

How many habitable planets are there in each planetary system? In our solar system there is one, Earth, and a possible second, Mars. So far, the search for life on Mars by the Viking probes has been discouraging (Chapter 5), but the barrenness of Mars is more directly due to its lack of an atmosphere than to its distance from the sun. If one assumes that many planets might have an earthlike greenhouse effect, there are an average of one or two planets in any ecosphere. Let us take $1\frac{1}{2}$ planets per ecosphere as an estimate.

If anything, $1\frac{1}{2}$ planets per ecosphere may even be too small a number. The arguments in the last paragraph presume that life must exist on the surface of a rocky planet. Jupiter, for example, is quite cold at the visible cloud surface, but deeper in the Jovian atmosphere it is warmer, and theoretical calculations indicate that water clouds exist. Maybe life evolved in those clouds. On the other hand, the optimistic estimates of the prevalence of the greenhouse effect on rocky planets may be biasing our value for n_p toward the optimistic side. Thus, $1\frac{1}{2}$ habitable planets per ecosphere seems to be a reasonable compromise.

To calculate n_L, the number of life sites per average star, we need to average over all stars, not just the good suns. The good suns, the F, G, and K main-sequence stars, comprise about one-sixth of all of the stars in the Milky Way (Table 13.1). With one-sixth of the stars as good suns, and $1\frac{1}{2}$ life sites per good sun, the total number n_L of life sites per average star is $\frac{1}{6} \times 1\frac{1}{2}$ or about 0.2. Pessimists might argue that binary stars should be discarded, and that the chance that a planet in an ecosphere might be a tiny one like Mercury would reduce the number of life sites. Optimists would take issue with the temperature limitations in this discussion as being too stringent, and suggest that two, three, or even four planets per star is more realistic. The present estimate of $n_L = 0.2$ may be half or double the true value. Other estimates of n_L by exobiologists range from 0.1 to 2.

★ 16.4 THE ORIGIN OF LIFE ★

With 0.2 good life sites per star, the Milky Way galaxy contains 4×10^{10} life sites at 2×10^{11} stars. Will life evolve on any of them? The third factor in Drake's equation, f_L, measures the fraction of life sites on which life in fact evolves. On one of these sites, the earth, life has evolved. We can analyze the origin of life on the earth and see if it is a natural process that would be likely to occur elsewhere. If we suspect that life does evolve under most conditions, then $f_L = 1$. But if the origin of life involves some highly unusual, extraordinary process, f_L might be a very small number, reducing the possible number of civilizations in the Galaxy. We thus ask how life evolved on our own planet.

The primeval earth, the earth before the appearance of life on its surface, had a rather different chemical environment than the one we have today. While the original atmosphere of hydrogen dissipated rapidly, a secondary atmosphere containing combinations of hydrogen with other chemical elements appeared as the result of outgassing by volcanic eruptions. The constituents of this atmosphere were methane (CH_4), ammonia (NH_3), and water (H_2O), combinations of hydrogen with the three most common reactive elements carbon, nitrogen, and oxygen. Several energy sources were available to assist in molecule building: Solar ultraviolet radiation, primitive thunderstorms, and volcanic eruptions are three possible ones.

The first step in the creation of life is *chemical evolution*, the development of the complex chemicals of life from the far simpler compounds initially present in the primeval atmosphere. Biology was revolutionized in the 1950s with the discovery that most biological processes are based on the reactions of a small number of chemicals. The most famous of these basic biological chemicals is the substance that carries the genetic code, deoxyribonucleic acid, or DNA. DNA, a huge, complex molecule, is built from some far simpler molecules: a sugar-phosphate backbone and four nucleic acids. Another significant constituent of all life is protein molecules, long chains of thousands of simpler compounds, the amino acids.

Chemical evolution has been simulated in a number of laboratories. Stanley Miller and Harold Urey, the pioneers, were the first to fill a reactor vessel with the chemicals of the primitive-earth atmosphere, supply some energy, and see what happened. In their initial experiments, performed in 1953, two amino acids and some compounds similar to nucleic acids were formed.

Subsequent work has refined the early work of Miller and Urey. A number of investigators have reproduced the processes of chemical evolution in a variety of primitive-earth environments that contained a variety of energy sources. The occurrence of chemical evolution, the production of biologically significant molecules, in all of these experiments shows that no special circumstances are required. The existence of biologically significant molecules in interstellar space (Chapter 10) demonstrates that, even in these hostile environments, the first stages of chemical evolution can occur. The analysis of several meteorites has shown that organic compounds exist in these objects. Chemical evolution will occur on all the life sites in the Galaxy.

But it is a long, long way from a collection of nucleic acids and amino acids to the first living cell. Simulations of chemical evolution have only progressed to the synthesis of the most primitive building blocks of life, and they are a far distance from making a little green bug in a test tube. The most complex chemicals synthesized in chemical-evolution experiments so far have been chains of a few amino acids. Proteins are long chains of amino acids, so the experiments are heading in the right direction. The path to life has no visible obstacle blocking the road, since the synthesis of more and more complex molecules has not been followed in the laboratory. Nature had a whole planet and a billion years of time to work with, and a scientist working in a laboratory has no such scope.

Perhaps the most difficult step to visualize along the path from chemicals to life is the production of the first living cell. How was it possible to create a separate unit that could control its chemical reactions in a reproducible way? Theoretical speculations abound, but no one has shown that this event must occur everywhere. It did occur on the earth within a billion years of the formation of our planet. The oldest fossil organisms are primitive blue-green algae, found in the rocks of the pre-Cambrian shields in northern Canada and southern Africa. If life were discovered on Mars, or anywhere in the solar system, it would confirm that the formation of living systems from chemicals is a relatively common occurrence. The actual existence of life in two independent places in the solar system would show that this process occurred twice in our own solar system, a microscopically small part of the Galaxy. Thus we set $f_L = 1$, optimistically arguing that where the conditions for life are right, life will arise. There is scientific evidence that chemical evolution will indeed occur everywhere, but we rely on faith to bridge the gap between chemical evolution and the formation of the first cell.

★ **16.5 INTELLIGENCE AND** ★
TECHNOLOGY

The emergence of the first living cells on the primitive earth marked the beginning of biological evolution, a process that has continued to the present day. Natural selection, the driving force behind the transformation of organisms into different sorts of organisms, was recognized a century ago by Charles Darwin. A particular organism, a bit more sophisticated than its neighbors, is better able to live in the environment provided by nature. This organism can grow and reproduce. It and its progeny will survive in greater numbers, and the organisms that did not change will die out. The process continues indefinitely, and it has been extensively documented on a number of islands, where the isolation of organisms from their mainland counterparts allows detailed study of their evolution. The chain from the primitive blue-green algae to the very sophisticated organisms of the present epoch is a long, tortuous, and unbroken one. Biologists studying fossils are gradually reconstructing the various steps of this evolutionary chain, one segment at a time.

One critical step in the evolutionary chain occupies a special place in the study of interstellar communication. A biologically sophisticated but unintelligent animal on some distant planet orbiting a faraway star is useless to us. We need an intelligent being with a radio telescope that we can communicate with. In what environments will intelligence evolve? Intelligence is, in its broadest sense, the ability to cope with problems creatively, and nature provides a wide variety of problems with which living beings must cope. There is good reason to believe that natural selection will encourage the development of creatures that are smart enough to build dwellings to keep out the cold rather than fly south for the winter. Human beings have populated the entire earth, from the frozen wastes of Antarctica to the tropical jungles of Africa. No other species has been able to adapt to such a variety of environments. Common sense (optimistically) dictates that this ability will evolve in most instances. Here I adopt a tempered optimistic value for f_l of $\frac{1}{2}$, supposing that half of all ecosystems will evolve an intelligent species in response to environmental pressures.

Intelligence, by itself, does not ensure the evolution of a partner at the other end of an interstellar communication link. The culture out there must be able to build equipment that can send and receive messages transmitted across the light years of interstellar space. Further, this culture must have the desire to engage in interstellar communication. One can only guess how many cultures would become technological and how many would have both the urge and capability to explore. Different terrestrial cultures have shown different technological inclinations. A great many cultures did not take the step of building very sophisticated machines in an effort to master the environment. The conventional wisdom of exobiology sets f_C, the fraction of intelligent cultures that will become able and willing to communicate, at 0.1, but this number is really only a guess.

★ **16.6 THE NUMBER OF TECHNICAL** ★
CIVILIZATIONS

☆ **LIFETIME OF A TECHNICAL** ☆
CIVILIZATION

Let's stop and review the progress toward determining N, the number of technical civilizations in the Milky Way galaxy. We have, so far, assigned numerical values to five of the components of Drake's equation (16.2):

$$N = 10 \times f_P \times n_L \times f_L \times f_I \times f_C \times L$$
$$= 10 \times 1 \times 0.2 \times 1 \times 0.5 \times 0.1 \times L \qquad (16.3)$$
$$= 0.1 \times L$$

The numerical values assigned to these five components are those discussed in the last four sections. Different writers on this subject use different numerical values, but the end result seems to be approximately the same: The number of technical civilizations in the Milky Way galaxy equals the lifetime of such a civilization in decades. This result agrees, within a factor of 10, with most of the estimates in the literature of exobiology.

One uncertainty remains: L, the lifetime of a technical civilization. For the other terms in the equation, our own experience with terrestrial ecosystems and cultures can provide a guide, but for L not even our own example can provide any insight. How long will our culture last? We have been capable of interstellar communication for approximately 10 years. If we blow ourselves up tomorrow, and if all other civilizations do likewise, then that span of one decade represents the average lifetime of a technical civilization. If this pessimistic estimate is true, how many civilizations are there in the Galaxy? We have $N = 0.1 \times 10 = 1$, one civilization in the entire Milky Way. That one civilization is us.

Depressing as the morning headlines may seem sometimes, we probably are not on the road to immediate extinction. Suppose we estimate that some fraction, say 1 percent, of all civilizations will manage to avoid self-destruction, cope with their technology without destroying their environment, and last a stellar lifetime (10^9 years or more). The average lifetime of a civilization is then obtained by totaling over 100 civilizations, 99 of which self-destruct and one of which survives:

$$L = \frac{99 \times 10^2 + 1 \times 10^9}{100} = 10^7 \text{ years}$$

With $L = 10^7$, the number of civilizations is $N = 0.1 \times 10^7$, or 10^6.

This last paragraph is sheer speculation, for you can insert your own estimate of chances for long-term survival and come up with some other estimate for L. Yet this speculation does show that if any significant fraction of the technological cultures that arise in the Milky Way can overcome their difficulties and last a stellar lifetime, communicative civilizations in the Galaxy will be numerous. With 10^6 civilizations in the Galaxy, 1 star in 20,000 will be inhabited, and the nearest civilization will be 50 pc away. In such a case, interstellar communication would be a reasonably optimistic proposition.

☆ REFLECTIONS AND OBJECTIONS ☆

The previous sections represent the consensus of those who have spent a certain amount of time and effort on the problems of exobiology, the problems of interstellar communication. Two serious questions regarding the validity of that analysis are often raised.

We have only one real example of a biological system, the ecosystem of the earth, to use as a basis for the analysis. Is it not possible that by using us as an example, many possible life sites in the Milky Way galaxy have been overlooked? Carl Sagan, the pre-eminent exobiologist of contemporary science, argues that various narrow-minded attitudes (he calls them "chauvinisms") may be narrowing the range of our search. Drake's equation is in itself planet-chauvinistic because it presumes life can only exist on planets. The definition of an ecosphere is water-chauvinistic because it assumes that a life site must possess liquid water as a basis for life. It is quite possible that liquid ammonia could serve both as the blood and as the biochemical basis of some hypothetical creature (Figure 16.4). This creature would find Jupiter or one of its satellites a very reasonable place to live and would be quite surprised to find organisms thriving on a hot planet like the earth.

The existence of only one example of a biological system can produce a false sense of optimism as well as a chauvinistic attitude. There are a number of significant gaps in the great evolutionary chain leading from protostar to communicative civilization. If any of the steps in the development of intelligent cultures is a difficult one, crossed on only one of each million planets, the fraction f in Drake's equation associated with that particular step will be very small and the number of civilizations will also be small. To me, the most significant possible gaps are the development of life from chemicals and the development of technology. While most scientists in the field of exobiology tend to minimize the significance of these gaps, specialists outside this subdiscipline do not all share the optimistic attitude characteristic of the conventional analysis presented in this book. But there is no reason why these gaps cannot be bridged by natural evolution.

FIGURE 16.4 Drawing by R. Grossman; © 1962 The New Yorker Magazine, Inc.

"Ammonia! Ammonia!"

Even granting the validity of the conventional analysis, another obstacle to interstellar communication is apparent. The conclusion from Drake's equation is that life is abundant if the average lifetime of a technical civilization is long, 10^7 years for example. An average civilization in the Milky Way will be one that has lived for an average lifetime, 10^7 years. Our civilization has been in the communicative stage for only 10 years, and so the cultures that we might contact will be far more advanced than we are. Is true communication possible between a 10^7-year-old civilization and a 10-year-old civilization? Even counting the 10^4 years we have spent in the precommunicative stage, these civilizations will still be hundreds of times older than we are. If life is abundant, if the number of civilizations in the Galaxy is between 10^4 and 10^6, we are one of the youngest (and perhaps the stupidest) of them all.

There may be problems with the conventional viewpoint that interstellar communication (sometimes known by the acronym CETI = Communication with ExtraTerrestrial Intelligence) is a realistic possibility. However, as long as at least some cultures overcome the self-destructive tendencies that emerging technology produces, there should be a large number, say 10^6, communicative civilizations in the Milky Way. The nearest such civilization would then be 50 pc, 150 light years, away. Let us push all fears and doubts aside and explore how, with present technology, we could establish communication with it.

<h2 style="text-align:center">★ 16.7 INTERSTELLAR ★
COMMUNICATION</h2>

The first decision that must be made in even considering concrete steps toward interstellar communication is the determination of the medium that might be used for such messages. How could one civilization best transmit information to another civilization tens or hundreds of parsecs away? If we are to try to detect a message, we must first guess the form of the message. It is useless to search for interstellar spacecraft if the other civilization is beaming radio signals to us.

In guessing how some culture we have never heard of, on a planet orbiting a distant star, might communicate with us, we can only appeal to the one common ground that the two cultures have. We both live in the same universe, governed by the same laws of physics, and we will be trying to establish our individual ends of the communication link with the same general type of materials. The overriding principle in each culture will be a *principle of economy*. All civilizations will seek the least expensive way to communicate in order to avoid wasted effort. Whether the expense is measured in dollars or deutschmarks or yen or rials is irrelevant; the principle of economy or minimization of effort operates even in a hypothetical society with no monetary medium of exchange. Why waste hours of work when the same job could be done in minutes? The remaining hours can then be spent listening for messages from more stars. This simple principle, trivial as it sounds, dictates radio communication as the preferred mode of interstellar contact.

Physical contact over interstellar distances, interstellar space flight, may be beloved by science fiction writers, but it is a very wasteful method of exchanging information between cultures. The distances between stars are unimaginably vast. A speed of 1 km/sec is almost exactly equal to 1 pc per *million* years. Our space probes, traveling at a few tens of kilometers per *second,* would take tens to hundreds of thousands of years to reach even the nearest star. It would take as much time for such a probe to leave the earth and return as has passed since the Stone Age.

Even using the best rocket permitted by the laws of physics, a rocket far beyond our present capabilities, interstellar space flight is a discouraging proposition. A 1000-ton rocket, capable of carrying 12 people, taking 10 years for the round trip to Alpha Centauri at near the speed of light, would consume 3×10^{31} ergs of energy in accelerating and decelerating twice, once at our end and once to slow down, stop, and turn around. Over the decade that the flight took, this one rocket would be consuming energy at 4000 times the rate at which the entire United States consumes energy for all purposes. One-tenth of the earth's surface would need to be covered with 100 percent efficient solar cells in order to store up enough energy in a decade to fuel this single rocket. A timely search of even a dozen or so nearby stars would require that the entire earth be covered with solar cells to supply energy. Searching the tens of thousands of stars that would need to be searched in order to make contact with even one civilization would either require 10,000 ships or 1 million years.[2]

In contrast, communication by electromagnetic waves is far more efficient, since traveling photons use far less energy than traveling objects. One telescope can search many stars at one time. Let us compare, for example, the needs in equipment and energy to search 10,000 stars in one decade. Such a search effort would be the kind that would be needed to uncover a civilization in a reasonable amount of time. Ten thousand rockets would be needed, versus only one radio telescope and a transmitting beacon. An interstellar rocket ship would be likely to cost at least as much as a telescope, so the equipment cost would be ten thousand times greater if space flight were used as the communications medium. Further, these rockets would use an inordinate amount of energy, about 3×10^{35} ergs. A similar search effort

[2] The magnitude of interstellar distances, the huge number of stars in the Galaxy, and the enormous energetic cost of interstellar space flight make it unlikely that large numbers of spaceships are flying around interstellar space and visiting insignificant places like the earth frequently, in spite of the large numbers of reports of unidentified flying objects (UFOs). I know of no UFO sighting where the evidence is strong enough to exclude the explanation that the sighting is due to natural atmospheric phenomena, airplanes, balloons, or some other prosaic occurrence. Readers interested in UFOs should be sure to read intelligently written books on both sides of the controversy. J. Allen Hynek, in *The UFO Experience: A Scientific Inquiry* (Chicago: Regnery, 1972), argues the case for an extraterrestrial explanation, and P. J. Klass, in *UFOs Explained* (New York: Random House, 1974), presents the case that UFOs are natural phenomena. The arguments of Erich von Daniken that ancient astronauts visited the earth are spurious and are based on overinterpretations of the evidence; see, for example, Barry Thiering and Edgar Castle (eds.), *Some Trust in Chariots* (New York: Popular Library, 1972).

by radio would require the transmitting civilization to use about 3×10^{24} ergs to operate its beacon, or 1/100,000,000,000 of the energy needed in space flight. Such an energy consumption is only 0.04 percent of the current American energy consumption, so it would not place an undue strain on us, not to mention a more advanced civilization.

Thus, energy considerations based on fundamental physical laws dictate that electromagnetic radiation is the best medium for interstellar communication. What region of the electromagnetic spectrum is best? Should we be looking for gamma rays or radio waves? Energy considerations argue for low-energy, low-frequency forms of radiation. Any intelligible signal will be made up of pulses of some sort, and a pulse must contain at least one photon if it is to be distinguished as a pulse. The lower the frequency of the radiation, the smaller the energy per pulse. Further, we seek a region of the electromagnetic spectrum where the messages we send will not have to compete with the background babble of the Milky Way itself. In optical and infrared frequencies, it would be difficult to distinguish between a signal from a civilization and the light of the star. Since stars do not emit large quantities of radio radiation, this confusion with the star does not arise.

But where should we look in the radio spectrum? Radio waves run from short wavelengths of about 1 cm to very long wavelengths of several kilometers. Again, the natural properties of the Milky Way galaxy, in which both we and those out there live, will guide us. The Milky Way itself emits copious quantities of low-frequency, long-wavelength radiation, making it difficult to detect low-frequency signals. The quietest part of the cosmic electromagnetic spectrum lies between wavelengths of 3 and 30 cm, or frequencies of 1 and 10 GHz (1 GHz = 10^9 Hz). This part of the spectrum is the most likely location for interstellar radio signals.

Whether the search range in the electromagnetic spectrum should be narrowed further is a matter of opinion. Searching all frequencies from 1 to 10 GHz is time consuming and costly, and a number of investigators have thought about the possible existence of preferred frequencies in this range. Philip Morrison and Giuseppe Cocconi, in the first paper ever published in the mainstream of the scientific literature on the topic of interstellar communication, suggested that the 21-cm line of hydrogen, the most common element in the universe, might be a logical wavelength for interstellar messages. A potential disadvantage of this wavelength is that the Milky Way emits a fair amount of this radiation, and, even though 21 cm is in a quiet region of the spectrum, the hydrogen itself is quite noisy and would tend to hide possible signals. Others have suggested that perhaps a frequency half or double that of the 21-cm wavelength would seem logical to an interstellar culture. The most currently fashionable suggestion of a region within the quiet 1- to 10-GHz band is the "water hole" between 18 and 21 cm. The hydroxyl radical, OH, emits quite strongly at 18 cm, and exobiologists guess that a creature requiring water for survival would be attracted to a region of the electromagnetic spectrum between two transitions of two constituents of its basic molecule ($H_2O = H + OH$). Where have desert travelers met for generations? At the water hole, of course. However, Richard Grossman's ammonia-based creature (Figure 16.4) would not find the water hole compelling, either in the desert or in the electromagnetic spectrum. Whether the water hole has any special significance is a matter of individual choice.

Interstellar communication by radio, the most economical method, is feasible given our present level of technology. More advanced civilizations would find it technologically even easier. The largest radio telescope in the world is the 1000-foot facility at Arecibo, Puerto Rico (Figure 16.5). With a power of 200 kilowatts, not much more than the power used by the strongest commercial radio stations, the Arecibo telescope could communicate with another similar telescope located anywhere in the Milky Way galaxy, transmitting about 1 pulse/sec, a rate similar to the rate used by the first transcontinental telegraphs.

Energetic and economic considerations dictate that radio contact is the preferred mode for interstellar communication. It is sufficiently easy technologically so that we are capable of it even now, only 70 years after the first radio signal was exchanged between two antennas on the earth. However, the establishment of interstellar contact by radio requires that some signals be emitted by one civilization, and that another civilization be capable of and interested in detecting them. Is it at all reasonable to expect that civilizations will be emitting detectable signals, either deliberately or inadvertently?

We ourselves are already broadcasting our presence to the universe. Television and FM radio stations transmit radio waves that are not blocked by the earth's atmosphere. While transmitting stations are mounted on high hills and try to direct their signals to receivers on the ground, some of the signal inevitably leaks out into space and travels through the universe. Using existing technology, we ourselves could detect our own leaking radiation at a distance of 20 pc. While other civilizations might be able to do a better job with larger telescopes than we now contemplate building, eventually our signals will be lost amid the background radiation of the Milky Way.

Because of the short range of inadvertent radio transmissions, extraterrestrial civilizations interested in contacting us will probably establish a transmitting beacon announcing their presence. Unless almost every good sun in the Milky Way harbors a communicative civilization, it would be fruitless to try to direct initial transmissions toward individual stars, because the chance that we would be listening to them at the same time that they would be transmitting to us is extremely small. A recent analysis by an interdisciplinary scientific team concluded that a relatively modest investment in a 1000-megawatt beacon, an investment that would cost us $80 million per year at current (industrial) power rates, would produce a signal that could be detected by stars in every direction to a range of 100 pc or more. If cultures interested in interstellar communication exist, it is not unreasonable to suspect that they would set up such a beacon.

The principal and possibly fatal difficulty in establishing interstellar contact is the large number of stars that must be searched in order to find one transmitting civilization. With the optimistic assumption, derived in section 16.6, that 10^6 communicative civilizations exist in the Galaxy, one good sun in 20,000 contains an active, flourishing civilization. We would have to search about 20,000 stars in order to have a better than even chance of finding the one that is active. Recall that the estimate of 10^6 civilizations in the Galaxy was an optimistic one, presuming that the gaps in the evolutionary

FIGURE 16.5   The 1000-ft telescope at Arecibo, Puerto Rico. This telescope could communicate with another of comparable size located anywhere in the Milky Way galaxy. (The Arecibo Observatory is part of the National Astronomy and Ionosphere Center which is operated by Cornell University under contract with the National Science Foundation.)

chain from protostar to intelligent, technological culture are easily crossed, and that at least some cultures last a stellar lifetime. With fewer civilizations, we would have to search more stars, for they would be more thoroughly hidden among the huge number of good suns in the Galaxy.

Assume, for the moment, that the optimistic estimate is valid. The length of time required for a search of 20,000 stars depends on the sophistication of the radio telescope used in the search. NASA collected a group of astronomers, electrical engineers, and other scientists in the summer of 1971, and this team designed a radio telescope that would be specifically built to search for signals from another civilization. Their system, dubbed Project Cyclops, consisted of a collection of radio antennas forming a giant multi-mirror telescope about 3 km in diameter. The cost of this system is high, about $10 to $20 billion. The system includes an automated search scheme that could search about 100,000 stars in less than a decade. If the optimistic estimate of the number of civilizations is correct, and if most of them have established transmitting beacons, Cyclops could find them. If the number of civilizations in the Galaxy is smaller, Cyclops could search for them for generations. Insterstellar communication is a long-term proposition, requiring a commitment of at least a decade, and quite possibly several lifetimes.

In spite of the difficulties of interstellar communication, a few token efforts in this direction have been made. The Pioneer 10 and Pioneer 11 spacecraft that recently flew by Jupiter each carried a plaque, etched in gold, on the side of the spacecraft. The plaque was a message from the earth, showing a man, a woman, a map of our location in the Galaxy, and sundry other information about our solar system. We were tossing a bottle into the vast ocean of interstellar space; Carl and Linda Sagan and Frank Drake of the Pioneer 10 team thought that we should include a message along with the bottle. It is extremely doubtful that the Pioneer craft will in fact ever be intercepted by another civilization, and the message was primarily intended to galvanize public consciousness on the subject on interstellar communication.[3]

Another message to space was deliberately transmitted at the ceremony celebrating the construction of a new and more accurate surface on the Arecibo telescope. This message was directed to the globular cluster M 13 (Figure 11.1), a large collection of about 30,000 stars. This message, too, showed a human being, an indication of our chemical makeup, and information about the solar system and the size of the human population.

Several attempts have been made to search for radio messages from nearby stars. The first search, named Project Ozma, was conducted by Frank Drake at the National Radio Astronomy Observatory in 1961. Two stars were observed for 400 hours. Since this pioneering effort, several radio astronomers have looked at a number of stars and found nothing. At the present time, five searches are in progress around the world. Recall that at least 20,000 stars need to be searched, even if an optimistic estimate of the abundance of life in the Galaxy is correct. It would be highly unlikely that we would find another civilization after searching the few hundred that have been examined so far. Cyclops remains on the drawing boards, but some project like it is needed if signals are to be found in the near future.

The magnitude of the task of interstellar communication should not be underestimated. The Cyclops sytem, costing tens of billions of dollars, just might work in a decade or so if life is as abundant as our optimistic estimates indicate. The Cyclops workers would have to be prepared to devote their entire lives to the project, a project that might do nothing more than accumulate negative result after negative result. Taxpayers would have to be prepared to support the project for decades or generations. Even if all the gaps in the evolution of cosmic matter from protostar to intelligent being can be easily crossed, life may be common in the universe but hidden among an enormous number of barren stars.

The recognition that life may be very common in the universe is perhaps the culmination of the Copernican revolution. Over 400 years ago, Copernicus started the intellectual movement that pushed the human race off center stage and into a small, insignificant corner of the universe. It is quite probable that other equally insignificant corners of the universe contain life forms similar to ours. Yet the insignificance of these life-containing parts of the universe, the enormous number of stars in the Milky Way galaxy, makes the establishment of contact between two cultures a formidable task. We know how to do it. The cost, in dollars and in human terms, is considerable. Optimists argue that we should go ahead and search anyway, while pessi-

[3] Carl Sagan, in *The Cosmic Connection* (New York: Doubleday, 1974), describes the motivations for and reactions to this message, the first one sent by human civilization.

mists point to the problems and to the likelihood of frustration. At some time, we will undertake the search. The only question is when in human history it shall be begun.

SUMMARY

The conventional, optimistic view argues that life is abundant in the universe, that the processes that led to life on our planet have been duplicated many times on other stars. An analysis of the various factors governing the abundance of extraterrestrial life offers provisional support to this view, although there may be some steps toward intelligent beings that seem simple to us but that in fact require special circumstances that only occur rarely. But all the right ingredients seem to be there: good suns, habitable planets, the chemicals that make living systems arise as a natural result of the workings of the evolution of stars. The actual discovery of life in the universe is within our technological grasp.

This chapter has covered:

Drake's equation, a framework for calculating the number of communicative civilizations in the Milky Way;

The methods used to discover planets around other stars;

The requirements that make a star a good sun, a possible center for a planetary system with a habitable planet;

The way in which life originated on the earth and an estimate of the likelihood that this process occurred elsewhere;

Some speculation on the sociological factors governing the abundance of extraterrestrial life; and

The methods that would be used to establish interstellar communication.

KEY CONCEPTS

Chemical evolution   Good sun   Natural selection
Ecosphere   Perturbation

REVIEW QUESTIONS

1. Summarize the astronomical, biological, and sociological factors that govern a calculation of the abundance of extraterrestrial life.

2. Why do we need Drake's equation in order to decide whether an interstellar communication system like Cyclops will work?

431

3. Suppose that a planetary system around a distant star contained no gas-giant planets. Would it be as easy to discover as one that did contain some massive gas giants? Explain.

4. Why is a low-mass, M-type main-sequence star not a good sun?

5. What is the distinction between chemical and biological evolution?

6. Make your own estimates for f_l and f_c, and justify them as well as you can.

7. What limitations on our analysis of the abundance of extraterrestrial life are produced by the existence of only one known example of a planet with life on it?

8. Why are radio waves better than space probes as a medium for interstellar communication?

FURTHER READING

Bracewell, R. *The Galactic Club: Intelligent Life in Outer Space*. San Francisco: Freeman, 1975.

NASA Summer Faculty Fellowship in Engineering Systems Design. *Project Cyclops: A Design Study of a System for Detecting Extraterrestrial Intelligent Life*. (NASA CR 114445, available free from Dr. John Billingham, NASA/Ames Research Center, Code LT, Moffett Field, CA 94035)

Ponnamperuma, C., and A. G. W. Cameron. *Interstellar Communication: Scientific Perspectives*. Boston: Houghton Mifflin, 1974.

Shklovsky, I. S., and Carl Sagan. *Intelligent Life in the Universe*. New York: Delta, 1966.

APPENDIX A   UNITS OF MEASUREMENT

Å: angstrom (10^{-8} cm; measures length)

AU: astronomical unit (1.496×10^{13} cm; length)

cm: centimeter (length)

°C: degrees Celsius (temperature)

erg: erg (energy)

eV: electron volt (1.60×10^{-12} erg; energy)

g: gram (mass)

g/cm³: grams per cubic centimeter (density; water has a density of 1 g/cm³)

Hz: hertz (frequency)

K: Kelvin (temperature)

km: kilometer (10^5 cm; length)

kpc: kiloparsec (10^3 pc; length)

m: meter (10^2 cm; length)

μ: micron or micrometer (10^{-4} cm; length)

MHz: megahertz (10^6 Hz; frequency)

Mpc: megaparsec (10^6 pc; length)

pc: parsec (3.085×10^{18} cm; length)

sec: second (time)

METRIC-ENGLISH CONVERSION SYSTEM

1 cm = 0.3937 in

1 m = 1.0936 yd

1 km = 0.6214 mi; about $\frac{5}{8}$ mi

1 g = 0.0353 oz

1 kg = 2.2046 lb

1 in = 2.54 cm

1 ft = 0.3048 m

1 yd = 0.9144 m

1 mi = 1.6093 km; about $\frac{8}{5}$ km

1 oz = 28.3495 g

1 lb = 0.4536 kg

APPENDIX B   CONSTELLATIONS

The 88 recognized constellations are listed here in alphabetical order, according to their Latin names. Also listed are the English names. The "Remarks" column notes the zodiacal constellations and lists any bright star or star group contained in the constellation. Constellations marked with an asterisk contain a bright star or star group and are those that should be learned first.

LATIN NAME	ENGLISH NAME	REMARKS
Andromeda	Andromeda	M 31 (nearest large galaxy)
Antlia	Air Pump	
Apus	Bird of Paradise	
Aquarius	Water Carrier	Zodiacal constellation
Aquila*	Eagle	Altair
Ara	Altar	
Aries	Ram	Zodiacal constellation
Auriga*	Charioteer	Capella
Bootes*	Herdsman	Arcturus
Caelum	Chisel	
Camelopardus	Giraffe	
Cancer	Crab	Zodiacal constellation
Canes Venatici	Hunting Dogs	
Canis Major*	Big Dog	Sirius
Canis Minor*	Little Dog	Procyon
Capricornus	Goat	Zodiacal constellation
Carina	Keel	Canopus (visible to southern observers)
Cassiopeia*	Cassiopeia	
Centaurus	Centaur	Alpha Centauri and Beta Centauri (bright stars visible to southern observers)
Cepheus	Cepheus	
Cetus	Whale	
Chamaeleon	Chameleon	
Circinus	Drawing Compass	
Columba	Dove	
Coma Berenices	Berenice's Hair	
Corona Australis	Southern Crown	
Corona Borealis	Northern Crown	
Corvus	Crow	
Crater	Cup	
Crux	Southern Cross	Acrux and Mimosa (bright stars visible to southern observers)
Cygnus*	Swan	Deneb
Delphinus	Dolphin	
Dorado	Swordfish	
Draco	Dragon	
Equuleus	Small Horse	

LATIN NAME	ENGLISH NAME	REMARKS
Eridanus	River	Achernar
Fornax	Furnace	
Gemini*	Twins	Castor and Pollux; zodiacal constellation
Grus	Crane	
Hercules	Hercules	
Horologium	Clock	
Hydra	Hydra	Alphard
Hydrus	Sea Serpent	
Indus	Indian	
Lacerta	Lizard	
Leo*	Lion	Regulus and Denebola; zodiacal constellation
Leo Minor	Little Lion	
Lepus	Hare	
Libra	Scales	Zodiacal constellation
Lupus	Wolf	
Lynx	Lynx	
Lyra*	Lyre	Vega
Mensa	Table	
Microscopium	Microscope	
Monoceros	Unicorn	
Musca	Fly	
Norma	Square	
Octans	Octant	
Ophiuchus	Ophiuchus	
Orion*	Orion	Betelgeuse and Rigel; Orion's belt, Orion nebula
Pavo	Peacock	
Pegasus*	Pegasus (Flying Horse)	Great Square
Perseus*	Perseus	
Phoenix	Phoenix	
Pictor	Easel	
Pisces	Fishes	Zodiacal constellation
Piscis Austrinus*	Southern Fish	Fomalhaut
Puppis	Stern	
Pyxis	Ship's Compass	
Reticulum	Net	
Sagitta	Arrow	
Sagittarius*	Archer	Zodiacal constellation
Scorpius*	Scorpion	Antares; zodiacal constellation
Sculptor	Sculptor	
Scutum	Shield	
Serpens	Serpent	
Sextans	Sextant	
Taurus*	Bull	Aldebaran, Pleiades; zodiacal constellation

LATIN NAME	ENGLISH NAME	REMARKS
Telescopium	Telescope	
Triangulum	Triangle	
Triangulum Australe	Southern Triangle	
Tucana	Toucan	
Ursa Major*	Great Bear	Big Dipper
Ursa Minor*	Little Bear	Polaris
Vela	Sails	
Virgo*	Virgin	Spica; zodiacal constellation
Volans	Flying Fish	
Vulpecula	Little Fox	

NAME	VALUE	DEFINED IN
Fundamental physical constants		
Speed of light	$c = 2.997925 \times 10^{10}$ cm/sec	Box 1.3
Gravitational	$G = 6.67 \times 10^{-8}$ dyne cm^2/g^2	Equation 3.1
Planck's constant	$h = 6.626 \times 10^{-27}$ erg/Hz	Box 3.1
Stefan-Boltzmann	$\sigma = 5.669 \times 10^{-5}$ erg/cm^2/sec/K^4	Box 3.1
Astronomical constants		
1 parsec	3.085678×10^{18} cm	Equation 9.2
1 astronomical unit	1.49598×10^{13} cm	Box 3.2
Solar radius	6.9599×10^{10} cm	
Solar mass	1.989×10^{33} g	
Solar luminosity	3.826×10^{33} erg/sec	
Earth radius	6378.164 km	
Earth mass	5.976×10^{27} g	
Conversion factors		
1 year	365.2422 days = 3.1556925×10^7 sec	
1 parsec	206,265 AU = 3.26 light years	
1 electron volt	1.6022×10^{-12} erg	

Powers of ten are further explained in Box 1.1.

Lists of various properties associated with different types of astronomical objects are given in Table 5.1 (Planetary Temperatures), Table 6.1 (Outer-Planet Properties), Table 7.1 (Planet Distances), Table 7.2 (Meteor Showers), Table 9.1 (Stars), Figure 10.10 (Interstellar Clouds), and Table 14.1 (Galaxies).

APPENDIX D   NUMERICAL PROBLEMS

CHAPTER 1

1. $(2 \times 10^7) \div (5 \times 10^4) =$ _____

2. $(6 \times 10^{-27}) \times (3 \times 10^{14}) =$ _____

3. $(6 \times 10^3)^4 =$ _____

4. An optical photon has a wavelength of 0.4 μ. Express this in centimeters and meters.

5. Express the following Fahrenheit temperatures in degrees Celsius and in Kelvins: 100°F, 0°F, −30°F, 10,000°F.

6. Devise a rule for converting Fahrenheit temperatures to Celsius.

7. Calculate the energy and frequency of the following types of radiation:

Green light (wavelength = 5000 Å)

Extreme-ultraviolet radiation (wavelength = 400 Å)

Radio waves (wavelength = 10 cm)

8. What is the wavelength of a 1-kiloelectron-volt (keV) x-ray photon?

CHAPTER 2

9. If you could measure the position of the vernal equinox with a probable error of 0.5°, how long would it take to verify the existence of precession?

10. The stars rise 4 minutes earlier each day. If the Pleiades rise at midnight on August 1, at what time do they rise on September 1? On December 1?

11. Use Figure 2.11 to calculate the declination of a star passing overhead at latitude 40°.

12. Use the result of problem 11 to show how navigators could determine their latitude by knowing the declination of a star passing overhead. Suppose a Polynesian in an outrigger canoe noticed that Arcturus, declination +19°42′ north, was overhead. What latitude would he be? In what direction would he sail to reach the island of Hawaii, at latitude 20° north? (The overhead-star method of navigation was, in fact, used by Polynesian navigators.)

CHAPTER 3

13. In the manner of Box 3.1, show that Newtonian gravity will keep a satellite in orbit with a mean distance from the earth of 42,250 km and a period of 24 hours. (This orbit is the one used by communications satellites.)

14. Apply Kepler's law as expressed in Box 3.3 to find the mass of the sun from the orbital characteristics of the earth. (Use numerical data from Appendix C.)

15. Kepler's third law can also be applied to satellite systems of planets. Use it to determine the masses of the following planets.

PLANET	SATELLITE	ORBIT RADIUS (km)	PERIOD (days)
a. Earth	Moon	384,000	27.32
b. Mars	Phobos	9,000	0.318
c. Jupiter	Io	422,000	1.769
d. Uranus	Oberon	586,000	13.463

16. Verify the statement in the appendix to Chapter 3 that the delivery table exerts much more gravitational force on a newborn baby than the planet Mars. Assume that the delivery table has a mass of 100 kg and is located 20 cm from the newborn baby. Mars has a mass of 0.108 Earth masses and never comes closer to Earth than 0.4 AU.

CHAPTER 4

17. Calculate the densities of the following planets.

PLANET OR SATELLITE	MASS (gm)	RADIUS (km)	TYPE OF OBJECT
a. Venus	4.871×10^{27}	6,052	Planet
b. Ganymede	1.49×10^{26}	2,635	Satellite of Jupiter
c. Saturn	5.690×10^{29}	60,000	Planet

18. Calculate the surface gravity of each of the objects listed in problem 17.

CHAPTER 5

19. Calculate the temperatures of the following objects using the simple theory of Box 5.1.

PLANET OR SATELLITE	ALBEDO	AVERAGE DISTANCE FROM SUN (astronomical units)
Jupiter	0.57	5.2
Io	0.63	5.2 (satellite of Jupiter)
Callisto	0.17	5.2 (satellite of Jupiter)
Pluto	0.5	39.4 (albedo estimated, assuming its surface is methane frost; see Chapter 6)

20. The energy units in Figure 5.2 are appropriate for the effect of the earth's atmosphere on its surface temperature. How hot is the earth's surface on the average? Compare your answer with the actual temperature listed in Table 5.1. (*Hint:* Use the temperature calculated in the absence of an atmosphere in Table 5.1 and the expression giving the total radiation emitted by the earth's surface in terms of the surface temperature. Compare the total amount of radiation emitted by the earth's surface in each of the two figures.)

21. If the earth were completely covered with ice, its albedo would be about 0.5. Calculate its surface temperature. Would the ice melt?

CHAPTER 6

22. Use the simple theory of Box 5.1 to calculate the temperatures of the Galilean satellites when Jupiter was at its brightest, with a luminosity of 0.001 times the luminosity of the sun. Assume that the primeval satellites were made of ice with an albedo of 0.6. The distances of the various satellites from Jupiter are: Io, 4.21×10^5 km; Europa, 6.71×10^5 km; Ganymede, 1.07×10^6 km; and Callisto, 1.88×10^6 km. If the melting point of ice is 273 K, which satellites will retain ice in their interiors and which will be rock?

23. If the average albedo of Pluto is 0.5, what is its temperature? Methane condenses at a temperature of 40 K in conditions that existed when Pluto formed. Is it reasonable that solid methane exists on Pluto's surface?

CHAPTER 7

24. Justify the statement on page 175 that an Apollo asteroid should collide with the earth every million years, assuming that there are 1000 Apollo asteroids, 1 in 10 collides with the earth, and their lifetime is 10^8 years.

25. The shield rocks of northern Canada cover about $1\frac{1}{2}$ percent of the earth's surface and can preserve impact structures for about 500 million years. How many craters 25 km and larger in diameter are expected to be there? About four such large structures have been discovered; is it correct to state that the number of discovered Apollo asteroids is roughly in agreement with geological evidence?

26. About 10 percent of the earth's surface is reasonably densely populated (75 percent is ocean, and about half the land area is desert or tundra). Using an average population density of 5 people/km², estimate the death rate from collision of Apollo asteroids with populated areas. (Assume the average Apollo asteroid 1 km in diameter produces a 25-km-diameter crater, and assume that the people in the crater are killed while all others survive.) Compare this death rate with some others, like the 40,000 to 50,000 people killed per year in the United States by automobile accidents.

CHAPTER 8

27. A helium nucleus has a mass of 4.0028 atomic mass units, and a hydrogen nucleus has a mass of 1.008 atomic mass units (1 atomic mass unit $= 1.6603 \times 10^{-24}$ g). One of the results of Einstein's theory of relativity is that mass and energy are equivalent, according to the famous formula $E = mc^2$ (where c is the speed of light). Calculate the mass difference between four hydrogen nuclei and one helium nucleus, and calculate the total energy produced when four hydrogen nuclei fuse to make a helium nucleus. This energy difference should be slightly larger than the energy given in section 8.5, because some of the energy produced is given off in the form of neutrinos and is not available for heating the solar core.

28. About 6×10^{10} neutrinos pass through a 1-cm² area of the earth's surface every second, according to theory. If the Davis neutrino experiment were to capture about one neutrino per day as the theory says it should, and if the chlorine tank is a cube 10 m on a side, what fraction of the neutrinos are trapped by the chlorine? If the earth is as good a neutrino trap as the chlorine tank is, how many neutrinos are stopped by the earth every second?

CHAPTER 9

29. Relaxing comfortably on board a yacht, you observe that the 1-watt navigation light is as bright as a 500-watt street light on the dock. If you are 10 m away from the navigation light at the bow, how far away is the street light? (*Warning:* Do not attempt to use this method to navigate your way to a dock at night. Eye estimates are sufficiently imprecise to make them dangerously inaccurate. Further, if it is foggy, the intensity of the street light will be cut down; you will think it is further away than it actually is, and you will crash into something.)

30. In rural Connecticut, I have noticed that a 100-watt incandescent light bulb burning in a house about 1.3 km away seems to be about as bright as the bright star Sirius. A 100-watt light bulb only emits 2.4 watts of light energy. Using the luminosity of Sirius given in Box 9.4, estimate the distance to Sirius. (Recall that Chapter 1 pointed out how easy it is to visualize the stars as little light bulbs a few kilometers away!)

31. Verify the relationship between absolute and apparent magnitude (equation B9.5) for as many of the stars in Table 9.1 as your patience can stand or your instructor wishes.

32. Verify the relationship among luminosity, and radius, and temperature (equation 9.3) by calculating the luminosity from the temperature and radius of as many of the stars in Table 9.1 as you can.

33. Some of the giant stars in Table 9.1 are too far away to have their distances measured; distance estimates, therefore, have been made. What would the luminosity of Betelgeuse be if it were 1000 pc away? Where would it appear in the H-R diagram were it at such a distance?

34. One day I heard the Metroliner go past Newark, Delaware, traveling between Philadelphia and Washington. Its whistle tone dropped in pitch a major third when it passed, so that the frequency of its whistle when it was traveling away from me was 80 percent of its frequency on approach. If the speed of sound is 1200 km/h, how fast was the Metroliner traveling?

35. The star Procyon is in fact a double star, with a white-dwarf companion. If the orbital separation is 15.8 AU and the orbital period is 40.6 years, calculate the combined mass of the two stars.

CHAPTER 10

36. Assume that each of the clouds listed in Figure 10.10 is a sphere of volume $\frac{4}{3}\pi R^3$. Calculate the radius of each of the clouds in the table.

37. Calculate the energy change in each of the following transitions:

a. The 21-cm line of neutral hydrogen atoms;

b. The 0.35-cm rotational transition of SiO;

c. The 8.1-μ vibrational transition of SiO;

d. The 6562-Å transition of neutral hydrogen.

Which of these transitions could be observed from the ground?

CHAPTER 11

38. Calculate the main-sequence lifetime of each of the following stars. Assume that a star leaves the main sequence when 10 percent of its hydrogen has been transformed into helium and that 4.19×10^{-5} ergs of energy are liberated by the fusion of four hydrogen atoms. Compare your approximate calculations with the exact values given in the table.

MASS (solar masses)	LUMINOSITY (solar luminosities)	ACTUAL MAIN-SEQUENCE LIFETIME (years)
1.0	1.0	1.0×10^{10}
1.5	5.4	2.0×10^9
3.0	94.9	2.2×10^8
9.0	4,450.0	2.1×10^7
15.0	20,200.0	1.1×10^7

39. Use the theory of Box 5.1 to calculate the earth's temperature in the following stages of stellar evolution.

a. 2 billion years from now, when the solar luminosity is 1.37 times the present solar luminosity;

b. At core-helium ignition, about 6 billion years from now, when the luminosity is 300 times the present luminosity;

c. When the sum becomes a white dwarf with a luminosity 0.01 times the present solar luminosity;

d. At the same time as **c**, but recognize that the earth will be covered with ice that has an albedo of about 0.5.

40. How far away is Delta Cephei, with a visual magnitude of approximately 3.5 and a period of 5.37 days? Use the period-luminosity relation of Figure 11.10.

CHAPTER 12

41. Calculate the density of Sirius B from the information given in section 12.2.

42. If the earth, with a mass of 6×10^{27} g, were to become a black hole, how large would its event horizon be?

CHAPTER 13

43. Suppose a Population II Cepheid variable were misidentified as a Population I Cepheid, its luminosity determined from its period, and its distance measured. What would be the error in the distance measurement? For definiteness, take a Cepheid with a period of 10 days and an apparent magnitude of $+15$, and use the period-luminosity relation shown in Figure 11.10 for your answer.

44. Calculate the proper motion of a star with a transverse velocity of 10 km/sec if it is (a) 10 pc away and (b) 100 pc away.

45. Kepler's third law is a relationship among orbital period, orbital radius, and the mass of the orbiting object. While it is strictly valid only for point masses, it can be applied to other objects to obtain a mass estimate. The sun's orbital period around the Galaxy is 250×10^6 years, and its orbital radius is 10,000 pc. What is the mass of the Galaxy?

CHAPTER 14

46. Assume that the main sequence in an elliptical galaxy consists of 10^6 G stars, 10^7 K stars, 10^8 relatively hot M stars, and 10^9 cool M stars. This galaxy also contains 10^5 red giants and 10^6 white dwarfs. What is the mass-to-light ratio of this galaxy? Use the data in Table D.1. (Such a population of stars is typical for an elliptical galaxy or the nucleus of a spiral galaxy.)

47. Calculate the mass-to-light ratio of a group of stars in a spiral arm, consisting of a main sequence with 10^8 cool M stars, 10^8 hotter M stars, 4×10^7 K stars, 2×10^7 G stars, 10^6 A stars, and 10^5 B stars, along with 10^8 white dwarfs and 10^5 red giants. This collection of stars is a simplified representation of the population in the local spiral arm.

TABLE D.1 DATA FOR PROBLEMS 46 AND 47

STAR TYPE		MASS	LUMINOSITY
Main sequence:	B	3	110
	A	1.8	1.5
	G	1	1
	K	0.6	0.1
	Hot M	0.4	0.01
	Cool M	0.15	0.0001
Red Giant		1.2	70
White Dwarf		0.5	0.01

Problems 46 and 47 are based on the information in Table 13.1 and on the analysis of the nucleus of M 31 contained in H. Spinrad and M. Peimbert, "The Stellar and Gaseous Content of Normal Galaxies as Derived from Their Integrated Spectra," in Allan Sandage, Mary Sandage, and Jerome Kristian (eds.), *Galaxies and the Universe* (Chicago: University of Chicago Press, 1975).

CHAPTER 15

48. Calculate the M/L ratio of a closed universe that has 1 percent of its mass in the form of elliptical galaxies with an M/L ratio of 15, 3 percent of its mass in the form of spiral galaxies with an M/L ratio of 5, and 96 percent of its mass in the form of invisible matter that contributes mass but no light.

49. What fraction of the deuterium present in interstellar space must *not* have been made in the big bang if the universe is to be closed? (You will need to refer to Figure 15.6.)

CHAPTER 16

50. Calculate the number of communicative civilizations in the Galaxy, given the following changes in the probabilities listed in the text.

  a. Life emerges from prebiological chemical evolution only 1 time in 10^4 rather than always.

  b. Only 1 in 100 ecosystems evolves an intelligent species.

  c. Organisms can use ammonia as a working fluid, so there are five habitable planets per solar system.

For these calculations assume that L, the lifetime of a communicative civilization, is 10^7 years.

51. A reasonable approximation of the distance to the nearest communicative civilization is

$$\text{Distance} = 2.5(N/10^{10})^{-1/3} \quad \text{in parsecs}$$

Assume that $L = 10^7$ years. Calculate the distance to the nearest civilization using the values for N described in the text and in parts **a** through **c** of problem 50.

APPENDIX E ACTIVITIES FOR AMATEUR ASTRONOMERS

Some of you, having taken an astronomy course, might wish to keep up your interest in astronomy. Astronomy is one of the few scientific fields in which an informed lay person can remain relatively close to the world of research. Listed below are a few addresses, sources of more information, and ideas on further activities that you may wish to pursue.

ARMCHAIR ASTRONOMY

For those who wish to know more about current research in astronomy, two magazines are devoted to surveying current astronomical research but require little or no background to understand their articles:

Astronomy, 411 E. Mason St., 6th floor, Milwaukee, WI 53202, currently $15 per year (12 issues).

Sky and Telescope, 49–50–51 Bay State Rd., Cambridge, MA 02138, currently $12 per year (12 issues).

Scientific American, Smithsonian, and *Natural History* all have occasional articles on astronomical topics. Most of these magazines have book-review columns that can guide you to recently published books describing current research in nonmathematical, nontechnical language. Four recent books in this line are:

Friedman, H. *The Amazing Universe.* Washington, D.C.: National Geographic Society, 1975.

Golden, F. *Quasars, Pulsars, and Black Holes.* New York: Scribners, 1976.

Motz, L. *The Universe: Its Beginning and End.* New York: Scribners, 1975.

Shipman, Harry L. *Black Holes, Quasars, and the Universe.* Boston: Houghton Mifflin, 1976.

ACTIVE AMATEUR ASTRONOMY

Many amateurs own their own telescopes and enjoy using them to observe the skies. *Sky and Telescope* and *Astronomy* both contain many advertisements from telescope makers; a buyer's guide to the different types of telescopes that are commercially available was published in *Astronomy*, October 1975, pp. 58–67.

Amateur astronomer's clubs are active in most areas of the country. Your local planetarium or university can probably provide the names of club officers, who should be contacted by anyone interested in joining. These clubs often have an observatory open to club members, discuss current astronomical research at meetings, and sometimes sponsor star parties at which constellations are pointed out to the general public.

A few nationwide organizations coordinate the activities of serious amateurs making systematic observations of various astronomical phenomena. Planetary and comet observations are coordinated by the Association of Lunar and Planetary Observers (ALPO), c/o Walter H. Haas, Box 3AZ, University Park, NM 88003. For meteor observations contact the American Meteor Society, c/o David D. Meisel, Box 213, Geneseo, NY 14454. Variable-star observations (see Chapter 11) are collected and published by the American Association of Variable Star Observers (AAVSO), 187 Concord Avenue, Cambridge, MA 02138. A number of amateurs time the occultation of stars by the moon to derive information about the moon's orbit, and the International Occultation Timing Association, c/o David W. Dunham, 4032 N. Ashland Avenue, Chicago, IL 60613 coordinates this activity.

ASTROPHOTOGRAPHY

With the increasing sophistication of commercially available cameras and films, a number of amateurs enjoy taking photographs of the skies. Their photographs are sometimes aesthetically quite satisfying; samples of amateur astrophotography are shown in Color Photo 9 (Comet West), Figure 2.1 (Sagittarius in the trees), Figure 3.2 (moonrise and saguaros), and Figure 2.4 (the constellation Orion). (Dennis di Cicco and Tom C. Cooper, the photographers of Color Photo 9 and Figure 3.2, are not strictly amateurs since they are on the staffs of *Sky and Telescope* and *Arizona Highways*, but the equipment they used is the same equipment used by amateurs.) *Astronomy* magazine contains much helpful information for amateur astrophotographers, and the Eastman Kodak Company distributes a booklet on photographing the stars that is available through Kodak or through most photo dealers. A standard 35-mm camera is all that is needed to start taking a few pictures of bright objects in the sky. Try framing the moon in an interesting collection of trees, buildings, or lights, and try several different exposure times ranging from a second or two to a minute. With a little practice, the results can be quite satisfying. Figures 2.1 and 2.4 were both taken with a standard 35-mm camera by Helen Moncure, an amateur astronomer, with a 58-mm lens (the usual lens you buy with one of these cameras) opened to f/1.4 with High Speed Ektachrome film, ASA 160, push-processed to ASA 400. Exposure times were 13 sec (Figure 2.1) and 20 sec (Figure 2.4).

APPENDIX F AN ASTRONOMICAL
CROSSWORD PUZZLE

WAY OUT BY THREBA JOHNSON/PUZZLES
EDITED BY WILL WENG

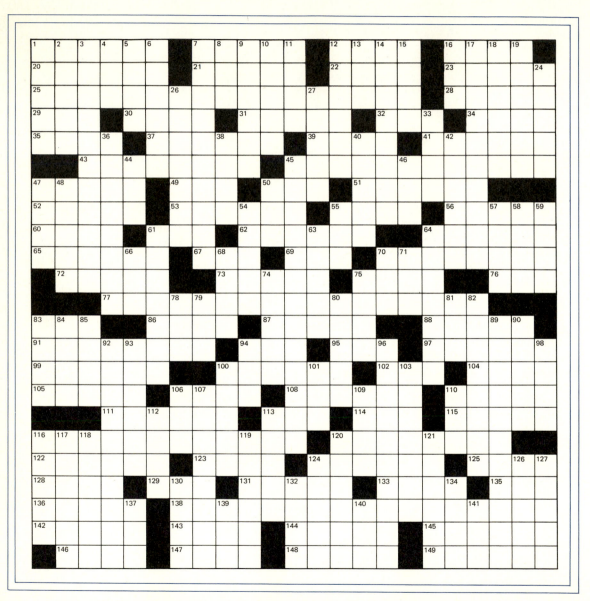

This puzzle appeared in the *New York Times Magazine* on Sunday, November 16, 1975. The solution is on page 452.
© 1975 by the New York Times Company. Reprinted by permission.

ACROSS

1 Constructs
7 Half a continuum
12 Other half of continuum
16 American ship initials
20 Eur. capital
21 Rural areas: Sp.
22 "_____ a man with seven . . ."
23 Changes, in music
25 Orderly view of cosmos
28 Mettle
29 Dress part
30 "_____ homo"
31 French thoughts
32 Little devil
34 Icelandic tale
35 Seed cover
37 Detective of fiction
39 German song
41 Rabbit ears
43 Bulldozer, for one
45 Two fathers of 25 Across
47 "_____ no questions . . ."
49 Diamond _____
50 Pick or wit
51 U.S. Japanese
52 E. Indian birds
53 "_____ last!" (finally)
55 Avesta translation
56 _____ pamby
60 Flightless bird
61 Color
62 Dip the boat out again
64 Be sociable
65 Scythe handle
67 College course: Abbr.
69 Snake
70 Ingredient of 92 Down
72 British auto parts
73 Scrooge et al., for short
75 Heavenly unit
76 Ways: Abbr.
77 Heaven on the go
83 Mouths

86 Tonic plant
87 Swagger
88 Kind of potato
91 Special move by Cornell eleven
94 Word of surprise
95 Bird of legend
97 Musical alley
99 Release, as a boat
100 Takes the helm
102 Choose
104 Bristle
105 French soldier
106 Dimensions: Abbr.
108 Black Sea city
110 Restless
111 _____ fideles
113 Pausing sounds
114 Times of day: Abbr.
115 Certain rays
116 Heavenly twist
120 Cooke of TV
122 Circus people
123 Lease
124 Class: Prefix
125 Night, in Norse myth
128 Rhodes, to Italians
129 Teaching degree
131 Word form for a Mideast land
133 Spanish lady
135 Fish
136 Adlai _____ Stevenson
138 Restless heaven
142 Peter's friend
143 Bone: Prefix
144 Farewell
145 Gem weights
146 Separate: Abbr.
147 Norms: Abbr.
148 Harvested, to poets
149 Oozes

DOWN

1 Turkish title
2 Absolute
3 Heavenly route
4 Baltic or North
5 Wavy, in heraldry
6 Soul
7 Heavenly slaves
8 Nicklaus's org.
9 More la-di-da

10 Betty and others
11 Town in Italy
12 Makes a connection
13 Nigerian river
14 Heavenly circle
15 Word study: Abbr.
16 Hesitant sounds
17 Heavenly monster
18 Movie lot
19 Shoe
24 Norse bard
26 Boston square
27 Spartan serf
33 Gets gold in a way
36 Father of 45 Down
38 Pell _____
40 Ford
42 Heavenly
44 Ship: Abbr.
45 Violent view of cosmos
46 Conjunction
47 Boats
48 _____ Saens
50 Compass point
54 Antelope
55 Kind of code
57 Baseball V.I.P.'s
58 Ink spot
59 Desires
61 Lose hope
63 Assyrian god
64 Island near Canaveral
66 Whammy
68 Yield
70 Educ. broadcasting
71 New Guinea port
74 Follow
75 Food: Prefix
78 Landon
79 Negative
80 Part of R.N.
81 _____ Lanka
82 Name linked with 7 and 12 Across
83 Down _____
84 Western city
85 Gazelle
89 Heavenly photo
90 Kind of yoga
92 Heavenly blows
93 Town near Paris
94 Peaks: Abbr.
96 Heavenly litter

98 Certain votes
100 Upright stone
101 Ways: Abbr.
103 Fondness
106 Telegram contents: Abbr.
107 Heavenly song
109 Store event
110 P.I. tree
112 Miss Lanchester
113 Mob-scene player
116 Cork or thumb
117 Farmer, at times
118 Antiseptic
119 Flavorings
120 Freedom from pain
121 Gins' companions
124 Nasty
126 Judgment
127 Lock
130 Heroic poem
132 Perfume: Var.
134 Of grandparents
137 Cheat
139 Inc., in Britain
140 Pol. party
141 Much: Prefix

449

APPENDIX G ASTRONOMICAL
COORDINATE SYSTEMS

Suppose you have just unwrapped your shiny new telescope and wish to point it at some object in the sky that you cannot see with the unaided eye. Many telescopes are mounted so that the dials or setting circles on the two axes around which the telescope rotates can be used to point the telescope at an object whose astronomical coordinates are known.

Navigators or surveyors wishing to describe the location of some place on the earth's surface assign two numbers, longitude and latitude, to that location. The latitude of a place is its distance north or south of the equator, measured as an angle in degrees. To determine the longitude of a place, draw an imaginary line or *meridian* through that place and through the North and South Poles. The angle between that meridian and the meridian passing through the Greenwich (pronounced *Gren-ich*) Observatory in England is the longitude of the place, measured in degrees west or east from the Greenwich meridian.

In the sky, *declination* (defined in Chapter 2) is the astronomer's equivalent of latitude. For longitude, astronomers are burdened (for historical reasons) with the coordinate of *right ascension*. This coordinate uses the vernal equinox as a reference point and is measured eastward, in units of hours, minutes, and seconds, where 24 hours corresponds to a full circle (Figure G.1).

When you point your telescope at some faint object, you don't care whether this faint object is east or west of the invisible vernal equinox. What matters to you is its relationship to your local meridian, the imaginary line in the sky passing from the north point on the horizon, through the North Celestial Pole, through the overhead point and southward. If an equatorially mounted telescope is set up correctly, so that one of its axes—the polar axis—is pointed toward the North Celestial Pole, one of the setting circles will measure declination and the other will measure the telescope's *local hour angle*, its distance east or west of the local meridian.

FIGURE G.1 On the earth (color), coordinates are measured in latitude and longitude. In the sky (black), coordinates are measured in right ascension and declination. Right ascension is measured eastward; sidereal hour angle, the navigator's coordinate, is measured westward. G marks the location of the Greenwich Observatory in England.

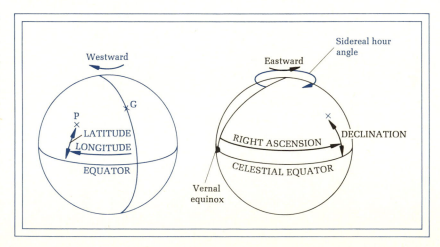

The relationship between local hour angle (LHA) and an object's fixed coordinate in the sky, its right ascension, depends on the local sidereal time (LST), the right ascension of an object on the meridian. The LST increases through the night, and while a sidereal day is 4 minutes shorter than a star day, the difference in rate between sidereal time and ordinary or civil time is generally small enough for most observers to neglect. Figure G.2 shows the relationship between LHA, LST, and right ascension. If the star is west of the meridian, evidently

$$\text{West LHA} = \text{LST} - \text{RA} \tag{G.1}$$

and if it is east of the meridian,

$$\text{East LHA} = \text{RA} - \text{LST} \tag{G.2}$$

The easiest way to determine LST, if you will need it during an observing session, is to point the telescope at some bright star and use equation G.1 to determine the local sidereal time. You can then assume that 1 minute of civil time, marked by your watch, equals 1 minute of sidereal time, and not be too far off. For precise work, you might wish to buy a sidereal clock.

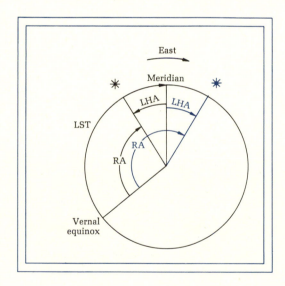

FIGURE G.2 Relationship between local hour angle (LHA), local sidereal time (LST), and right ascension (RA) for west hour angles (black) and east hour angles (color).

APPENDIX H   PRONUNCIATION GUIDE TO SELECTED ASTRONOMICAL NAMES

Aldebaran: al-*debb*-a-ran

Altair: *al*-tayr, al-*tah*-ir, al-*tayr*

Antares: an-*tay*-reez

Arcturus: arc-*too*-rus

Betelgeuse: beetle juice, bet-el-*jooz*

Bootes: bo-*oh*-teez

Cepheid (variable): *sef*-ee-id

Fomalhaut: *foam*-al-hot

Mizar: *meye*-zar

Orion: O'Ryan

Pleiades: *plee*-a-deez

Procyon: *pro*-si-on

Regulus: *reg*-you-luss

Rigel: *Rye*-jel

Sirius: *sir*-i-us

Spica: *spy*-ka

Uranus: you-*rain*-us, *you*-run-us

Vega: *vee*-ga, *vay*-ga

SOLUTION TO THE CROSSWORD PUZZLE OF APPENDIX F

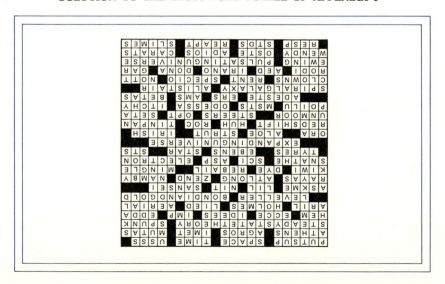

Further information on items in this glossary, and information on items not included, can be obtained by finding the item in question in the index and referring to the appropriate page in the text.

Absorption lines: deficiencies of photons of particular wavelengths in the spectrum of an object, generally caused by the removal of photons from the light beam by absorbing atoms located between the light source and the observer

Accretion disk: the disk of gas surrounding a neutron star or black hole that, by its swirling action, becomes hot enough to emit x rays

Albedo: the fraction of light reflected by a body

Angstrom unit: 10^{-8} cm

Arroyo: the bed of a stream that is generally dry; found in terrestrial deserts and on Mars

Asterism: recognized group of stars that is not a whole constellation (the Big Dipper, for example, is an asterism, part of the constellation of Ursa Major)

Astrobleme: an impact crater on the earth

Astronomical unit: the average distance from the earth to the sun

Carbonaceous chondrite: a class of meteorite, one of the most primitive solar-system objects

Cataclysmic variables: binary stars that erupt at time intervals ranging from weeks to months

Celestial sphere (pole, equator): the imaginary sphere in the sky that is useful for visualizing celestial motions

Cepheid variable: a variable star with a well-defined relation between period of variation and luminosity; useful as a distance indicator

Chromosphere: a region of the sun's outer atmosphere, located just above the surface layer seen in visible light

Constellation: a grouping of stars with well-defined boundaries established by the International Astronomical Union in 1930; most constellations were first delineated in ancient times

Convection: a mode of energy transport in which heat is transferred from one place to another by the motion of hot gas

Corona: the outermost layer of the sun's atmosphere

Crab nebula: the remnant of a supernova that erupted in 1054 A.D.

Cygnus X-1: the best stellar black-hole candidate

Degeneracy pressure: pressure exerted by particles that are compressed so that they touch; resistance to compression is provided by the tendency of the particles themselves to resist being squeezed, not by the motion of the particles

Density: mass per unit volume

Deuterium: a form of hydrogen containing a nucleus of one proton and one neutron; ordinary hydrogen has a simple proton for a nucleus

Doppler effect: the shift in frequency or wavelength of any sort of wave caused by the motion of the wave source relative to the observer

Drake's equation: an expression used to estimate the number of technical civilizations in the Milky Way galaxy

Eclipse: the obscuration of the light of the sun or moon

Ecliptic: the path through the sky traced out by the sun in the course of a year

Ejecta blanket: material surrounding an impact crater, thrown out by the force of the impact

Electron volt: a unit of energy equal to 1.6×10^{-12} ergs

Emission line: a concentration of photons at a particular wavelength, generally produced by atoms moving from a higher to a lower energy level, emitting photons of particular energies

Equinox: one of the two points in the sky where the ecliptic intersects the celestial equator; the time of year when the sun is located at one of these two points

Erg: a unit of energy, equal to the energy of motion of a 2-g insect crawling along at a speed of 1 cm/sec

Event horizon: the boundary of a black hole, the location beyond which nothing can escape from the hole without exceeding the speed of light

Flare star: a star that irregularly increases in brightness in a way similar to a solar flare

Flux: the amount of energy crossing 1 cm² of telescope mirror per second; a quantitative way of measuring the brightness of an astronomical object

Globular cluster: a collection of about 10^4 to 10^5 stars with a marked central condensation, generally found in the halo regions of the Milky Way galaxy

Greenhouse effect: the action of a planetary atmosphere in trapping infrared radiation, increasing the temperature of the planetary surface

H I, H II regions: regions of the interstellar medium containing neutral (H I) or ionized (H II) hydrogen

Highlands: regions of the lunar surface containing many mountains and craters; regions that were not filled by lava flows about 3 to 4 billion years ago

H-R (Hertzsprung-Russell) diagram: a graphical plot used to classify stars, in which temperature or a temperature indicator is plotted on the horizontal axis and luminosity or a luminosity indicator is plotted on the vertical axis

Hubble's law (Hubble constant, Hubble time): a mathematical expression indicating how the expansion of the universe affects the motion of distant galaxies. The velocity of recession v equals Hubble's constant H times the distance D, or $v = HD$. If the galaxies have always been receding from each other at the same speed, the Hubble time is the time that passed since the big bang. With the currently accepted value of 50 km/sec/Mpc for the Hubble constant, the Hubble time is 20 billion years.

Inferior planet: a planet with an orbit lying inside Earth's orbit (Venus, Mercury)

Kelvin: degree Celsius above absolute zero

Kepler's laws: laws describing the motions of bodies around each other under the influence of gravity; Kepler's third law can be used to measure the masses of binary stars

Luminosity: the energy radiated by an astronomical object; one can discuss the luminosity in a particular segment of the electromagnetic spectrum (visual luminosity, infrared luminosity) or the total luminosity of an object

Magnetosphere: the volume of space, surrounding a planet, in which the planetary magnetic field governs the motion of charged particles

Magnitude: another way of measuring the brightness of an astronomical object; magnitude $= -2.5 \log (\text{flux}) + \text{a constant}$

Main-sequence star: a type of star that is fusing hydrogen at its center

Maria: regions on the moon, dark in color, where lava flows filled impact basins between 3 and 4 billion years ago

Maser: the action of a collection of atoms in an upper energy state amplifying a pulse of radiation by falling to their lower energy state upon being stimulated by a passing photon

Meteor: the vapor trail of a particle burning up in the earth's atmosphere

Meteorite: a rocky particle that falls to the ground after colliding with the earth's atmosphere

Meteoroid: a name applied to any solid interplanetary particle

Micron: 10^{-6} m

Mira variable: a red-giant star that varies in brightness over a period of several hundred days

Neutrino: a particle that travels at the speed of light and interacts only very weakly with matter

Neutron star: a ball of neutrons, 10 km or so in diameter, with a mass of $\frac{1}{2}$ to 1 solar mass, roughly; neutron-star matter is extremely dense, packed at billions of tons per cubic inch

Nova: a star that eruptively increases its brightness by 1 million times or so

Open cluster: a cluster of stars found in the galactic disk

Parallax: a method of measuring the distances to stars; relies on triangulation with the earth's orbit as a baseline

Parsec: a unit of distance measurement, 3.085×10^{18} cm

Photon: a packet of electromagnetic radiation

Photosphere: the layer of the sun that is seen as visible light

Planetary nebula: a cloud of gas surrounding an old, hot star

Populations I, II: two groupings of stars, of differing ages; Population II stars were formed early in the evolution of the Milky Way galaxy, while Population I stars are younger

Precession of the equinoxes: the shift, relative to the stars, of the positions of the equinoxes, solstices, celestial equator, and celestial pole, caused by the shifting orientation of the earth's axis in space

Proper motion: the angular motion of a star across the sky

Pulsar: a rotating neutron star that emits pulses of radio radiation periodically

Radial velocity: a star's motion toward or away from the earth

Radiation belt: a volume of space surrounding a planet and containing high-speed charged particles

Recombination: the union of an ion and an electron

Red-giant stars: stars in the upper right corner of the H-R diagram, which are tens to hundreds of times larger than the sun; in general, these stars do not fuse hydrogen in their centers

Reflecting, refracting telescopes: a telescope uses an optical element to collect light and concentrate it at a focus; in a reflector, this optical element is a mirror; in a refractor, this element is a lens

Retrograde motion: the apparent backward motion of a planet that occurs when the earth passes that planet in its orbit

Rille: an irregular depression on the lunar surface, probably volcanic in origin

Solar cycle: the 11- (or 22-) year cycle that governs all forms of solar activity: sunspots, flares, coronal size, and so on (see Chapter 8)

Solar flare: an eruptive outburst on the surface of the sun, resulting in an abrupt increase in the brightness of a small part of the solar surface and often accompanied by bursts of high-speed particles

Solar wind: the continuous outflow of the corona into the outer solar system

Solstice: the point along the ecliptic that is farthest from the celestial equator; the time of year when the sun is at that point

Sunspot: a cooler region of the solar surface

Superior planet: a planet with an orbit lying outside Earth's (Mars, Jupiter, Saturn, Uranus, Neptune, Pluto)

Supernova: the final explosion of a star, resulting in a tremendous increase in brightness in which the star outshines the entire galaxy of which it is a part, increasing its brightness 1 billion times or more

Synchrotron radiation: radiation produced by high-speed charged particles spiraling in a magnetic field

Tectonics: geological processes involving the movement of large blocks of the earth's crust

21 cm: wavelength of photons emitted by hydrogen atoms flipping the orientations of electron and nuclear spins

White-dwarf stars: stars that pack the mass of the sun into a space no larger than the earth

Zodiac: a band of constellations, near the ecliptic, in which planets are generally found

Zodiacal lights: emission from interplanetary dust particles that are illuminated by sunlight

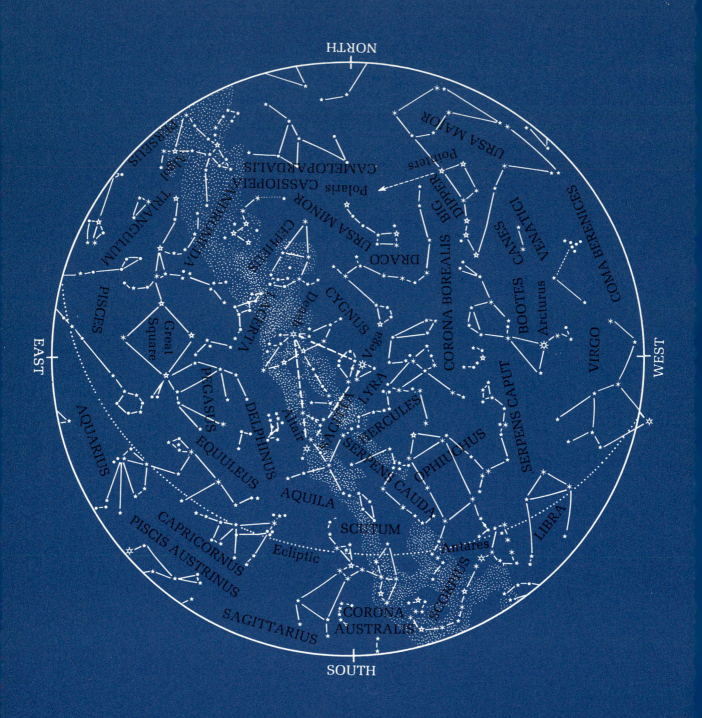

The late summer sky, on August 16 at 9 P.M.; September 1 at 8 P.M.; or September 15 at 7 P.M., standard time.

Adapted from H. A. Rey, *The Stars*, published by Houghton Mifflin Company. Copyright © 1952, 1962, 1967, 1970 by H. A. Rey.